Assessments and Conservation of Biological Diversity From Coral Reefs to the Deep Sea

Assessments and Conservation of Biological Diversity From Coral Reefs to the Deep Sea

Uncovering Buried Treasures and the Value of the Benthos

Jose Victor Lopez
Nova Southeastern University's Halmos College of Natural Sciences and Oceanography (NSU HCNSO), Dania Beach, FL, United Sates

Academic Press is an imprint of Elsevier
125 London Wall, London EC2Y 5AS, United Kingdom
525 B Street, Suite 1650, San Diego, CA 92101, United States
50 Hampshire Street, 5th Floor, Cambridge, MA 02139, United States
The Boulevard, Langford Lane, Kidlington, Oxford OX5 1GB, United Kingdom

Copyright © 2024 Elsevier Inc. All rights reserved.

No part of this publication may be reproduced or transmitted in any form or by any means, electronic or mechanical, including photocopying, recording, or any information storage and retrieval system, without permission in writing from the publisher. Details on how to seek permission, further information about the Publisher's permissions policies and our arrangements with organizations such as the Copyright Clearance Center and the Copyright Licensing Agency, can be found at our website: www.elsevier.com/permissions.

This book and the individual contributions contained in it are protected under copyright by the Publisher (other than as may be noted herein).

Notices
Knowledge and best practice in this field are constantly changing. As new research and experience broaden our understanding, changes in research methods, professional practices, or medical treatment may become necessary.

Practitioners and researchers must always rely on their own experience and knowledge in evaluating and using any information, methods, compounds, or experiments described herein. In using such information or methods they should be mindful of their own safety and the safety of others, including parties for whom they have a professional responsibility.

To the fullest extent of the law, neither the Publisher nor the authors, contributors, or editors, assume any liability for any injury and/or damage to persons or property as a matter of products liability, negligence or otherwise, or from any use or operation of any methods, products, instructions, or ideas contained in the material herein.

ISBN 978-0-12-824112-7

For information on all Academic Press publications
visit our website at https://www.elsevier.com/books-and-journals

Publisher: Nikki P. Levy
Acquisitions Editor: Simonetta Harrison
Editorial Project Manager: Catherine Costello
Production Project Manager: Neena S. Maheen
Cover Designer: Mark Rogers

Typeset by STRAIVE, India

Contents

About the author ix

Foreword to "Assessments and conservation of biological diversity from coral reefs
to the deep sea: Uncovering buried treasures and the value of the benthos" xi

Preface xv

Acknowledgments xvii

CHAPTER 1   The seabed—Where life began and still evolves 1

Introduction 1

Setting the place—Biogeographical regions and abiotic components
of the seafloor 3

Expeditions to the deep blue 5

Pelagic-benthic connections (and vice versa) 10

Connectivity within benthic species 18

How common is benthic continuity and cosmopolitanism? 21

Higher diversity of benthic macrofauna 26

Porifera 28

Crustacea 31

Echinodermata 32

Cnidaria, Mollusca, and other phyla 33

Marine microbial diversity 37

Generation of biodiversity 40

Marine symbioses—Getting to know each other better 42

Corals provide the structure for many benthic ecosystems 45

Possible origins of biodiversity 49

References 51

CHAPTER 2   Multiple approaches to understanding the benthos 75

Living in the era of big science 75

Big experiments on the seafloor are difficult 77

Informative maps show a way 81

Genetic and genomic maps 82

Leading wedge technologies for benthic assessments 86

Submersibles and remotely operated vehicles (ROVs) are leading
wedges that visit the ocean floor 87

Benthic monitoring—All eyes on the sea 92

Underwater soundscapes, landscapes, and unexpected sources of innovation 95

The rise of artificial intelligence (AI) 97

v

vi   Contents

Biotechnologies applied to the seafloor..99
Bioprospecting for new natural products and ideas 101
Secondary metabolites from marine microbes..105
Molecular ecology, conservation genomics, and genome sequencing..................107
References...116

CHAPTER 3   Diversity hotspots on the benthos—Case studies highlight hidden treasures 131

Continuing a Census of Marine Life..133
Cold water extremes at Arctic, Antarctic, and deep sea habitats..........................134
Profiles of multiple coral reefs..136
A resident's view of the Florida Reef Tract 137
The twilight zones 146
The Clarion-Clipperton Fracture Zone (CCZ): One jewel
 of the deep sea..148
Hydrothermal vents and seeps 150
The continuum of biodiversity to their genetic sources 150
Biodiversity includes an array of unique genes and functions..............................152
References...157

CHAPTER 4   Threats to benthic biodiversity...................................... 169

What kind of "information age" do we live in?..170
Ecosystem stability and disturbance 172
Climate change continues to heat up 173
Habitat loss and destructive extraction methods 177
Pollution and eutrophication 178
Seagrass habitats..180
Marine diseases that afflict the benthos..180
References...184

CHAPTER 5   Possible solutions for the conservation of benthic habitats and organisms.. 191

Exploration, education, engagement, and optimism..191
Biodiversity policies from governmental and nongovernmental organizations.....197
 UN biodiversity Goals A to D 199
Art within science, and science with an art...201
Local energy engages and educates 203
Coral nurseries ex situ and in situ..206
Marine-protected areas—"Wildness is the preservation of the world" 207
Deepwater *Oculina* coral reefs..208
Biodiversity databases: Pooling of knowledge or a tangled web?........................212

Contents **vii**

Education and engagement (for one and all)—"The Gathering of Tribes"...........218
The present and future (blue) bioeconomy...221
Coda 225
References...227

Index 237

About the author

Jose V. Lopez is a full professor at Nova Southeastern University in Dania Beach, Florida, and has been conducting research in molecular genetics, microbiology, and marine biology since the early 1990s. He has previously held positions at the National Cancer Institute, Smithsonian Tropical Research Institute, and Harbor Branch Oceanographic Institution. At the latter, Dr. Lopez dove to 3000 ft on the Johnson Sea-Link submersible and has also logged more than 300 SCUBA dives over his career. He co-founded the genomics not-for-profit, the Global Invertebrate Genomics Alliance (GIGA), to assist in bioinformatics training and whole genome sequencing of nonmodel invertebrates. He has written approximately 100 peer-reviewed articles and multiple book chapters and remains an avid student of biodiversity genomics and an advocate for the conservation of marine ecosystems.

Foreword to "Assessments and conservation of biological diversity from coral reefs to the deep sea: Uncovering buried treasures and the value of the benthos"

By Jose V. Lopez, PhD

The story of magnificent slime across deep time

It is a cosmic reality that most of the history of life was written on the benthos or seafloor, the original primordial laboratories for evolution. From the precursors of bacteria at about 4.2 billion years ago to the pre-Cambrian explosion about 600 million years ago, almost all the evolutionary action was on the bottom of the sea. Sea floor residents include about 19 of the 33 metazoan phyla, with most species yet to be discovered. A major theme of this book is the origins of benthic biodiversity, which the author attributes to habitat heterogeneity—many different types of living space. Indeed, from the tide pools to the abyss, there are niches described here in fascinating varieties.

The seafloor is the largest continuous habitat on the planet, with more than 70% of the Earth's surface, with an average depth of 14,000 ft. Yet it remains one of the least known areas on Earth. From what little we have seen (less than 20% of it is mapped), it gets strange, especially at the deeper end. Here, the *macrobenthos*—the largest class of animals—is defined as organisms larger than 1 mm (0.04 in.). However, titans also tread the bottom. The Japanese spider crab can reach up to 3.7 m (12.1 ft) from claw to claw. Giant barrel sponges can reach 2 m in height and over 2000 years in age, the redwoods of the sea. Evolutionary wonders dwell there as well. The crustacean class Malacostraca harbors over 40,000 species of shrimp, crabs, lobsters, crayfish, krill, and mantis shrimp. The slug-like nudibranchs include over 3000 species, many with amazing coloration, dubbed the flowers of the sea. Some sea slugs practice kleptoplasty, wherein they consume algae but "steal" their chloroplasts as energy-producing prisoners. Giant tube worms (*Riftia pachyptila*) have a partnership with symbiotic bacteria which allows both organisms to thrive at deep-sea vents. The venomous cone snails (genus *Conus*) can be picked up in shallow waters by unsuspecting beach-combers, sometimes with lethal consequences. Over 700 species exist, one of which (the magician's cone snail) has provided a painkiller ziconotide. Many other pharmaceuticals await discovery in the vast array of undescribed marine bacteria. Wonder drugs from marine microbes are already proving effective against cancer and other ailments, and one of the author's home institutions is pioneering these efforts; Harbor Branch Oceanographic Institution (HBOI) was founded in 1971 by J. Seward Johnson Sr. as a research organization dedicated to ocean science, exploration, and conservation.

Once below the sunlit zone (about 200 m depth), much nutrition comes from above, as a biological gravitational pump feeds the deep communities with organic particles, dissolved nutrients, whale falls, and marine snow and its undesirable replacement—microplastics. However, nutrients can also flow in the opposite direction, perhaps the most spectacular example being thermal vents that release energy and minerals into the water column. This back and forth is known as benthic-pelagic coupling (BPC), and the filter-feeding oyster is a classic example. The Atlantic *Crassostrea virginica* was estimated to clear the waters of Chesapeake Bay (60 trillion liters) in a matter of days back in colonial times. About 1.7 billion individual oysters (1.05 billion lbs) were reportedly harvested in 1891–1892 alone, dwindling to only a few million pounds in the 1990s.

An understanding of the ocean and its biodiversity requires mapping genomes as well as reefs, seamounts, and abyssal plains. Genetic studies have brought exciting new frontiers to the study of benthos; globe-spanning species such as the ghost worm *Stygocapitella subterranean* turns out to be hiding at least 8 cryptic species in the Northern hemisphere alone. Lobsters and corals produce millions of eggs and maintain connections across whole ocean basins. Genetic surveys are revealing the silent majority of marine bacteria, with thousands of new variants ripe with pharmaceutical possibilities.

This book is exceedingly timely, as many major research advances (summarized here) arrived only in the last 30 years, and the research horizon sparkles with possibilities. Indeed, we may be approaching a golden age of deep-sea exploration. Imagine automated underwater vehicles that roam the ocean and collect information for months or years without a guiding human hand. This volume couples such research advances with conservation concerns, providing a benefit to professional and lay-reader alike. To assess the impact of deep-sea mining, the DISCOL (Disturbance and Recolonization) project tracked plowed study sites. Notably, no recovery was evident 25 years after the plow. The High Seas Treaty was passed at the UN in March 2023, but remains to be finalized. The National Coral Reef Monitoring Program (NCRMP) in the United States was founded in 2000 to implement standardized protocols, so the scientist on a Pacific island can compare results with a Caribbean colleague. Nova Southeastern University Coral Reef Restoration, Assessment & Monitoring (CRRAM) laboratory has been conducting long-term monitoring of the South Florida reef track (FRT) for almost 2 decades. The Census of Marine Life ran from 2000 to 2010, including 14 areas of special concern, from the High Arctic to the equatorial coral reefs, with inventories of marine life that are continually updated. Video documentaries also have a role in keeping the public engaged in conservation issues. Voice of the Sea (https://seagrant.soest.hawaii.edu/about-voice-of-the-sea/) in Hawaii reaches audiences across the wide Pacific, injecting wonder and enthusiasm into conservation messages. National Geographic Pristine Seas Initiative (https://www.nationalgeographic.org/projects/pristine-seas/) documents the marvels where no people occur, with the goal of promoting Marine Protected Areas all over the planet.

Here, it is my pleasure to introduce author Jose "Joe" Lopez, who I have known since before the turn of the century. I first met Joe when we both participated in regular Conservation Genomics courses sporadically offered by our common sponsor, Stephen J. O'Brien. Then, Joe was sorting out evolutionary lineages in corals. This is a notoriously difficult task due to their morphological plasticity, but Joe brought genetics to bear on the issue with great success. He subsequently taught at Florida Atlantic University's Harbor Branch Oceanographic Institution while delving into the mysteries of sponges and marine microbes, again using the molecular toolkit. His career blossomed after arriving at Nova Southeastern University in 2007, and he now has around 100 publications on a variety of marine organisms. With accomplishments in genomics, ecology, invertebrate biology, microbiomes, and even a few fish publications, Joe has a combination of expertise that is truly unique, making him uniquely qualified to write this book. My favorite Lopez paper (so far, as they keep coming) is on marine invertebrate conservation genetics (*Annual Review of Animal Biosciences* 7, 2019). However, he also has a zinger on colonizing new planets with select microbes (*FEMS Microbiology Ecology*, 95, 2019). When Joe came up with the idea of covering all the bottom habitats of the ocean, it seemed like a fantastic dream. Now, after reading the five chapters herein, the dream is a masterful synthesis.

It is an invaluable resource for marine biologists, conservationists, and those who crave to understand the biological diversity of seafloor ecosystems. The technically daunting aspect of getting to the seafloor is especially poignant in the wake of the tragic 2023 loss of tourist submersible *Titan* with five passengers while diving to view the *Titanic*. Even after my 35 years as a professional marine biologist, this book expanded my scientific universe. It also gave me hope, in the post-COVID world, that people can make the changes necessary for a sustainable world. Hope comes from the 35-year effort to ban ozone-depleting substances from the atmosphere, a global triumph. Hope springs from the Convention on International Trade in Endangered Species of Wild Flora and Fauna (CITES), which has drastically reduced endangered species commerce through international agreement. From citizen science to ocean optimism to global treaties, the conservation solutions are complex but feasible. Joe artfully mixes in anecdotes from his own career, and the experiences of storied researchers. Herein you will find quotes and quips from scientists, poets, visionaries, and rock stars, as the author seeks to "weave a thread of wonder" through the description of the bottom of the sea. He has succeeded magnificently.

Brian W. Bowen
Hawaii Institute of Marine Biology, University of Hawaii, Honolulu, HI, United States

Preface

The central topic of this book is biological diversity (= biodiversity), which many know both technically and viscerally as the variation of life forms or species on the planet and specific habitats. The second focus is how this biodiversity thrives or struggles on the seafloor, which encompasses a plethora of possibilities and landscapes. Heterogeneity stands as one operative word when describing the many benthic seascapes, as well as some views of this book. However, combining both biodiversity and benthos into the major focus also created a major task and challenge. As with any large endeavor, we must measure and steel ourselves, and with this being my first book, I was initially hesitant to take on the task. The broad scale of the ocean benthos is daunting in itself, and my own career path into molecular, genetic, and cellular scales, albeit with a marine focus, was limited in experiences for some of the methods and experiences described herein. However, after more contemplation, I thought of an approach that melds marine ecology and evolution as part of the context to address the main scope of benthic conservation and assessment. I concluded that a combination of perspectives—from the molecular, organismal, ecological to the evolutionary—could possibly yield something useful. This was also another learning experience that I viewed as both privilege and opportunity to think and grow more, all while wearing the hat of the proverbial professional student. In this regard, I can invoke Frank Lloyd Wright's saying, "an expert is someone who has stopped thinking." During the process of researching and writing for this book, I definitely had to think more. Whether I will be successful in conveying the vital messages will be for others to decide. Although I will admit that I could not cover all relevant aspects of benthic biodiversity assessment and conservation in this volume, I can say that novel angles to their study will be approached and tested herein. I will have learned much in the process.

Almost every book, good or bad, has a few nuggets of gold, kernels of truth worthy of reading. For this reason, I also utilize past eloquence in the form of salient quotations at important intersections in the text. If an idea has already been pithily crafted from the vast library of human thought, I will unabashedly borrow it. Sometimes the juxtaposition of concepts may not always be purely scientific or technical in origin and can appear unorthodox. Indeed my first mentor, J. Herbert Taylor, a US National Academy member, once substituted the term "DNA" for every time "the Word" or "light" appeared in the gospel verse of John 1:1-5. I do not think it was meant to be tongue in cheek, as the reworked piece was published as supplemental to a peer-reviewed article. The recombination, reuse, and synthesis of old or new ideas is how we can thwart the pitfalls of our own invention of artificial intelligence (AI). By definition, AI cannot escape its inherent formula or algorithm, to which our reasoning is not bound. Moreover, a reference book should have many kernels, which I hope to convey to the best of my ability, not just as kernels but as bushels of facts in the form of scientific evidence and examples. Yet, even with the best or most knowledge-filled books (e.g., encyclopedias, textbooks, monographs), we will often skim over the bulk of pages and find that hidden gem of information which we were seeking or may be a fleeting phrase on the page. The modern Internet (or "World Wide Web" as first coined in the early 1990s) allows us to cover more content faster. Also, the most useful and enduring facts are often repeated, akin to the commonality of riffs among different folk tunes.

XV

xvi   Preface

Most of what I have tried to do in this volume is compile some of the brilliant work of many marine scientists and my colleagues who have spent whole careers and long hours underwater, assessing, observing, and helping conserve their chosen habitat of the benthos. The accolades also go to the ocean engineers, policymakers, environmental managers, students, faculty, and, more recently, social scientists and artists who, like me, may consider themselves beneficial interlopers, observing and synthesizing data for the greater good. Astute readers will also notice that because of my limited expertise and experience in various ocean engineering or physical oceanography subjects and methods, I may skim over some of those and defer to more in-depth reviews in the literature if they exist. I hope these are more helpful to find these collected instead of risking any potential misrepresentation.

I cannot provide answers to all of the longstanding questions regarding benthic biodiversity, such as their origins or how deep-sea diversity is maintained. However, with the juxtaposition of diverse models, ideas, and competing hypotheses, I attempt to review and compile here new syntheses and approaches to advance the protection of benthic ecosystems, large and small, shallow or deep. This possibility helped sustain my drive to complete this book. In the end, we are united in hoping to protect the fascinating life forms that inhabit the mostly unseen.

Acknowledgments

The author is grateful to many friends and colleagues for their helpful comments and feedback to improve this treatise. I thank Dr. Gonzalo Giribet for his assistance with invertebrate taxonomy and Dr Maria Ablan-Lagman and Caitlin Crisostomo for information and in situ photos from the Coral Triangle, respectively. I thank Dr. Brian Walker and his lab for numerous enjoyable dives to visit our own benthic communities for a proper first-hand view. Dr. Walker has also generously provided figures and video of the local Florida Reef Tract, which helps underscore local research and activity. I thank Keri Baker, Lisa Ferrara, and Joana Fernandez-Nunez, our erudite librarians for their assistance in finding benthic multimedia examples and organizing resources on NSUWorks. I am also grateful to my home institution, Nova Southeastern University, Halmos College of Arts and Science, and the Guy Harvey Oceanographic Center for allowing me the time and space to develop the ideas for this book. This includes administration, colleagues, and students who have provided feedback. I especially thank Amy Doyle, John Reed, and Brian Walker for final proofing of the galleys.

This book is dedicated to my family and all of the students who have the curiosity to dive into the challenging unknown. This includes young Victoria, who embodies a love and passion for the ocean. Also not in the least am I indebted to Amy, who was my sustenance and spark for helping me finish this project.

CHAPTER

The seabed—Where life began and still evolves

1

All trades, all callings, become picturesque by the water's side, or on the water. The soil, the slovenliness is washed out of every calling by its touch. All river-crafts, sea-crafts, are picturesque, are poetical. Their very slang is poetry.
Fuller, Margaret. Summer on the Lakes, in 1843 (Fuller, 1991).

Introduction

This book is not about poetry but rather about finding the poetry in natural phenomena. With the ocean as the prime subject, however, poetry may actually be easy to find. The mechanisms that sustain long-standing habitats such as the ocean bottom in particular and the deep evolutionary history of its many benthic creatures do have a certain poetic existence if not license. Intermingling the arts and the sciences has benefits. When practiced throughout this book, it is with the primary intention of stimulating thought, while conveying the scientific messages and data more effectively. This is especially true when it comes to a topic such as ocean seabed or benthic habitats, a seemingly endless expanse. As mentioned earlier, one of the goals of this book is to assess biological diversity on the seafloor. Thus, focus will be on the *benthos*, defined since Greek antiquity as "the assemblage of organisms inhabiting the seafloor."

Encyclopedia Britannica defines *benthos* as all of the organisms at the seafloor, which can be broadly categorized into the following:

(A) *Macrobenthos*—organisms larger than 1 mm (0.04 in.), dominated by polychaete worms, pelecypods, anthozoans, echinoderms, sponges, ascidians, and crustaceans.
(B) *Meiobenthos*—organisms between 0.1 and 1 mm in size, include polychaetes, pelecypods, copepods, ostracods, cumaceans, nematodes, turbellarians, and foraminiferans.
(C) *Microbenthos*—organisms smaller than 0.1 mm, include eubacteria and archaeabacteria (two out of the three major domains of life), diatoms, ciliates, amoeba, and flagellates. This listing and classification are not exhaustive. Perhaps these categories could have provided the muses, mermaids, and Venuses of antiquity, even if only in some ancestral mariner's imagination.

As we all intuitively know, our planet has a plentiful supply of water. This essential substance is literally the stuff of life. Although we can only infer that life began in the oceans, biologists hypothesize that from a biochemical and ontogenetic perspective, earthly life is totally dependent on water, which is a simple compound with two hydrogen and one oxygen atom, in great abundance in the oceans of our

Assessments and Conservation of Biological Diversity From Coral Reefs to the Deep Sea. https://doi.org/10.1016/B978-0-12-824112-7.00002-9
Copyright © 2024 Elsevier Inc. All rights reserved.

1

2   **Chapter 1** The seabed—Where life began and still evolves

planet. (As we increase our gaze to the stars with better telescopes, we also realize that water may be more common than expected in the cosmos, e.g., comets contain water and perhaps the moon as well (Smidt, 2018).) We can continue waxing more about water, its function as a universal solvent within biological cells, or as a vehicle of nutrients and the next generations' seed (gametes), conveyor of cold and hot temperatures at mesoscales, and its presence above and weighing upon our chosen habitat of the benthos, the subject of this book. Yet although the oceans are far flung, they are not infinite, especially when viewed from space satellites above the earth. The oceans have clear boundaries, so we intrinsically know beyond the horizons we may enjoy at sunset that the water has a limit and more precisely a bottom.

The goals of this book are to provide a current assessment of benthic habitats in the modern age. The habitats addressed will be limited to marine areas. However, saltwater covers approximately 361,000,000 km^2 (70.8%) of the planet, and the oceans have an average global depth of 14,000 ft. Thus, *heterogeneity* of benthic habitats is the current norm but addressing each possible iteration and causes for variation goes beyond the scope of this book and this sole author. For example, although temperatures may be more stable below 600 m across most basins, we can also conceive of dynamic broad and localized flow patterns (currents, eddies) at depth, similar to weather reports, which are broadcast to us each day for the continents. For example, Gage (1997) showed the effects of strong underwater currents on bivalve and polychaete biodiversity at the Nova Scotian continental rise in the North West Atlantic. These movements are sometimes referred to as "benthic storms." On continental margins, benthic habitat heterogeneity is well known and can include rocky intertidal pools of the New England or Costa Rican coasts along with the steep walls of a Puerto Rican or Bay Island coral fore reef, Monterey Canyon or the abyssal plains below the South Pacific Gyre (SPG), which is the largest oligotrophic ocean region, covering ~10% of Earth's surface, occurring from 4000 and 6000 m just above the hadal continental slopes. Therefore, most benthic biodiversity measurements will be focused locally as generalizations to other similar types of habitats cannot be easily transferred. Each coral reef, in the Pacific Coral Triangle, Indian Ocean, or off the coast of Andros Island or Bermuda or Belize in the Caribbean, may be generated by common processes (subsidence, accretion) or maintained only in oligotrophic clear waters. However, each reef also has unique properties and abiotic factors dictated by their specific geographic locale. Thus, extrapolations for biodiversity estimates are rare and only used with caution. Heterogeneity and patchiness of benthic seafloor habitats make their comprehensive assessments difficult.

We will review and highlight the concept of biodiversity and its conservation in the context of specific benthic marine habitats, ecology, and methodologies. This will be providing unique perspectives on biodiversity. In this first chapter, I will broadly describe some of the habitats of benthic organisms, then attempt to parade and describe a fraction of the enormous diversity of organismal groups that inhabit the bottom of the seas. I mostly focus on higher taxonomic levels, at the species level and up. These representative taxa will help conversely reflect the physical nature of habitats found across multiple oceanic basins. Population level data and studies will be discussed in the context of other topics—connectivity or new technologies etc. Next, I will describe several processes and mechanisms, such as benthic-pelagic coupling, evolution, and symbiosis, which likely have led to some of the diversity at all relevant levels—molecular, organismal, and ecosystem. In Chapter 2, a unique approach for addressing benthic biodiversity assessments will be constructed around a trifold "Big Science" framework of Big Maps, the Big Experiment, or leading Technological cutting edges. This framework will introduce relevant scientific technologies that have been used traditionally or recently to characterize benthic habitats and organisms. In Chapter 3, specific diversity hotspots are literally visited and highlighted to further support the methods and quantitative assessments. Chapter 4 briefly summarizes the most pressing threats that endanger some of the most unique benthic habitats previously listed and described. The last Chapter 5 then provides an array of possible approaches and remedies to many of the conservation dilemmas exposed in the previous chapters. Sufficient answers to the problems may or may not be fully completed but will be left open to the reader.

Setting the place—Biogeographical regions and abiotic components of the seafloor

The benthic habitat can be viewed as one of the original primordial laboratories for organic evolution on the planet (Martin et al., 2008). As long as sources of energy, building blocks of carbon, nitrogen, hydrogen, oxygen (electron acceptors) are available, the possibilities of forming and sustaining life exist. We enjoy hearing origin stories. Many initial conversations often begin with "where were you born or where do you hail from?" Perhaps because few ever eyewitness the true origins of cherished people, places, or ideas, the impossibility to verify and thus refute that great tales of imagination about the ocean were once weaved.

Before diving into the biology, we should describe some of the different physical settings of biodiversity in the seabed. Consider that most land area to the square foot or meter, with the obvious exception of Antarctica, has been trodden, visited, mapped, or photographed either in person or via high-definition satellites. Yet, we know very little about the basic nature of the ocean seabed. This is also why arguments against spending large amounts of funds and resources for trying to settle people on other planets *while* ignoring our own planet's biodiversity decline, have a sound basis (Lopez et al., 2019b; Vieira et al., 2020).

The benthos can be seen as an interface. It is a surface underwater, and all surfaces are interfaces between two different states or media. Yet the seafloor is also the largest continuous habitat and remains one of the least known on the planet. Similar to landforms, the highly variable topography, which includes canyons, flat plains, tall mountains, mounds, volcanoes, pockmarks, crevices, and the deepest trenches, can hold a large variety of life (Ramirez-Llodra et al., 2010). The heterogeneity can arise within small areas. Think about the extremes of vertical relief where tectonic plates collide—e.g., Himalayas and Sierras, or habitats on the millimeter scale for bacteria and meiofauna.

As is well known from marine biology textbooks (Levinton, 2017), marine zones can first be categorized by the depth. These are shown in Fig. 1.1. Multiple experts have described such zones and provinces over many years. For the basis of this book, we will rely on the UNESCO 2009 benthic zone descriptions currently accepted by most oceanographers. These were mapped by multinational groups.

Because they are nearest to civilization and commercial ports, the most studied and well-known benthic zones are coastal and shelf habitats from 0 to 300 m. These are termed the littoral and neritic zone (or **sublittoral** zone) for the ocean that precedes the drop-off to the continental shelf. Although not as large an area as the abyssal plains of the ocean, coastal areas can contain wide shelves that host diverse habitats. For example, the continental shelves off of Argentina, or below the South China sea off Malaysian coastlines, and Antarctic shelves under the Ross and Weddell seas are wide. Also, shallow coral reefs occur in the nearshore littoral zones. Above 200 m, coastal shelves are in the epipelagic and photic zones, so primary productivity by phytoplankton is high and benefits from

4 Chapter 1 The seabed—Where life began and still evolves

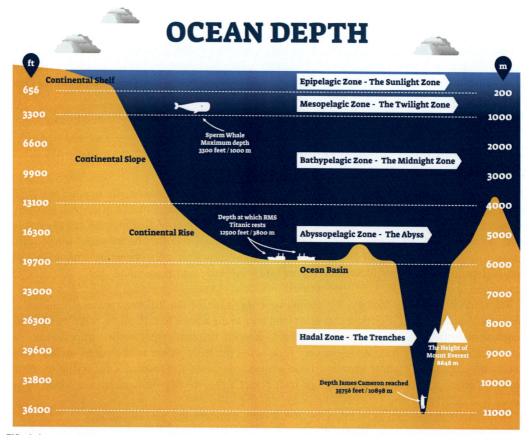

FIG. 1.1
Vertical depth zones of the ocean. Pelagic depth definitions roughly correspond to the same prefixes for benthic zones discussed in the text.

This is a Shutterstock image and does not require perms in the permission log.

any nutrient runoff from rivers and the land. Below the shelf regions, the mesopelagic overlaps with the upper **bathyal** zone from 200 to 800 m followed by the lower bathyal zone from 800 to 4000 m. No sunlight penetrates below 1000 m or the aphotic zone. Essentially the **bathyal** marine ecologic realm roughly follows continental margins and extends down from the edge of the continental shelf to the various depths at which the water temperature will dip to 4°C (39° F). This slope can be abrupt or gradual, including canyons, but the ultimate destination of all modes reaches the abyssal depths. The **abyssal** zone ranges from about 4000 to 6500 m and is considered the largest environment for Earth life, covering approximately 300,000,000 km^2 (115,000,000 mile^2). This represents about 53% of the planet's surface and 83% of the area of oceans and seas (Menard and Smith, 1966). With such a vast expanse, estimates of the number of abyssal species can range from 500,000 to tens of millions. Lastly, the ultra-abyssal and **hadal** zone covers all depths and trenches greater than 6500 m. This greatest depth zone derives its name from Greek mythology and god Hades, who ruled his vast mostly "unseen" realm of underworld (Jamieson, 2018).

Questions about how the abyssal biodiversity is generated will be addressed throughout this book. The physical complications and hazards of living in the deep have produced unique organisms from natural selection. Although the habitat is large, resident deep organisms must also contend with no radiant sunlight, an average temperature (of $\sim4°C$), and a low supply of organic matter (i.e., $1–10\,mmol\,C\,m^{-2}\,yr^{-1}$). Moreover, hydrostatic pressure increases by about 1 atm (approximately 14.7 pounds per square inch at sea level) every 10.3 m. Thus, at the abyssal depths, pressures can reach 75 MPa (11,000 psi) 200 and 600 atm, while the hadal includes the deepest trenches with pressures at 1100 standard atmospheres (110 MPa; 16,000 psi). Wolff (1970) asserted that the fauna of the hadal zone has high degrees of endemism, on the same order of magnitude as that of the abyssal zone but which probably exceeds that of the bathyal zone. The various descriptions throughout this book help support this contention, but overall, we can conclude that insufficient data currently exist. Jamieson (2015, 2018) gives a more recent perspective on hadal science.

Multiple models and ecological factors affect the distribution and relative abundances of benthic organisms at these deeper depths. For example, Rex et al. (2005) list several factors that can affect deep-sea biodiversity, which we will also revisit throughout this book: environmental stability (Sanders et al., 1965; Sanders, 1968), food availability and biotic interactions (Rex, 1981), sediment grain-size heterogeneity (Etter and Grassle, 1992), metapopulation dynamics and dispersal (Etter and Caswell, 1994), boundary constraints (Pineda and Caswell, 1998), topography (Vetter and Dayton, 1998), hydrodynamics (Gage, 1997), bottom-water oxygen concentration (Levin and Gage, 1998), and gravity-driven sediment failure (Levin et al., 1994). In a global-scale analysis based on 116 deep-sea sites, Danovaro et al. (2008) placed biodiversity parameters such as biomass and bacterial productivity at the forefront, indicating their importance for maintaining deep-sea ecosystem functions (Fig. 1.1).

The above mentioned characterizations provide a frame of reference and should not distract us that seafloor biogeography is as much vertical as horizontal. Some of the most thrilling scuba diving can occur along vertical walls of fore reefs, as we must suspend ourselves above the benthos. The depths to the bottom can span hundreds of meters, and the depth gradients themselves act as biogeographic barriers and delineate the zones. The noticeable change in habitats and fauna can be seen on various submersible ascents after deep dives on the DSV *Alvin* or other undersea vehicles. This becomes especially salient after spending hours in the relative darkness of the seafloor and gradually returning back up to the photic zone, starting with the twilight zone around 200 m. Vertical relief is often counted as a metric for coral presence (Lester et al., 2020). Moreover, organisms that recruit on vertical substrates may be less likely affected by sedimentation factors, now recognized as one major benthic stressor (Gleason, 1998; Jones et al., 2019; Rushmore et al., 2021). The future of exploration is not horizontal but rather up and down.

Expeditions to the deep blue

Before describing some of the specific assessment issues and taxa on the benthos, I will take a slight historical digression to discuss how we arrived at our current stage of knowledge of depth and biogeographic zones mentioned earlier. The answer requires many weeks and years of developing sea legs on

6   Chapter 1 The seabed—Where life began and still evolves

sometimes long cruises out at sea. Although most of my practical experience has been with benthic habitats, I had a rewarding foray into pelagic research with the DEEPEND/RESTORE project in the Gulf of Mexico (http://www.deependconsortium.org). This project was spearheaded by fish ecologist, Tracey Sutton, who initiated and continues to lead a stalwart and diverse group of scientists and students to characterize multiple pelagic organisms from the shallow to the bathypelagic zones (~1500 m) in the Gulf of Mexico. This project stemmed from the tragic Deepwater Horizon oil spill (DWHOS) from the deep Macando oil well on April 20, 2010. This accident killed 11 oil rig workers and released over 210 million gallons (4.9 million barrels) of crude oil in one of the worst ecological accidents in US history. Gulf of Mexico Research Initiative (GOMRI) provided funding and support for an array of projects to determine the accident's effects to the marine environment after this tragedy. DWHOS created one proverbial silver lining on the dark cloud via a massive influx of research funding to study the impact and aftermath of the disaster. This work may be best represented by the formation of the GOMRI (Milligan et al., 2019; Eklund et al., 2019). More about GOMRI's educational value and other restorative measures will be elaborated later. DEEPEND continued on as DEEPEND/RESTORE generating a large volume of pelagic data (Cook et al., 2020; Sutton et al., 2020). The project also sparked more questions and thinking on the tangible difficulties of addressing the concept of *benthic-pelagic coupling* (BPC).

DEEPEND required multiple cruises to meet its goals of characterizing the pelagic realms of northern Gulf after the DWHOS. Besides the impetus of scientific research, there may also be a conserved section of DNA in our genomes coding for behavior, such as the human inclination to explore the unknown. In many individuals, this cannot be easily suppressed. Perhaps this a good point to interject the lyrics of a song called "Wand'rin' Star" (written by Alan Lerner and Frederick Lowe) comically sung by Lee Marvin in 1969 for the movie version of the musical, *Paint Your Wagon* (https://www.youtube.com/watch?v=-jYk5u9vKfA). The setting is the American west, but the vague reference to Odysseus allows application to old salty sailors and captains who also navigate the ocean by the stars:

I was born under a wandrin' star
I was born under a wandrin' star
Wheels are made for rollin'
Mules are made to pack
I've never seen a sight that didn't look better looking back
I was born under a wandrin' star
Mud can make you prisoner, and the plains can bake you dry
Snow can burn your eyes, but only people make you cry
Home is made for comin' from, for dreams of goin' to
Which with any luck will never come true
I was born under a wandrin' star
I was born under a wandrin' star
Do I know where hell is?
Hell is in hello
Heaven is goodbye for ever, it's time for me to go
I was born under a wandrin' star
A wandrin' wandrin' star
(Mud can make you prisoner, and the plains can bake you dry)

> (Snow can burn your eyes, but only people make you cry)
> (Home is made for comin' from, for dreams of goin' to)
> (Which with any luck will never come true)
> (I was born under a wandrin' star)
> (I was born under a wandrin' star)
> When I get to heaven tie me to a tree
> Or I'll begin to roam, and soon you know where I will be
> I was born under a wandrin' star
> A wandrin' wandrin' star

The song reached the number 1 spot in Ireland and the United Kingdom but is again meant to highlight the penchant to explore and sometimes understand.

Looking back further, but with more serious intentions, the *HMS Challenger* voyage probably also echoed with sea shanties, while carrying out scientific expeditions from 1872 to 1876 to characterize the world oceans (Aitken and Foulc, 2019) and in the process formally began the discipline of oceanography. Sanctioned by Britain's Royal Society, the ship was transformed into a floating laboratory, which included specimen jars, filled with alcohol for sample preservation, microscopes, and chemical equipment. This allowed Chief scientist Sir Charles Wyville Thomson of the University of Edinburgh, along with a team of naturalists, to identify and catalog common and novel marine organisms in the water column and at the bottom. Traveling up to 68,890 nautical miles (79,280 miles; 127,580 km), the *HMS Challenger* expedition surveyed ocean bottoms with special dredging and trawling equipment (which replaced the ship's cannons). After 492 deep-sea soundings (with one down to 26,850 ft.; 8184 m), 133 bottom dredges, 151 open water trawls, about 4700 new marine species were recorded. The full report of the expedition is described in the riveting *Report of the Scientific Results of the Exploring Voyage of H.M.S. Challenger during the years 1873–76* (Thomson, 1878).

In the spirit of *Challenger*, but 130 years later, geneticist Craig Venter funded and retrofitted his own personal 95-ft sloop called the Sorcerer II for expeditions to sample and help "sequence the oceans" (Venter et al., 2004). At the time, this was a truly innovative and audacious enterprise. However, Venter had already been accustomed to blazing trails, such as formulating the "shotgun" cloning approach and the parallel company called Celera Inc. to provide the second whole human genome sequence (Venter et al., 2001), patenting cDNA sequences and inventing a synthetic bacterial genome complete with his own watermarks in amino acid code (Venter, 2014; Gibson et al., 2010; Hutchison III et al., 2016). After 2 years of collecting water samples and circumnavigating the globe at mid-latitudes, Sorcerer returned to port, and the era of high-throughput DNA sequencing of marine microbiota was enjoined. The effort yielded 7.7 million sequencing reads (6.3 bp—an impressive volume of sequence data at the time), finding that most differentially abundant genes were related to phosphate transport and utilization as 1874 variants of the ubiquitous light-absorbing proteorhodopsin gene (Rusch et al., 2007).

Moreover, the European Union (EU)-funded *Tara* Oceans multinational consortium expanded upon the Venter group's efforts. *Tara* Oceans outfitted a 110-ft research schooner *Tara* in 2009, with a primary goal to profile both bacterial and eukaryotic planktonic microbes across the world's oceans using state-of-the-art methods, including molecules. Tara sampled at 210 sites and depths up to 2000 m in all the major oceanic regions. More Tara cruises followed and are discussed in Chapter 2. The initial datasets on planktonic microbiomes have been invaluable, greatly expanding our knowledge of pelagic

8 **Chapter 1** The seabed—Where life began and still evolves

denizens with large caches of metagenomics data, which scientists still pore through daily (Sunagawa et al., 2015; De Vargas et al., 2015; Brum et al., 2015; Pesant et al., 2015). These studies also advanced analytical methods of large metagenomic datasets, such as describing interactomes among present phytoplankton taxa, which included parasites, viruses, and grazers. A major conclusion by Lima-Mendez et al. (2015) was that biotic interactions determined by cooccurrence analyses has as much or even greater influence than abiotic environmental factors in structuring the local community (Lima-Mendez et al., 2015). The R/V *Falkor* operated by the Schmidt Ocean Institute also adds to the fleet of research ships and has visited Walvis Ridge (WR) and Rio Grande Rise (RGR) hotspot twins in the South Atlantic Ocean and Shatsky Rise in the Pacific Ocean to obtain magnetic bathymetry and seismic data (Thoram et al., 2022).

These research cruises set inspiring examples, complete with their own stories and driven personalities, which compel many of us to follow. At their respective stages, these were monumental efforts. We view these oceanic quests as instructive for humanity. All have in common the journey out, away from the comfort zone of warm hearth and home, and into the often cold and uncomfortable unknown. However, we have to keep in mind that most of these large-scale expeditions have mostly focused on the pelagic for their collections. More benthic expeditions would always be welcome. Few have reprised the scale of *Challenger* expedition to systematically sample multiple benthos on a large global or even single oceanic scale. Until this can be accomplished economically, our data will remain patchy. For example, the Danish ship *Galathea* trawled Challenger sites in the early 1950s at site 716 at depths of around 3570 m in the East Pacific. The expedition collected 2100 specimens representing 132 different species of benthic metazoans, a considerable finding at the time (Wolff, 1961).

Therefore, without sensational breaking news, such as the synchronized double landings of SpaceX's Falcon heavy boosters, sending a new probe to Venus or a new landing on Mars, the lay public may not care too much or at all about the average depth of the ocean or what organisms live on its seabed. However, as described earlier, the idea of mysteries has long resided in the minds of many oceanographers, marine scientists, and explorers. Therefore, whenever a new seafloor site is discovered and charted, could we not collectively summon the media for coverage and fanfare amounting to at least a quarter of the fraction given to extraterrestrial escapades (e.g., other worldly, but so far lifeless sites)? Perhaps we can take a lesson from guides on how to find and appreciate objects that can be awe-inspiring in our everyday lives (Keltner, 2023). From a biased viewpoint, the benthos likely possesses countless phenomena, which await display and appreciation from the lay public.

This is the great irony, as we live on a planet dominated by water, with five oceans, and yet only a relative few of us venture or are comfortable in aquatic environments. Many people cannot swim, nor have sailed on a boat, or even viewed the ocean firsthand. Perhaps this number could increase if we include the amphibious, as most of humans and their civilized constructs do favor and flock toward coastlines (Neumann et al., 2015; United Nations, 2017). Because few benthic locales have been named, are difficult to visit and even to pinpoint on oceanic maps, we cannot blame anyone for not hearing about these particular sites or knowing their location—e.g., Campeche Escarpment, Miami Terrace, Kamehameha Basin, the Menard or Eltanin Fracture zone systems, the Chatham Rise, Campbell Plateau, Wharton Basin, Cocos Keeling Rise, Ninety East Ridge, Mascarene, Tasmin and Natal Basins, Andrew Bain Fracture zone, etc. In addition, multiple deep-sea trenches delineate some of the aforementioned regions and dip deeply into the earth's crust—the Middle America Trench, Peru Chilean Trench, Java Trench, and of course, the Marianas Trench. All of these names have a touch of mystery about them, perhaps because they still are. Many of these sites have an area as large as small- to middle-sized countries. Yet very little biodiversity assessments exist for most of these locales. Methods such as monitoring and visualizations by submersible or AUV/ROVs will be discussed later and could be used to characterize some of the aforementioned sites. Images, especially when detailed in situ, of organisms and habitats are very useful. However, these literally only scratch the surface of most benthic ecosystems, and this effort only marks a beginning. There are few substitutes to physically holding a specimen and having the time to measure and fully analyze organisms for biodiversity.

One study of a deep-sea trawl of Tasman Basin, formed 90–52 MYA, was introduced and written so eloquently that I have to quote an excerpt in the following. This represents impressions of other divers or viewers when encountering many other countless unexplored regions of the benthic habitats for the first time (Keene et al., 2008):

> The eastern margin of the basin adjacent to the Dampier Ridge also lacks fans and is of a similar depth to the conjugate western margin. What appears to be a rise 100 km wide occurs where the seabed shoals into an embayment in the Dampier Ridge between 28°30′ and 29°30′S The seismic profile across this rise, along with a multibeam line further south, show it to have irregular topography on the order of 10's of meters interspersed with flat lying sediments. Small rises occur elsewhere along the base of the Dampier Ridge and are formed by the accumulation of sediment debris redeposited from the slopes possibly by less subsidence in the underlying crust. The basin is deep (~4800 m) along the base of the Dampier Ridge where the seamounts are close. There is evidence for erosion by bottom currents in these areas (Baker et al., 2008). Distinct erosion cha nnels occur at the western foot of the Dampier Ridge where the abyssal seabed between the Dampier Ridge and the Taupo and Barcoo Seamounts is only 50 km wide (Jenkins, 1984). Moats 6–30 km wide and 18–332 m deep also occur at the foot of Recorder Seamount (west side), Britannia Seamount (east side), the Monawai Ridge and an un-named seamount (Jenkins, 1984).

I will continue to expound on the relative deficiencies of wider benthic knowledge such as the aforementioned, the collective debts humanity owes to the services of largest ecosystem on the planet (Ramirez-Llodra et al., 2010, 2011), and the various scientific efforts and societal campaigns to remedy the situation (mostly in Chapter 5). Filling the gaps will be a major interconnecting thread throughout this book. Perhaps we can find evidence that even commercial cruise lines can play some positive roles in increasing and adding to ocean appreciation, supporting outreach, education, research, or bringing more people out to sea. However, the casual cruiser also carries the many creature comforts of *terra firma* with them, with an extra flare of entertainment and sensation, and so they are either in a mall-like cocoon or distracted while sailing. In one notorious commercial cruising misstep in the late 1990s, one of the largest cruise lines, Royal Caribbean (RC), which flew under the Liberian flag, was caught polluting US waters. RC had been illegally releasing tons of waste oil and chemicals in their wake. For this the company was fined a record $18 m fine (£11.4 m) (Martinson, 1999). Another silver lining to this debacle was that as part of their penalties, Royal Caribbean had to install two working atmospheric and ocean laboratories onto to their aptly named 142,000-ton Royal Caribbean flagship, *Explorer of the Seas*. The laboratories were jointly constructed and operated by Royal Caribbean International, University of Miami's Rosenstiel School of Marine and Atmospheric Science (RSMAS), the National Oceanic and Atmospheric Administration (NOAA), Atlantic Oceanographic and Meteorological Laboratory (AOML), the National Science Foundation, and the U.S. Office of Naval Research and would collect data almost every time they went out on to sea (Williams et al., 2002). For the *Explorer*

10   **Chapter 1** The seabed—Where life began and still evolves

of the Seas, only two major routes were travailed during the periods of data collection and recording—either an easterly path to Puerto Rico or a southern route to the Cayman Islands. This would effectively convert the cruise ship into a quasi-research vessel, ship-of-opportunity, sanctioned by the aforementioned partners, which would plow the same tract regularly and frequently. Large amounts of oceanographic data were recorded, justifying a "big science" outlook, and eventually some of these data reached peer review publications. For example, *Explorer's data* displayed significant seasonal variation in carbon dioxide levels that correlated with temperature changes, which allowed further calibrations of carbon dioxide levels with satellites (Marris, 2004; Olsen et al., 2004). Then, Beal et al. (2008) reported between May 2001 and May 2006 that the *Explorer recorded* mean transport of the Florida Current was lower (31.0 ± 4.0 standard deviations—Sv) than that typically inferred by cable voltages (32.4 ± 3.2 Sv) at 27°N over the same period. Therefore, this assuaged doubts that valid robust, scientific data could flow from a good time!

While not an RSMAS or a NOAA employee at the time, the author experienced the benefits of this partnership as one of array of scientists who were offered to serve on a 7-day RC cruise. The primary duties per day were to provide a tour of both NOAA laboratories to interested passengers. This was a fairly easy and pleasant exercise, learning the instruments and process along the way. At the end of the week, seminars were given to the same public who became interested in the guest scientist specific research focus. I lectured on our current focus on *Axinella corrugata* sponges at the time. The RC Explorer saga became an unexpected turn of events, turning lemons in lemonade, and this was fortunate. Working on the surface of oceans can be much easier than working below the waves. I will return to the work of more bona fide research vessels and instruments in the next chapter when we discuss the application of diverse technologies for benthic research and assessments.

Pelagic-benthic connections (and vice versa)

As mentioned, the pelagic oceans compose the largest potential habitat on earth due to their sheer volume, but this does not readily translate into more variable niches. Rather it is the benthos that underlies this volume and truly encompasses more than 70% of the earth's total surface area. Vertical relief and habitat heterogeneity likely contribute to the greater complexity of benthic habitats. For example, greater topographical relief and rugosity, which can reflect more coral cover or fish refugia, are known to have predictive power for biodiversity on reefs. Although macro- and mesoscale features have profound influences on ocean and benthic diversity (Johnston et al., 2019), we will not dwell upon them extensively in this book. When large physical phenomena impinge upon specific benthic subjects, the effects will be invoked. For example, we know the Gulf stream off of Florida or the Loop Current of the Gulf of Mexico or the Humboldt Current along the western coast of South America will influence many different types of biological interactions and distributions. What is less known is how pelagic water column properties fully impact or potentially affect benthic biodiversity distributions.

Regarding the concept of benthic-pelagic coupling (BPC), can the adage "as above, so below" apply? This depends. The water column obviously overlays the benthos that lies beneath, and thus, pelagic processes implicitly connect to the benthic ones. The two habitats are in constant contact. Due to the action of simple gravity, particulate organic carbon (POC) or material (POM) of any substantial weight dropped at the ocean surface, if not affected by strong currents or consumed on the way down, will eventually find its way to or near the benthos. However, this "biological gravitational pump

(BGP) is an important mechanism, while the physical 'solubility' pump acts to sink colder waters at higher latitudes, that help drive deep water currents and carbon storage in the ocean. The BGP has been studied at selected sites using a range of particle interception techniques such as sediment traps (deep-moored, surface-tethered free-drifting, or neutrally buoyant)" (Boyd et al., 2019). Studying the BPC has long been considered patchy at best, likely correlated to the heterogeneity of benthic habitats on the bottom. As mentioned earlier, obtaining samples of the benthos can be difficult and nontrivial especially as one goes deeper. Therefore, specific qualifications and criteria exist to accurately portray and quantify pelagic to benthic connections. As we shall notice again and again, one size does not fit all with respect to many biodiversity assessments, including BPC. Understanding the processes that link pelagic to benthic habitats can be illustrated by reviewing a wide range of examples. In general, many couplings could be viewed as either transitory or constant, up or down, though this will also depend on the locale and situation.

It is well accepted that the ocean plays a major role in carbon cycling for the whole planet. Because of the benthos' overall lower productivity, energy in the form of particulate organic materials (POM) or POC would be welcome in almost any form. The quality and quantity of organic matter descending to the sea floor as potential food depend on the original sources of primary production, phytoplankton sinking rate, zooplankton grazing rate, mixed layer depth, overall water column depth and proximity to land runoff sources (Wassman, 1984; DeVries, 2022). Moreover, carbon/nitrogen (C/N) and stable carbon isotope (^{13}C) ratios can indicate amounts of particulate that reach the bottom of various habitats (Grebmeier et al., 1988). Evidence indicates that deep cold-water corals appear dependent on the primary productivity of surface waters reaching the sea floor via transport and hydrodynamics and for survival (Duineveld et al., 2004; Van der Kaaden et al., 2021). However, estimates indicate that only 0.5%–2% net primary production ever reaches the abyssal bottom, which varies geographically according to levels of primary production and export efficiency in overlying depths; this essentially delineates abyssal and hadal ecosystems as extremely "food limited" (Smith et al., 2008).

BPC may not be one of the most commonly addressed or integrated topics in oceanographic literature, although it is important and growing. BPC is complex, encompassing both physical and biological parameters, possibly even inchoate aspects of life itself. As knowledge accumulates about both types of realms, as well as disparate disciplines such as ecology and oceanography, the combination and synergy will likely drive research momentum, making connections paramount (Lignell et al., 1993; Lovvorn et al., 2005; Rodil et al., 2020). The phenomenon of BPC also should be seen as one common denominator among geographically disparate benthic habitats at relatively the same depth. For example, a coral reef atoll in the South Pacific may share similar BPC variables with Curacao or Cuban reefs in the Caribbean. Also, BPC would be expected to be stronger in littoral zone of shelves compared with deeper zones. Not enough data exist to support this contention yet, but perhaps should be explored further. BPC studies have traditionally focused on the sinking deposition of POM to the benthos, bio-resuspension, and inorganic nutrient release from the sediments (Smetacek, 1985; Graf and Rosenberg, 1997; Raffaelli et al., 2003). Multiple studies support the notion of looking more closely to functional group compositions of the benthos. The groupings can greatly affect ecosystem activities in the context of global biogeochemical cycles (Snelgrove et al., 2018). Fig. 1.2 portrays several parameters of BPC along either physical or biological routes in an effort to show biogeochemical processes interacting with biodiversity on the seafloor. More data and modeling like this could provide a unifying bridge between various types of habitats and biogeographic zones (Crisp et al., 2011; Donovan et al., 2018).

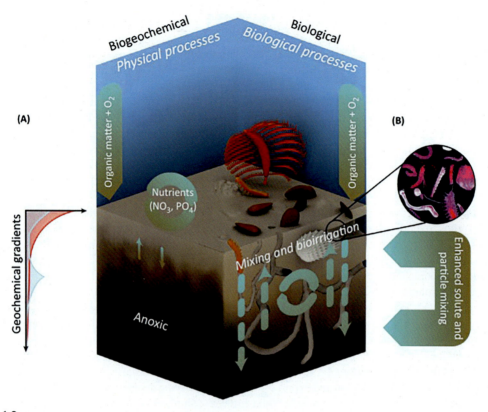

FIG. 1.2

Summary of contrasting geochemical (A) and biological views (B) of organic matter decomposition, illustrating differences in emphasis on the predominant processes and in the relative complexities of the two perspectives.

Reproduced with permission from Snelgrove, P.V., Soetaert, K., Solan, M., Thrush, S., Wei, C.L., Danovaro, R., et al., 2018. Global carbon cycling on a heterogeneous seafloor. Trends Ecol. Evol. 33(2), 96–105.

Perhaps the most obvious example of the coupling can be ascertained through the descent and presence of particulate sediments from the surface waters to the benthos as *marine snow*. These particles, a form of sedimentation, are defined as "mostly organic debris, such as decomposed organisms, fecal matter, and sand, that drift downward through the ocean from the upper levels, serving as a food source for a variety of organisms inhabiting the deeper regions where sunlight does not reach." Alldredge et al. (1998) and others have shown that POC, particulate organic nitrogen (PON), dry mass, and chlorophyll *a* are proportional to the size of marine snow particles. Sedimentation, silt, and turbidity can represent significant problems in shallow waters for some sessile organisms, but conversely viewed as a boon and an avenue for nourishment to organisms at greater depths. For example, the movement of sediments by disturbances or currents at depth can disperse minerals, dissolved organic matter (DOM), or microbial communities. Biodiversity within sediments likely depends on its major physical parameters, which enable the particulates to represent its own ecosystem—composition, sediment grain size, layering, oxygenation, etc. As a grain of sand can be coated with scores of microbes, natural sediments, such as mud and fine sand, harbor diverse microbial biofilms. D'hondt et al. (2015) have found that aerobic microbial communities found in deeper sediments indicate the presence of oxygen "throughout the entire sediment sequence in 15%–44% of the Pacific and 9%–37% of the global sea floor. Subduction of the sediment and basalt from these regions is a source of oxidized material to the mantle." Although much shallower, in the author's laboratory, a recent descriptive study compared sediment microbial communities and their habitats (e.g., "microbiomes") within a sheltered major port (Port Everglades) and adjacent natural reefs beyond the port. Dredging in the port was by hopper and clamshell methods but only for short periods of time. Over a 2-year period, the microbiomes offshore displayed more dynamic, shifts in spatial community structure ostensibly due to greater wave action and currents offshore (Krausfeldt and Lopez et al., 2023).

Multiple studies in deeper waters address various physical and biological processes that influence BPC. In general, deeper waters (>750 m) did not exhibit unequivocal benthic-pelagic coupling or community-level trophic diversity because sinking organic matter most likely can get intercepted before reaching the seabed (Stasko et al., 2018). For biological processes, marine microorganisms remain central to coupling mechanisms (Acinas et al., 2021). Tanet et al. (2020) interestingly posit that ubiquitous bioluminescent bacteria in the water column or within symbionts transfer to sinking marine snow and fecal pellets. However, transfer to the seabed may be short circuited by coprophagy, causing variability in carbon transfer. Other studies point to physical processes such as midwater mixed layer pumps (physical) and carbon dioxide solubility that can modulate overall carbon sequestration (Boyd et al., 2019; Buesseler and Lampitt, 2008; Dall'Olmo et al., 2016; National Academies of Sciences, Engineering, and Medicine (NASEM), 2021). Photosynthesis, which captures CO_2 from the atmosphere and converts the gas to usable organic carbon at the surface, composes the primary biological carbon pump. A second pump (or sink) and coupler consists of detrital carbon or sinking particles (larger than 0.5 mm of diameter) known as marine snow, are a combination of phytodetritus, living and dead organisms, and fecal pellets (from zooplankton and fish) from the surface layer to the dark ocean and its sediments by various pathways (Siegel et al., 2016). Various studies have indicated nitrification through ammonia and nitrite oxidation as the main sources of energy for carbon fixation in the dark ocean (Alonso-Sáez et al., 2012; Könneke et al., 2005; Pachiadaki et al., 2017; Wuchter et al., 2006). Marine snow (POC), rich in carbon and nutrients, and its surrounding solute plumes are hot spots of microbial activity in aquatic systems (Alldredge and Gotschalk, 1990; Alldredge and Silver, 1988; DeLong et al., 1993). However, seasonality of particle flux into the deep sea has been observed several times and probably linked to variable detrital loads (Billett et al., 1983; Thiel and Schriever, 1989; Durden et al., 2019).

The effects of turbidity on reef health or the responses to sedimentation by the reef community, depending on how we want to view the phenomenon, appear variable. The effects will depend on the total amount of turbidity, which blocks irradiation for primary production and overall composition of the reefs in question. Even counterintuitively, in some instances, the decreased irradiance may enhance coral survival by reducing the effects of warming sea surface temperatures (Sully and van Woesik, 2020). For example, a study of submerged patch reefs (Eve's Garden, Anemone Garden and Siwa Reef) in the Miri-Sibuti Coral Reef National Park (MSCRNP) at about 5–15 m depth in Borneo was conducted due to expected higher sedimentation levels created by increased local logging and deforestation (Browne et al., 2019). The nearby Baram River (10 km north of the reef complex) was recorded to discharge $2.4 \times 10^{10} \, \text{kg yr}^{-1}$ of sediments. The MSCRNP is not species rich, but still showed healthy coral coverage ranging from 22% to 39% of encrusting forms of *Diploastrea*, *Porites*,

14 **Chapter 1** The seabed—Where life began and still evolves

Montipora, Favites, Dipsastrea, and *Pachyseris* corals and some bleaching stress during the study period from October 2016 to October 2017. More importantly, however, were the documented signs of resilience, indicating that this community had stabilized in the face of chronic turbidity.

These studies have implications for biodiversity occurring in high turbidity and sedimenting environments. It is likely that microbial and meiofaunal components at these sites may affect overall oxygen profiles. Sediment biota generally includes bacteria, meiofauna (metazoa+Foraminifera) and macrofaunal, so these should be less negatively affected by new sedimentation. More discussion of meiofauna will follow later.

Deep-sea sediments act as a huge reservoir of organic carbon with an estimated $\sim$13% of the total global bacteria living in the upper 10 cm (Jørgensen and Boetius, 2007). However, molecular-based metagenomic data of deep sediments remain scarce compared with coastal habitats. Lower layers hold both aerobic and anaerobic bacteria. In a recent study, the core microbiomes that were characterized at the Mariana Trench along the abyssal-hadal transition zone were studied (i.e., $\sim$5455–6707 m) (Jing et al., 2022). *Woeseia, Gemmatimonas,* and *Nitrosopumilus* appeared as the dominant prokaryotic genera within the core microbiome at all depths.

Sedimentation should also be viewed as one other type of BPC and occurs more in coastal areas or where wave action and perturbation is high (Viles and Spencer, 2014; Krausfeldt and Lopez et al., 2023). Sedimentation could also occur at greater depths with strong currents or turbidity, but with less evidence due to limited monitoring capacities. Nonetheless, how natural (riverine plumes) or human activities (suspended sediments via coastal dredging) can generate large sedimentary plumes is well documented (Delandmeter et al., 2015). Many examples of the former exist around the world, showing how nutrients are introduced to delta or coastal areas. They vary based on the size and location of the river as well as season (Cooley et al., 2007). For example, most rivers deposit sediments at the river mouth. In contrast, Amazon River empties into a turbulent Atlantic Ocean, which can disperse most of the sediment. Nonetheless, the Amazon produces a very large plume composed of tons of DOM and inorganic particulates that create a breeding ground for phytoplankton, which absorbs $\sim$5.8 g C m^{-2} yr^{-1} atmospheric CO_2 (0.014 Pg C yr^{-1} over $15 < S < 35$ across approximately 2 million square kilometers) (Körtzinger, 2003).

The extent of the Amazon River plume on pCO$_2$ can reach 3000 km east to 25°W during peak autumn surface flow of the North Equatorial Countercurrent (Lefèvre et al., 1998), and large amounts of sediments likely reach the seabed. In contrast, in the Congo River plume, net community production (NCP) enhanced the mixing-driven air-sea CO_2 gradient (Cooley and Yager, 2006). Coles et al. (2013) combined seasonal CO_2 flux estimates with satellite-based plume areas to estimate the total annual plume CO_2 uptake that includes observed variability. Freshwater plumes, such as those at the Mississippi river, can travel hundreds of kilometers offshore with likely widespread sedimentation effects (Payandeh et al., 2021). Overall river plumes contribute to BPC in variable ways.

Lastly, because most of the abyssal benthos may be carbon- and nitrogen-limited, the occasional falling of carcasses should be discussed as a BPC. Early spectacular examples of whale falls were discovered in the Monterey canyons from 382 to 1820 m by Robert Vrijenhoek's group using the Monterey Bay Aquarium Research Institute's (MBARI) ROVs *Tiburon, Ventana,* or *Doc Ricketts* (Lundsten et al., 2010). Whale falls have revealed new, albeit temporary ecosystems, which "pop up" when large morsels fall from the surface waters (Smith et al., 2015). Although deep-sea scavengers such as octopus, hagfish, and sharks may be quick to find a freshly sunken carcass, some can take decades to be fully consumed and decompose. Indeed, after the relatively quick digestion of soft tissues,

Pelagic-benthic connections (and vice versa) **15**

harder whale bone decomposition requires action of bone-consuming microbes and metazoans, including the worm *Osedax* (Rouse et al., 2004). The latter is an important deep-sea, symbiotic host, and requires sulfate-reducing bacteria for the initial burrowing into the bones. Once established, the worms appear as red plumes on top of the bare bones. *Osedax* worms rely on chemosynthetic bacteria held in their trophosomes for energy, as *Osedax* lacks a digestive tract. Goffredi et al. (2005, 2007), Fujiwara et al. (2019), and Osman and Weinnig (2022) have identified the symbiont as an Oceanospirillales that can produce enzymes to hydrolyze collagen and cholesterol, which sustain the worm. It appears that larvae may be suspended abyssal waters until a carcass appears. Thus, living on the ocean seabed requires a fair amount of opportunism for survival. This also provides another powerful example for the importance of symbiosis across multiple habitats and situations, as well as the patchiness of food in the deep sea.

Perhaps at this juncture, we can imagine the invention of novel underwater technologies, such as miniaturized, inexpensive trackers that are also nonpolluting and biodegradable. Their primary utility would be to follow and report back to land-based researchers the movement of small particles mimicking POC at depth and illustrate the tracks of midwater currents, eddies, and deeper convection patterns. In this context, phytoplankton composes the largest source of organic material that can deposit to the ocean floor (Tamelander et al., 2008). If they could emote, benthic inhabitants would show a great deal of gratitude to various plankton—diatoms and radioloaria and other productive surface dwellers that may eventually sink to the benthos. For example, abyssal sediment in waters shallower than 4000m in equatorial to temperate regions are composed primarily of the calcareous shells of foraminiferan zooplankton and of phytoplankton such as coccolithophores. Some of these may be naturally benthic residents, as most foraminiferans. One interesting foraminiferan is the xenophyophore, whose name means "bearer of foreign bodies." Occurring at abyssal-to-hadal depths (3500–10,600m), the xenophyophores display some of the largest unicellular organisms with multiple nuclei, reaching up to 20cm (8in.) in diameter (Gooday et al., 2011).

Microplastics represent a recent, synthetic yet undesirable substitute of marine snow. Carbon is in the composition of many of the microplastics, but often in an unusable form for benthic organisms. Therefore, microplastics are more of an accumulating pollutant than a beneficial nutrient supplement (Egbeocha et al., 2018; Shen et al., 2020; Lim, 2021). At this stage, it is still probably too early to know the full effects of microplastic dispersal and distribution across the water column and benthic habitats. Understandably, we will likely observe the effects on shallow areas first. However, awareness and studies of this unique pollutant continue to increase for good reason (Foley et al., 2018; Trestrail et al., 2020). We will return to this issue more in Chapter 4.

In recent years, sessile marine sponges have been shown to have large effects on nutrient cycling. Seminal studies (de Goeij et al., 2013, 2017) elucidated the DOM–sponge–fauna pathway. In this and a follow-up paper, the authors point out that "DOM represents the largest source of organic matter (>90% of total) on reefs and is operationally defined as organic matter passing a fine filter GF/F with $\pm 0.7\,\mu m$ pore size or polycarbonate filter with $0.2\,\mu m$ pore size." In addition, the exact role of microbes living within sponges was set up as a target for future studies. Into this area, Freeman et al. (2021) have reviewed how sponge microbial symbionts contribute to host ecological divergence and convergence, nutritional specificity and plasticity, and allopatric and sympatric speciation.

The transfer of coral-derived DOM to detritovores in the Sponge Loop hypothesis fits into the wider scope of BPC (Rix et al., 2018). For example, marine sponges certainly represent a taxon that could provide a coupling of pelagic and benthic processes via biomineralization. This is firstly accomplished

16 **Chapter 1** The seabed—Where life began and still evolves

by their ability to convert free silica into spicules for their skeletons. Sclerocytes perform these functions. Maldonado et al. (2021) have been spearheading quantitative studies to demonstrate these conversions. Secondly, sponge directly filters water and converts DOM into solid benthic biomass available for other reef fauna. A 1 kg sponge can filter up to 24,000 L of seawater per day removing POM and bacterioplankton (Vogel, 1977).

Biomineralization is a trait reflected by other benthic organisms, such as bivalves and other shelled molluscs, and which is a direct example of BPC. The creation of crystal arrangements, organic matrix components (proteins) from calcium carbonate crystal structures has variable origins (Addadi and Weiner, 2014). For example, molluscan proteins p10 and pfN23 in the nacre and chitin have been shown to be important in crystallizing the prismatic layer of the pearl oyster, *Pinctada imbricata fucata* (Bean et al., 2020).

Oyster (Phylum Mollusca) bed reefs represent an easily perceptible example of BPC. The connection is based on the typical oyster lifestyle of filtering water for plankton, bacteria, dissolved organic matter, nitrates, ammonia, and phosphates. Individual oysters have filtration rates of up to 190 L (50 US gal) per day though this will vary with species. The filtering activity of large populations of species such as Atlantic *Crassostrea virginica* was estimated to clear the waters of Chesapeake Bay in a matter of days back in colonial times. In James Michener's historical novel, *Chesapeake,* the narrator described how indigenous protagonists cultivated individual oysters, which reached sizes of a foot or more. Oysters in this habitat were also considered an essential foundation species (Baird and Ulanowicz, 1989).

However, because of habitat loss, pollution, disease, and the oyster's popularity at the dinner table, the loss of oyster reefs eventually occurred in Chesapeake Bay (as well as across multiple locations) and is well documented (Jackson et al., 2001). For example, as oyster harvesting has continually escalated after American colonialism, productivity would never fully return to their peaks (Kirby, 2004; Reeder-Myers et al., 2022). About 1.7 billion individual oysters (1.05 billion lbs.) were reportedly harvested in 1891–1892 alone, when this fishery was truly a staple of indigenous as well as colonial settlements. By 1930, the Chesapeake oyster harvest declined significantly to less than 40 million pounds, less than 20 million pounds in the 1980s, and only a few million pounds in the 1990s (Kennedy, 2018). Carranza et al. (2009) indicated that perhaps 85% of the 19th-century oyster reef fisheries have been lost by the early 21st century.

It is unfortunate that modern societies have not embraced more sustainable practices for many fisheries or agriculture, as espoused by Reeder-Myers et al. (2022) and other authors (Graeber and Wengrow, 2021) because this allowed oyster beds to persist for 10,000 years or more, albeit with much smaller human populations. These early societies also had their own commerce and effective systems to manage their natural resources. The evidence of this bygone resource can be seen in the piles of used shells or middens scattered along coastlines around the world. We can envy what BPC and resilience the earlier societies may have witnessed. Lastly, a model is being developed using oysters as indicator species to assess hydrodynamic variables (water depth, speed, salinity, temperature, and vertical mixing) and overall coastal ecosystem health (Ahn and Ronan, 2020).

In a recent study, Griffiths et al. (2017) focused on the Baltic Sea for one example of coupling, which could provide lessons in the wider context, as current knowledge on the exchange of inorganic nutrients and organic material between benthic and pelagic habitats was summarized. The study reiterated the impact of excess nutrient flows in the Baltic Sea, which relate to oxygen usage, long considered a vital environmental regulator across the sediment-water interface (Conley et al., 2009; Norkko et al., 2015).

Nutrients can also flow in the opposite direction of sinking marine snow, i.e., upward to the surface in a few situations. For example, hot smokers at thermal vents release energy and minerals into the water column at spreading deep ocean ridges. This is a case of productivity starting from the bottom, and moving upward and around, as hydrogen sulfides, methane, and carbon dioxide gases support the diverse symbiotic chemosynthesis of the microbes that live in various invertebrates such as the vestimentiferan *Riftia* worms. We will discuss symbiosis in more detail later.

Buoyancy provides the physical lift for some materials and a temporary departure of biomass from the seafloor. Coral mass spawning events fit this bill. Nutrient availability after mass spawning events has been recorded to induce up to a fourfold increase in pelagic production 1 day after spawning. The spawning material can be incorporated into the sediments of coral reef lagoons either directly or via sedimentation (advective-driven percolation) (Glud et al., 2008). For example, upon release from a gravid broadcasting coral colony, hermaphroditic egg-sperm bundles slowly float to the surface where they burst. The sperm are then free to swim and fertilize coral eggs from other colonies, buoyed by a rich lipid interior. Spawning often occurs only once a year, either in August or September in the northern hemisphere, and is synchronized by sea surface temperature and moon phases. This event can initiate multiple chains of trophic interactions, such as fish feeding on the coral spawn, then defecation of labile carbon pellets to the reef sediments. Alternatively, phytoplankton blooms and benthic microalgae (BMA) are stimulated by the sudden nutrient releases of the primary spawning event. For example, Eyre et al. (2008) carefully studied coral spawning events as large episodic infusions of nutrients from the benthos to pelagic. To assess the effects of coral spawning on total nitrogen (TN) and phosphorous (TP) cycling in a coral reef ecosystem, the authors measured diurnal changes in water column nitrogen (N) and phosphorus (P), light and dark benthic N and P fluxes, and denitrification immediately before and daily for 7 days after coral spawning on a Heron Island reef flat, Australia. They found that the input of N and P via the deposition of coral spawn and associated phytodetritus resulted in large changes to N cycling in the sediments, but only small changes to P cycling. Most of the TN consisted of dissolved organic nitrogen (DON) (average 71%) and particulate nitrogen (PN) (average 20%). Ammonium (NH_4), nitrate (NO^3), and nitrite (NO2) concentrations were mostly below 0.60, 0.40, and 0.20 mmol/L respectively, and combined (dissolved inorganic nitrogen, DIN) only made up a small proportion of the TN (average 9%). Dark N2 fluxes were generally the largest of the N fluxes, making up on average 54% of the total net benthic dissolved N flux (on the basis of dark N2 fluxes only). The rapid response of BMA and denitrification to the increased supply of N postmajor spawning illustrate that N was a limiting factor for both coupled nitrification-denitrification and ammonification. Alarmingly, recent evidence shows some coral disruption in spawning synchrony (Shlesinger and Loya, 2019; Fogarty and Marhaver, 2019).

Lastly, we cannot leave the topic of BPC without mentioning global thermohaline circulation (THC) (Via and Thomas, 2006). Global circulation patterns are complex and affected by factors such as tidal forces and wind, as well as salinity, air and water temperature. The THC phenomenon was once thought to drive a mesoscale global ocean conveyor belt (OCB), a term coined by climate scientist Wallace Smith Broecker to describe an earlier model of oceanic circulation whereby high-latitude colder waters descended down continental slopes, to incessantly flow toward the equator (Stommel, 1958). This *overturning* begins with downward flow of cold arctic waters, which also cause a counterflow "at the surface in an opposite direction," thus creating a continuous conveyor belt-like motion (Warren and Wunsch, 1981; Broecker, 1987; Via and Thomas, 2006). This model also provided transport and specific structures of massive water bodies in the ocean, which could in turn affect biogeographical partitioning. In more recent years, the conveyor belt paradigm has been modified, as most of

18 **Chapter 1** The seabed—Where life began and still evolves

the subpolar-to-subtropical exchange in the North Atlantic occurs along interior pathways while the OCB breaks up into eddies at 11°S (Dengler et al., 2004; Bower et al., 2009). New analyses also showed that it is more accurate to refer to the OCB as a meridional overturning circulation (MOC) (Lozier, 2010).

Thus, THC and MOC phenomena have distinct, multiple effects on physical oceanic processes, which in turn likely affect benthic ecosystems. Due to the sheer size and multiplicity of these water masses, integrating specific BPC effects is not easy. However, physical oceanographers also understand that THC can bring large amounts of organic materials to the seabed with the sinking colder waters, while also deep ocean upwelling conversely brings nutrients back to shallower pelagic waters in some coastal zones. Future oceanographers and marine biologists could find models to integrate THC into benthic biogeographic mapping, which could also be digestible for the wider public.

Connectivity within benthic species

With the description of some of the physical processes, which link both vertical and horizontal zones, and which can delineate where some benthic organisms will live, we can also see how biogeographic regions can be drawn. A logical next step would be to review in more detail how the biological processes of connectivity could affect BPC and help resident benthic organisms thrive. We should also revisit some first principles when discussing biodiversity and connectivity. These can be found in the realm of evolutionary biology and origins (Futuyma, 2017). For example, diversity is necessary in order to help drive the mechanism of natural selection to enable the rise of more adaptive phenotypes.

Intuitively, most oceanic waters are physically connected, and we have just described better known cases of massive water movements. Thus, describing potential biological barriers in marine waters requires physical data, its careful analyses and nuance. Maintaining genetic connectivity between living organisms, from the individual to the population scale, is a vital, yet sometimes subtle principle. Individuals belong in populations, whether they may be dense or widely dispersed. However, when small, isolated populations lose contact with larger populations, genetic consequences can arise with detrimental effects such as excessive inbreeding and consequent loss of genetic diversity in the "founder population" (Mayr, 1970). This original population may eventually also go through a genetic bottleneck during environmental stress and near extinction events. In the sea, genetic connections between populations would seem inherently easier, especially when an organism can swim, or its life history involves wide dispersal to different locations. Sessile organisms without or limited dispersal mechanisms could show greater negative effects of isolation, e.g., loss of genetic variation (e.g., heterozygosity) and connectivity. In this context, along with growth and reproduction, organismal dispersal capabilities have been considered as one of the most important life history traits affecting population structure, gene flow, and hence species evolution (Mora and Sale, 2002).

The concept of *gene flow* stems from population genetics theory and refers to the transfer and cross talk of genes (alleles) within a single species and their inherent variability from one population to another (Hellberg, 2009). The transfer typically occurs via random interbreeding or panmixis (the ideal situation). In addition, migration between different populations can prevent differentiation and increase homogeneity within a species. Genes obtain variation by random mutations, and these generate allelic forms in outcrossing eukaryotes. If some alleles are adaptive, they represent useful variability, which can provide advantages to species. Organisms that can mate and exchange gametes and their genetic material recombine into new configurations (haplotypes) in progeny and successive generations. By contrast, single-celled asexual organisms such as bacteria and archaea simply divide by binary fission and thus are not obligated to find a mate and recombine genes. This can lead to the phenomenon of "Muller's Ratchet," whereby haploid organisms cannot recombine genomes to eject potentially deleterious genetic mutations. These then may eventually accumulate and increase genetic load (deleterious) on the organism (Andersson and Hughes, 1996).

Gene flow can be blocked or diminished by various factors, leading to potential population subdivisions, followed by possible incipient speciation. Factors that could halt or slow gene flow between individual organisms (or gametes) include limited dispersal abilities, changes in currents, fecundity, or the sudden or gradual erection of physical barriers (natural or manmade) between populations. This physical separation may ultimately lead to allopatric speciation.

Endemism, the state of being native to or limited to a certain region, is a feature related to this discussion of connectivity. Endemism implies the halt of connectivity from an ancestral lineage, subsequent isolation, and successful speciation. Ecological metrics for endemism include species richness or alpha diversity as measured by the Shannon or Simpson Indices (Samadi et al., 2006; Fedor and Zvaríková, 2019). Areas that display a high number of endemic marine species are considered valuable and worthy of conservation, but the processes that lead to this abundance should be queried as to how this could occur, e.g., barriers to gene flow or inherent low mobility. We will return to the concept of endemism especially with respect to classifying some benthic sites as biodiversity "hotspots." Low dispersal can lead to the phenomenon of "species flocks," which are a group of closely related taxa, which have remained and evolved within a proximal area to the point of endemism (Bowen et al., 2020).

Many benthic invertebrate adults have a sessile (immobile) lifestyle (sponges, corals, crinoids). But this should not be mistaken as sloth. Diverse interactions and genetic implications can arise from apparently stationary organisms, which concomitantly may be expected to have significant population structures. However, some species can also continue to disperse genes from their fixed sites and have positive gene flow if they "broadcast" gametes, as in massive spawning events or produce swimming larvae that can travel reasonable distances that connect populations (Palumbi, 1992). Barriers also include behaviors, which lead to premating isolation, such as disparate timing of gamete release (Levitan et al., 2011). Alternatively, gamete incompatibilities (e.g., nonrecognition of species-specific receptors on sperm and eggs) preventing fertilization, limited hybridization or hybridizations yielding unfit progeny would encompass postmating barriers to gene flow (Palumbi, 2009).

Establishing forest corridors that allow plants and animals to connect with each other has been practiced for some time in terrestrial systems after the recognition of an "island-like" dilemma in some nature preserves (Diamond, 1975; Hubbell, 1979), similar ideas to address and preserve connectivity were applied to protecting endangered coastal habitats. Roberts (1997) promoted the idea of connectivity for conservation purposes and with regard to protected coral reef preserves and their inherent patchiness. The ideas have been applied widely to establishing marine protected areas worldwide and with other keystone fauna to protect (fish aggregation zones). On an ecological and management scale, demographic connectivity represents the frequent (i.e., weeks to years) exchange of individuals within a metapopulation. This is a fundamental ecological process affecting the management of marine fisheries and protected areas (Kritzer and Sale, 2004).

O'Leary and Roberts (2018) continued this discussion and also recognized the difficulties of linking midwater and benthic processes along the vertical axis. Many examples of vertical connectivity of species that can transit between shallow and deeper habitats exist, but there are equally other

20 **Chapter 1** The seabed—Where life began and still evolves

examples of failure when it would be expected (Van Oppen et al., 2011). O'Leary and Roberts (2018) also noted the double-edged sword of "high levels of open ocean connectivity suggest that there will be greater difficulties in assigning benefits to particular MPAs in this environment." We will revisit MPA strategies in more details in subsequent chapters.

Several examples of studying connectivity appear in the literature. For example, Morrison et al. (2011) measured genetic connectivity with nine microsatellite DNA markers in *Lophelia pertusa,* the dominant framework-forming coral on the continental slope in the North Atlantic Ocean. Samples in this study had wide geographic ranges from the western N. Atlantic, Eastern North Atlantic Ocean (ENAO) Gulf of Mexico (GOM), to seamounts off of New England (9000 km) as well as depth ranges from 140 to 1679 m. The timing of spawning appears to show that this species has the capacity to reproduce across this range. The results indicated reduced gene flow causing significant structuring of *L. pertusa* populations, which exhibited patterns akin to isolation by distance (IBD) and possible vicariance events or regional adaptation. New England seamounts showed the lowest heterozygosities and allelic richness of the compared groups. The extent of differentiation across most sites was unexpected, especially between the proximal GOM and southeastern US (SEUS) regions, and which yielded the most samples. Vicariance events due to past glaciation and climactic events were used to explain this variation.

Connectivity in relation to the important Caribbean spiny lobster (*Panularis argus*) fishery was described by Kough et al. (2013). This genus can release over a million eggs at the time of spawning. The authors used connectivity modeling to estimate the connectivity between different marine regions and assess the potential impact of management actions on the dispersal of lobster larvae. For example, by applying Lagrangian Stochastic, Physical Oceanographic, GIS-based Benthic models, and larval biology, a view of *P. argus* larval connectivity emerged and indicated local recruitment and the value of MPAs in larval transport. Populations in the Bahamas, Cuba, Nicaragua, and Venezuela mostly self-recruited. The authors were surprised to see that the models pointed to lobster larvae utilizing open ocean, pelagic nurseries as suitable habitat for larval development. In this context, areas such as the Caribbean basin had high mean flows, which could thwart connective corridors.

Although most fish are highly mobile, debates about how far and wide some "reef fish" species can disperse and connect across vast ocean biogeographic provinces has continued (Horne et al., 2008). Several extensive reviews on this topic are available, and so I will only highlight some of key and current take-home messages (Mora and Sale, 2002; Floeter et al., 2008). For example, some reef fish species can display significant population structure and limited gene flow, because they tend to spawn on or near the substrate, exhibit parental care strategies, or lack long-range planktonic larvae. Floeter and colleagues carried out an extensive survey of 2605 reef fish species across 25 areas of the Atlantic and southern Africa. The western Atlantic (WA) and southwest Indian Ocean (SWIO) displayed the highest levels of species richness, though the WA also had the highest levels of endemism. As seen in other reef fish studies, philopatry (self-recruitment to natal sites) contributed to local species richness. Broad patterns of fish diversity were found to be affected by several barriers to dispersal—e.g., major historical (closer of the Tethys sea 15–20 million years ago [Ma]) or tectonic events (closure of the Isthmus of Panama about 3 million years ago) and softer barriers such as the mid-Atlantic 3500 km expanse, freshwater outflows from the Orinoco and Amazon rivers, which divide Brazilian and Caribbean reef habitats, and the cold Benguela current, which isolates tropical Atlantic reef fauna from those in the southern Indian Ocean beginning after the late Pliocene/early Pleistocene (~2 Ma).

Lastly, the laboratory of Brian Bowen has spent much of the effort studying the connectivity of various reef organisms, which includes many reef fish species (Rocha et al., 2005; Craig et al., 2007; Eble et al., 2011; Tenggardjaja et al., 2014). This group of fishes generally encompasses species that live in shallow (< 100 m) tropical/subtropical waters, has generally short distance dispersing larvae, and consistently associates with benthic substrates such as coral pavement, algal, or rocky reefs. These are the groups that help define benthic biogeographical zones and regions, due to the high levels of endemism. In a study of the threespot chromis, *Chromis verater*, which inhabits shallow and deeper (mesophotic) reefs, Tenggardjaja et al. (2014) hypothesized exchange between shallow and mesophotic reefs based on the abundance of juveniles at depth. Mesophotic reefs exist in "the twilight," between sunlit (above 200 m) and aphotic zones (below 1000 m). Using well-established mitochondrial DNA sequences, cytochrome *b* and the hypervariable control region (CR) as markers, high levels of genetic connectivity were found between shallow and mesophotic populations of *C. verater* in the Hawaiian Archipelago. These results have implications for the "deep reef refugia" hypothesis, whereby species can shelter in relatively more stable environments during periods of greater disturbance or storms at shallower depths (Hoegh-Guldberg et al., 2008). The *C. verater* findings represent one positive example of the capacity to replenish populations vertically.

How common is benthic continuity and cosmopolitanism?

If we ask which the most widely distributed taxa on the planet are, one could list several, plausible candidates. On land, apex predators such as lions (*Panthera leo*) historically roamed widely through Africa, most of Eurasia, and North America during the Late Pleistocene (100–10 kyr ago) and would still be in those places except that they coincided and encountered humans (Barnett et al., 2006). Across the oceans, apex predators (sharks, dolphins) and many pelagic taxa would certainly qualify as widespread (phytoplankton or swimming crustaceans, fish, or cetaceans). These organisms have the inherent advantage of wide dispersal in their life histories. For example, the far-ranging migrations of megafauna such as baleen whales (fin, humpback, blue) and great white sharks (*Carcharodon carcharias*) are well known (Bonfil et al., 2005; Hoelzel, 1998; Hucke-Gaete et al., 2018). The moray eels (family Muraenidae) exemplify this capacity as well as any species, with high gene flow between populations far flung across two-thirds of the planet from Africa to Central America (Reece et al., 2011). However, if we narrowed down the question to which natively benthic group is the most widely distributed, examples are less common. To be *cosmopolitan*, a species or genus should have the capacity to spread and establish widely, which contrasts sharply with endemism. Cosmopolitanism entails the possession of sufficiently high genetic diversity and heritable phenotypic plasticity, which would allow the species to successfully respond to environmental variation and heterogeneity across benthic habitats. In this context, could a continuum of some benthic taxa be possible (natural or human-facilitated invasions), as considerable barriers to gene flow exist (Hutchings and Kupriyanova, 2018)? To continue this line of enquiry, we could ask "which taxa are found broadly across wide oceanic expanses, both shallow and deep, represent a continuum and could be used in comparative or biogeographical analyses"? Several ubiquitous intertidal mussels such as *Mytilis galloprovincialis* or *Mytilis edilis* have been described as cosmopolitan (McDonald et al., 1991; Murgarella et al., 2016). Shallow-water ascidians *Ciona intestinalis* and Caribbean *Didemnum cineraceum* may also qualify with six other ascidian species as possible cosmopolitan, although dispersal by cargo shipping was recognized as a factor

22 **Chapter 1** The seabed—Where life began and still evolves

(Monniot and Monniot, 1994). Zeppilli et al. (2011) also described new species of deep-water nematodes in the genus *Manganonema* (Nematoda: Monhysterida) from different Atlantic and Mediterranean regions at depths ranging from 567 to 4997 m. *Manganonema* has been found inhabiting all oceans (Atlantic, Indian, Pacific, Arctic, and Antarctic) and thought to be transported via turbidity currents and benthic storms over long distances. Deep sea hydrozoans in the 247 species-rich *Stylaster* genus appear fairly widespread in the western hemisphere and could be tested for cosmpolitanism (Cairns, 2011). Benthic species may show large-scale population structures that resemble pelagic species patterns. Even with no evidence of gene flow between isolated populations, benthic species may yet be useful for assessing the health and status of their respective habitats. However, very few benthic taxa or species are truly cosmopolitan and "found everywhere." This deficiency can limit the comparative approaches between regions. We will discuss more possible candidate benthic taxa, such as some meiofaunal species (amphipods, nematodes), macrofauna (deep-sea corals sponges; benthic fishes), citing specific examples again in later chapters in the context of their presence at specific diversity hotspots or as potential remedies to biodiversity problems and assessments. Even with or without genetic connectivity, closely related species, guilds, and complexes (Knowlton, 1993; Lopez et al., 1999b; Bowen et al., 2020) could still be considered as effective resources and targets for benthic assessments. Species may yet be advanced through closer genetic scrutiny, identifying the best indicator taxa, their amenability for comparisons and the various genetic landscapes that may still connect them (e.g., conserved sequences, runs of homozygosity). The genetic characterizations could then be integrated with physiological and ecological parameters to assess or model responses to specific environmental variables (Darling and Carlton, 2018; Rainbow and Phillips, 1993; Monaco et al., 2019).

At this juncture, we can touch upon *meiofauna*. By definition, these are organisms that form the distinct size class between 1 mm and 1 μm, making up many species invisible to the naked eye. This group encompasses many extant invertebrates, which tend to thrive as inconspicuous, having small body sizes, barely visible at less than 1–2 mm, such as many amphipod and annelid species. Marine benthic sediments are replete with meiofanua and should represent the next frontier of biological research, which encompasses not only genomics and ecology but also basic taxonomy—how to classify new species and organisms. Only recently has greater attention turned to characterizing meiofauna in various sediment habitats. This group of animals helps decompose and recycle nutrients. In general, the majority of meiofauna are concentrated in the top layers of benthic sediments (Zeng et al., 2018); however, many meiofauna organisms do not have a pelagic state. This static condition makes many meiofauna excellent candidates to reflect local environmental conditions (Alves et al., 2015). Moreover, this feature creates many opportunities for characterizing endemism. By their very nature, the diminutive meiofauna will have low biomass, so we would not expect them to have major effects on ecosystems. But by now we should learn that judgments based on cell or organismal size alone should be taken with caution. For example, single-cell algae or phytoplankton affect the atmosphere by producing at least 25% of of the planet's oxygen. This rivals the primary production derived from terrestrial plants. And what ecosystem services (tangible benefits for provided by organisms living in a community) could meiofauna actually carry out? As mentioned previously, besides elemental recycling, meiofauna can produce marine natural products (Avila and Angulo-Preckler, 2021).

Meiofaunal species would possibly make good candidates to elevate as "cosmopolitan" indicators across various benthic habitats. We should keep meiofaunal studies and specific taxa in the forefront with regard to biodiversity assessments. Moreover, taxonomic and parataxonomic identifications of meiofaunal could itself provide work and job security (in the rhetorical sense) and elevate mundane yet ubiquitous marine sediments and their biodiversity as diversity hotspots. To this end, novel courses such as "Biodiversity and Integrative Taxonomy of Invertebrates" taught by Kevin Kocot and Gustav Paulay at Friday Harbor Laboratories provide valuable training opportunities for the next generation of experts (https://fhl.uw.edu/courses/course-descriptions/course/integrative-biodiversity-and-taxonomy-of-invertebrates/). The array of fauna that could comprise typical marine meiofaunal residents include up to 19 of 33 metazoan phyla: Platyhelminthes, Nemertea, Gnathosomulida, Kinorhyncha, Loricifera, Nematoda, Rotifera, Gastrotricha, Priapulida, Sipuncula, Annelida, Arthropoda, Tardigrada, Mollusca, Bryozoa, Brachiopoda, Chordata, and Echindermata (see Table 1.1). Ongoing

Table 1.1 (a) Examples of extant benthic single-celled taxa and (b) extant metazoan phyla.

Phylum	First evolutionary occurrence (MYA)	Approximate number of species
(a)		
Chromista	~350[a]	50,000
Coccolithophores	230	200
Foraminifera	1150	10,000
(b)		
Xenacoelomorpha	No fossils known	400
Annelida	518	20,000
Arthropoda	555	1,200,000
Brachiopoda	529	440
Bryozoa	480	6300
Cephalochordata		35
Chaetognatha	535	130
Chordata	539	61,000
Cnidaria	560	13,300
Ctenophora	525	130
Cycliophora	No fossils known	2 (+2 undescribed)
Dicyemida	No fossils known	120
Echinodermata	500[b]	7300
Entoprocta	No fossils known	400
Gastrotricha	No fossils known	440
Gnathostomulida	n/a	100
Hemichordata	Also Burgess Shale[c]	130
Kinorhyncha	270	300–341
Loricifera	late cambrian	38
Micrognathozoa	No fossils known	1 (+1–2 undescribed)
Mollusca	541–485.4	73,000
Nematoda	396[b]	27,000
Nematomorpha	100	365
Nemertea	n/a[d]	1300
Onychophora	300[e]	200

Continued

24 Chapter 1 The seabed—Where life began and still evolves

Table 1.1 (a) Examples of extant benthic single-celled taxa and (b) extant metazoan phyla—cont'd

Phylum	First evolutionary occurrence (MYA)	Approximate number of species
Phoronida	No confirmed fossils	15
Placozoa	550+	4
Platyhelminthes	50[f]	30,000
Porifera	580	9300
Priapulida	Many from Burgess Shale	22
Rotifera	Cenozoic	3350
Tardigrada	145 - 66	1300
Urochordata	3000	510
Vertebrate/Craniata	400	66,178

[a]*Cavalier-Smith and Chao (2006).*
[b]*Poinar et al. (2008).*
[c]*Caron et al. (2013).*
[d]*Schram (1973).*
[e]*Garwood et al. (2016).*
[f]*Dentzien-Dias et al. (2013).*

efforts to characterize meiofaunal genome sequences have targeted at least five phyla, which include Annelida, Nemertea, Nematoda, Gastrotricha, and Mollusca, or the Spiralia clade (Laumer et al., 2015; Sevigny et al., 2021). Kocot et al. (2019) have applied these techniques to resolve the systematics of meiofaunal shell-less, worm-shaped aplacophoran molluscs (Solenogastres and Caudofoveata), and an enigmatic new species *Apodomenia enigmatica* sp. nov., which may be an intermediate between the two previously mentioned clades. As observed in many other habitats, ecological networks between meiofaunal species may be closely connected, for example, via predation and competition. Therefore, removal or addition of any one species could tip the balances to change dominance and community structures in a particular geographical location. Relatively little data exist on the ecology and biogeography of deep-sea meio- or even many macrofauna (Sinniger et al., 2016). For example, several bryozoan species are ubiquitous, early colonizers of some benthic polar habitats and can act as ecosystem engineers and indicators (Barnes and Griffiths, 2008; Santagata et al., 2018). Juveniles of other phyla may also compose some meiofaunal habitats, and these could be a focus. It would advance biodiversity studies if we could turn more attention and obtain resources to study deep benthic meiofauna.

In addition, similar to many microbiome abundance profiles portrayed in rank abundance analyses and the previously described "rare biosphere" from Sogin et al. (2006) in the context of microbiomes, many meiofaunal habitats appear to also be dominated by a relative few species. The long "tail" of less abundant species trails the more abundant subset, appearing mostly as single or doubletons, whose immediate role in the ecosystem functions is mostly unknown. Determining the functional roles of these low abundance taxa in any habitat may remain a major question and goal for metagenomics, microbiological, and meiofaunal ecologists for many years (Lopez, 2019).

Cerca et al. (2020) recently questioned whether the interstitial meiofaunal annelid (ghost worm) *Stygocapitella subterranea* Knöllner, 1934 (Parergodrilidae, Orbiniida) was truly cosmopolitan. After generating novel species-delimitation data (haplotype networks, GMYC, bPTP, maximum likelihood, posterior probability, and morphological) from four coastlines in the Northern hemisphere, eight new cryptic species appear to have been revealed. In another deep-water coral example, *Paragorgia arborea* population genetics was characterized in the Northern Pacific (NP), Southern Hemisphere, and Northern Atlantic (NA) basins. Genetic variation based on DNA sequencing the internal transcribed spacer 2 (ITS2) and five mitochondrial genes in this species correlated with geographic location at the basin-scale level, but not with depth. The hypothesis of an NP origin and youngest populations occurring in the NA followed a Miocene ocean circulation model (Herrera et al., 2012). A subsequent study has indicated that variation in *P. arborea* larval recruits off of Nova Scotia stem from differences in larval supply, fine-scale hydrodynamics, and postsettlement processes (Guy and Metaxas, 2022). Another taxon that could fit a benthic cosmopolitan distribution pattern, even with low dispersal capacities, is the feather duster worms (Sabellida, Annelida). These organisms can be fairly conspicuous on multiple coral reefs. Recently, however, a deep-sea lineage has been described that also displays a symbiosis with methanotrophic bacteria (Goffredi et al., 2020).

Without requiring sunlight, azooxanthellate Cnidarian genera, such as *Lophelia, Oculina, Stylaster, Madrepora, Desmophyllum,* and *Solenosmilia,* abound in deeper waters. For example, Cairns (2007) has provided several comprehensive assessments of deep-water corals: "among the 706 azooxanthellate Scleractinia, certain generalizations can be made about their morphology, bathymetry, and family affiliation (update of Cairns, 2004). Most azooxanthellates (519 species, 73.5%) are solitary in habit, and the remainder (187, 26.5%) are colonial. Thus, in the minority is *L. pertusa* (Linnaeus, 1758), the redoubtable deep-water bank framework coral (Rogers, 1999). Only 32 (4.7%) azooxanthellate Scleractinia species have been found beneath 2000 m, with the deepest occurring at 6328 m (Keller, 1976)." Furthermore, deep-sea currents may facilitate faunal gene flow across benthic biogeographic zones.

This relative abundance approach for assessing benthic biodiversity can entail species distribution modeling (SDM) (Dolan et al., 2008; Araújo et al., 2019; Randin et al., 2020). SDM can also help predict future species distributions in the wake of climate change and other stressors (Dambach and Rödder, 2011). However, much of SDM has been applied to only a few taxa such as cod and marine mammals and few benthic taxa except for stony corals (Howell et al., 2011; Zibrowius, 1980). Overall SDM has not yet proved effective to conservation efforts, since few SDM "outputs of the modeling have been linked specifically to management decisions" (Marshall et al., 2014). Although harder to locate and study, future characterization of solitary, deep-sea corals and other fauna could provide more data to fuel future SDMs, differentiate them from ecological niche models (ENMs) and add more data of vertical positions of taxa to improve benthic assessments (Roberts et al., 2006; Melo-Merino et al., 2020).

Before fully delving into the concept of biological diversity (biodiversity), the term must be fully defined. As is probably well known by now, biodiversity refers to the variety of life on Earth at all its levels, from genes to ecosystems, and can encompass the evolutionary, ecological, and cultural processes that sustain life (Duffy et al., 2007; Hector and Bagchi, 2007). The increased effort to measure species diversity and abundances rose more accurately to prominence after publication of E.O. Wilson's seminal eponymous book (Wilson, 1992). "Biodiversity" became a buzzword soon after, which helped realize Wilson's overarching goal of raising awareness about the concept and the urgency of protecting life among the lay public (Wilson, 1987). A flurry of other biodiversity books published, but few addressed marine biodiversity per se except for a few such as *Biodiversity II: understanding and protecting our biological resources* (Reaka-Kudla et al., 1997) or the early biodiversity assessment by UNEP (2000).

26 **Chapter 1** The seabed—Where life began and still evolves

Unfortunately, our planet's biodiversity is rapidly decreasing as the Anthropocene fully unfolds, which may include a likely sixth mass extinction event (Ceballos et al., 2015; Darling et al., 2019). The first scenario is clearly apparent due to the massive changes in landscape and various ecosystem functions wrought by human activities. Complete support for an ongoing, sixth mass extinction, such as accurate rates of past and present species loss, remains debatable. Nonetheless, it can likely be agreed that wholesale or even partial loss of unique habitats, which can host diverse organisms, will cause concomitant collapses in species richness. There is also a growing awareness that significant losses of biodiversity impoverish the human condition (Darling et al., 2019). And yet, we can also lament that we still do not fully understand and integrate all of the levels of biodiversity into much of our research. The late songwriter John Lennon once sang "You don't know what you got/Until you lose it," and this can ring all too true for too many locales in the sea.

Higher diversity of benthic macrofauna

Besides the various single-celled microorganisms, we can now continue the description of benthic life in the eukaryotic domain. What are some of adaptations of organisms that live in the diverse benthic habitats? How variable is the macrofauna across vertical and horizontal spaces—worms, molluscs, fish, or crustaceans, etc.? Also, how did they arise? What role do they play in ecosystems function or their trophic position in the local food chain?

If we aim to conserve biodiversity, we should try to know precisely what components we aim to conserve, and many ecologists and taxonomists would agree that with regard to the benthos, collectively we still have much to learn. The primary reasons for studying the seafloor are to characterize and understand as many organisms as possible that may inhabit it. This basically comes down to one of the most basic quests of many biologists—we are actively searching for signs of "life." When we posit that an organism, perhaps related to some creature known living closer to surface, may exist at extreme depths, or within aphotic ocean zones, then we also can try to determine how these organisms adapted to extremely high pressures or lack of sunlight's productivity. We can also ask how organisms adapt to the average ocean depth of 4250 m, where pressure is roughly 420 atm. In this largest and coolest of habitats, scarcity of nutrients will generally predominate as a factor driving adaptations such as slower metabolism and bioluminescence. Interestingly, however, fewer sessile organisms possess the ability to emit light compared with pelagic organisms (Martini et al., 2019). This could perhaps be attributed to a lower adaptive advantage for calling attention to oneself when there is no ability to evade by swimming away. Many of the animals that live in the darkness at these unseen depths tend to be small, cryptic, and perhaps "uncharismatic" when compared to some of the terrestrial mascots representing poster ready iconic megafauna assisted by the light of the day. Compare the photogenecity of typical anglerfish (*Melanocetus johnsonii*) to a cheetah (*Acinonyx jubatus*). In the darkness of three thousand meters depth of water, there are fewer eyes to impress. The increasing appreciation of the unique forms and functions that are only found in aquatic systems and after hundreds of million years of evolution. Czekanski-Moir and Rundell (2020) provide some pathways to increasing awareness of the benthos and habitats.

Thus, the first nucleation points of assessing biodiversity are finding, identifying, and characterizing the enormity and breadth of life at various levels. Then specific strategies for protecting and conserving these organisms will follow. Table 1.1 shows metazoan diversity at the phylum level, which is the starting point for describing Linnean hierarchy and biodiversity. Among named taxa, there are estimated totals of 61 eukaryotic phyla, 266 eukaryotic classes, 1253 orders, 9330 eukaryotic families, 140,000–200,000 eukaryotic genera, and around 1.9 million eukaryotic species. For some of these taxonomic groups, detailed monographs and keys for their identification may exist.

We emphasize here that the invertebrates (animals without backbones) account for the majority of animal phyla in Table 1.1 and approximately 95% of metazoan diversity (Zhang, 2011). For the invertebrates, we can point the reader to the many elaborate taxonomic descriptions in extensive monographs and comprehensive invertebrate textbooks (Giribet and Edgecombe, 2020; Brusca et al., 2022). Also, supplementary material is available in GIGA's first white paper (https://tinyurl.com/2zwbky5h). Insects represent the lion's share of the species and invertebrate diversity and were one of the primary inspirations for the movement to understand general biodiversity (e.g., E.O Wilson and Terry Irwin's work will be discussed later).

The breadth of marine biodiversity can be realistically viewed as just short of enormous and likely rivaling terrestrial diversity. The marine realm holds representatives of all but two animal phyla, and many of these are not found in terrestrial or limnic ecosystems. Aquatic invertebrates encompass a large portion of phyletic breadth and basically understudied taxa. However, for the vast majority, these specimens may not have been revisited, or full descriptions may be missing after the initial discovery. Besides the cnidaria, a plethora of fascinating and less well-known phyla occur on the seafloor and deep oceans. Interesting to marine taxonomic and organismal specialists for decades are the invertebrates. This group of animals makes up an artificial or paraphyletic clade and can be defined by several criteria—such as lacking a spine or backbone derived from a notochord (Brusca et al., 2022). According to several phylogenetic reconstructions, invertebrates can be clustered into three major groups: (i) early-diverging metazoan lineages, such as Porifera, Ctenophora, and Cnidaria; (ii) the protostome: Spiralia (Platyhelminthes, Mollusca, Annelida) and Ecdysozoa (Nematoda, Arthropoda); and (iii) the deuterostome clades comprising Echinodermata, Hemichordata, and Chordata.

In terms of biomass and number of species, invertebrates represent a clear majority of living animals in freshwater and marine ecosystems where they often play fundamental ecological roles (Stratmann et al., 2020; Anthony et al., 2023). Moreover, as a group, invertebrate organisms have emerged as targets for biodiversity enterprises (GIGA (Community of Scientists), 2014). Firstly, invertebrates encompass a large proportion of the world's biodiversity: of the approximately 1.9 million species described globally, invertebrates represent 79% of these (Appeltans et al., 2012). Wildlife, including many benthic invertebrates, provides a wide range of previously mentioned *ecosystem services,* which we will discuss throughout most of the chapters of this book (Freeman et al., 2014; Grabowski et al., 2012). Corals reefs, for example, halt or moderate coastal erosion, contain various fisheries, which help feed millions of people and also provide platforms for recreation that sustain local economies (Pascal et al., 2016). A central point is that there is no shortage of benthic species to be characterized, and at the current rate, humanity may run out of qualified taxonomists before species can be fully discovered or described.

Each species on the planet, and especially those living in unique habitats, has a story to tell through its life history and evolution. As prime examples, multiple freshwater and marine invertebrates contribute to ecotoxicology (such as oysters), or ex situ as model organisms in the laboratory such as the pond snail, *Lymnaea stagnalis* (Amorim et al., 2019), rotifer (Gribble and Mark Welch, 2017). Some coral species have served as marine lab rats in the development of quantitative response assays for the effects of different pollutants and chemicals, which can reach reefs (Renegar and Turner, 2021; Turner et al., 2021). We also do not discount the importance of other benthic taxa, such as protozoans, which are important prey for benthic macrofauna (Stoecker and Capuzzo, 1990).

28   **Chapter 1** The seabed—Where life began and still evolves

With the various examples of benthic biodiversity that will be provided in this assessment, we can correctly surmise that for biologists in general, the benthos represents a nearly endless treasure chest of likely new species. Almost at any site, from shallow nearshore coastlines to the hadal depths, life exists and awaits full characterizations.

The phenotypic diversity of the invertebrates may be attributed to their deep evolutionary lineages—many phyla represent the earliest evolving groups of animals that appeared in the fossil record during the Cambrian Period starting around 538–580 million years ago (Maloof et al., 2010; Table 1.1). Besides the rise of most animal phyla, complex morphology already appeared evident even in the trilobites of the Burgess shale. Much of the eukaryotic explosion of diversity was likely driven by growing levels of free oxygen, which was not present in the earliest Earth's atmosphere (Parfrey et al., 2011). Before the availability of oxygen, protozoa and other microbes had to likely contend with anaerobic sediments or environments, requiring the ability to utilize sulfides or other inorganic minerals.

Porifera

To start the parade of common benthic macrofauna, I will describe one phylum I have been familiar with—the Porifera. This is a suitable taxon to begin for several reasons. Sponges are benthic, sessile animals attached to the seafloor except as juveniles in some species. Testifying to their evolutionary and ecological success, Porifera represent one of the most diverse phyla with over 10,000 species, which can occur in freshwater habitats and every oceanic basin from shallow to abyssal (Hooper and Van Soest, 2002; Hooper et al., 2002; Van Soest et al., 2012; Pawlik and McMurray, 2020). Sponges also appeared in discussions about BPC earlier and will be discussed later in the context of marine disease.

Sponges can be found in almost every aquatic habitat, both freshwater and marine. Sponges occur in rivers and ponds and as discussed down to mesophotic and abyssal depths (Van Soest et al., 2012; Andradi-Brown et al., 2016; Hawkes et al., 2019). From another ecological aspect, sponges have recently been considered to be important recyclers of carbon and nutrients to the benefit of coral reefs. Sponges thus help compose the "Sponge Loop" (de Goeij et al., 2013). In this role, sponges cycle a large proportion of DOM derived from coral and algal exudates and make them available to other organisms on the reef as sponge detritus or POM (Rix et al., 2018). This also helped explain why reefs host a large diversity of organisms and biomass within oligotrophic waters. Others alternatives pathways for the fate of DOM have been proposed (Pawlik et al., 2016). Ongoing studies by Cara Fiore, Chris Freeman, Cole Easson, and others are now underway to fill in gaps on whether and how much sponges may process DOM. As sponges are so ubiquitous on many benthic habitats, one would presume that they could provide much needed biomass for potential predators. This would be even more valuable on oligotrophic waters of shallow reefs. However, some sponge species do not present appetizing morsels for would-be consumers, because of likely biochemical deterrents (Pawlik et al., 2015). The chemical ecology of sponges and other benthic organisms has been previously described, helping explain the radiation of diversity of biochemically equipped organisms at specific points in space and time, which provide a scope of their effects on the ecosystem (Paul and Ritson-Williams, 2008; Paul et al., 2019; Puglisi et al., 2019).

Several sponge species are also known to be involved in bioerosion, especially of their reef co-residents, especially stony corals. Bioerosion represents an anathema to the process of biomineralization. Many studies have detailed the action of boring and escavating sponge species (SchoÈnberg, 2002; Carballo et al., 2013; Chaves-Fonnegra et al., 2015). For example, the excavating sponge *Cliona delitrix* has been negatively impacting already stressed boulder corals on the Florida reef tract. At deeper regions, *Lophelia* coral erosion is the result of sponges, foraminifera, bryozoans, polychaetes, sipunculids, molluscs, and various microborers (Reed, 2002; Rogers, 1999).

The sponge phylum evokes interest in several ways. Taxonomically, sponges comprise at least 8000 species, but only a relative handful of specialists around the world can identify sponges to the species level. Because they are sessile, sponges are not as highly animated as other animals. They spend most of their time pumping water through incurrent pores (ostia), into internal canals, and out larger oscula. This is because sponges are filter-feeding organic matter from the seawater. If they can find a stable substrate to attach, and there is sufficient particulate organic matter in the water, sponges are able to survive. Species vary in their capacity to filter seawater and in their associated microbial communities, leading to diverse metabolic strategies that often coexist in one habitat. While it is well established that sponges are important in processing DOM, an important reservoir of reduced carbon compounds, and transferring this energy to benthic food webs, there has been limited work to understand the consequences of sponge processing on the composition of coral reef DOM and on pelagic food webs.

I have had the privilege to work with some very bright and interesting sponge biologists throughout the years. For a few years, I was part of a team called the Porifera Tree of Life (or PorToL), led by Robert Thacker. The team was able to resolve the phylogeny of a few difficult subclasses within Demospongiae: Keratosa, Myxospongiae, Haploscleromorpha, and Heteroscleromorpha (Redmond et al., 2013). Sponge classification has never been easy because of the simple sponge bauplan and relatively lack of "characters" that could be used to differentiate species. Nonetheless, sponge biologist Henry Rieswig did joyfully build a career describing the Hexactinellids or "glass sponges." This major class of sponge is mostly found at very deep waters and produces very ornate spicule shapes, which Reiswig documented in detail (Reiswig, 2002).

One phylogenetic story worth retelling involves the deep roots of the metazoan tree. For many years, the Porifera were considered as the undisputed primary ancestors of all animals (metazoans). This is understandable considering the very simple body form, which lacks true tissues. Rather the matrix of proteinaceous collagen fibers and siliceous spicules is called mesohyl. Initially, there were questions of whether sponges were animals at all. This debate reaches back to Linnaeus (1758). However, the animal nature of Porifera was eventually confirmed (Grant, 1826; Hooper and Van Soest, 2002).

Porifera has long been considered the earliest branching metazoan by several criteria, such as fossil and molecular phylogenetics (Bergquist, 1978). However, over the last decade and half, there have been active questions and stimulating conversations about what constitutes the earliest bifurcations at the base of the animal tree. The initial questions arose from an analysis of new large-scale next-generation sequencing datasets of invertebrate taxa generated by Dunn et al. (2008). Sponges' basal ancestral position went relatively undisputed until this study utilized deep DNA sequencing data, which converged on a novel branching order at the deepest animal roots. Although published at the relative start of the less costly and faster "next generation" era of massively parallel DNA sequencing, this paper described 150 genes in the form of expressed sequence tags (ESTs) from 77 taxa (71 metazoans). Using the tree building method called maximum likelihood with specific models of sequence evolution (WAG and bayesian (CAT11 and WAG)) and 21,152 characters and 55.5% of missing data in their super matrix of concatenated genes, the authors recovered the phylum Ctenophora as the earliest branching lineage relative to the other animal taxa for the first time. This was surprising as the 150 or so species of comb jellies possess more complex nervous, mesoderm-derived muscular, and epithelia than sponges. Again, sponges actually lack true tissues but have differentiated cells. The Dunn et al. (2008) study was followed closely by whole genome sequences from the phylum Ctenophore

30   Chapter 1 The seabed—Where life began and still evolves

(Ryan et al., 2013), which contended that sponges lost several derived traits such as nerves and muscles, to support the ctenophores placement. Together the hypotheses opened up lively scientific debate on which lineage is the earliest ancestor to animals (Edgecombe et al., 2011; Whelan et al., 2015; Pisani et al., 2015; Simion et al., 2017). Trying to resolve the correct branching with new data probably remains active. The debate revolves around the question of which taxon—ctenophore or sponge—represents the earliest branching and thus sister clade to the rest of the animals. Sponges (Phylum Porifera) are benthic residents while ctenophores can swim in the water column, thus escaping the focus of this book. Critiques of the new metazoan phylogeny revolved around the use of too few nonbilaterian species, too rapidly evolving genes, and inadequate evolutionary models (Nei and Kumar, 2000).

I apologize for the sudden spike in the jargon of evolutionary tree building methods. Nonetheless, phylogenetic placement of the basal ctenophores was later supported by more targeted gene and whole genome sequencing and the most parsimonious explanation for gene gain/loss seen across major animal clades (Ryan et al., 2010; Moroz et al., 2014). However, the case for the basal ctenophore phylogeny was also soon challenged by several studies with new data and reanalyses.

For example, in an effort to minimize the systematic bias of long-branch attraction often afflicting the reconstruction of trees with deep and ancient branches (Lartillot et al., 2007), Philippe et al. (2011) applied stringent parameters upon discarding unreliably aligned positions (producing a super-matrix of 18,463 positions), used the site heterogenous CAT model (Lartillot and Philippe, 2004), and less missing data (35.6%) to reconstruct a tree with the ctenophores appearing closer to Cnidaria and Bilateria. In another study, Pisani et al. (2015) reanalyzed several published datasets comprising concatenated alignments of many genes and a data matrix with more than 23,000 genes in different animal species to conclude that finding Ctenophore as sister taxa to the metazoa may be artifactual. A consensus metazoan phylogeny had been drawn with sponges as the sister clade to all animals (Telford et al., 2015), while novel phylogenomic approaches are continually proposed to help settle the controversy (Li et al., 2021; Redmond and McLysaght, 2021; Juravel et al., 2023). For example, a characterization has recently emerged displaying conserved syntenic blocks on chromosomes, which unites ctenophores and unicellular relatives to metazoa and contrasts with conserved blocks shared only among sponges, bilaterians, and cnidarians (Schultz et al., 2023). These data can keep the debate over the first metazoan phylum active at the time of writing.

With this somewhat extended introduction to the phylum, one genus of Porifera called *Xestospongia* can be singled out for now, as some its species represent of one of the most iconic sponges due to its barrel shape and sometimes very large sizes. The genus *Xestospongia* occurs from the intertidal to 1800 m depth (Hooper and Van Soest, 2002). *Xestospongia muta* has been called "the redwood of the reef" because of the massive sizes they can reach and their long life spans (McMurray et al., 2008). This species has been known to grow over 2 m in height and 3 m in diameter. *X. muta* longevity also has been dated to 2300 years, which is around the same age as the California redwood trees (McMurray et al., 2008). *X. muta* occurs in most reefs throughout the Caribbean, with some population densities as high as 0.28 sponge/m^2 (McMurray et al., 2010), and thus, it has key roles in providing ecosystem services that are used by the reef system as a whole (Diaz and Rutzler, 2001). Other *Xestospongia* genera occur in other basins, and a new *Xestospongia kapne* sp. nov. has been discovered from deep-sea coral mounds of the Campos Basin in the SW Atlantic (Carvalho et al., 2016).

Similar to the large *Xestospongia*, the Japanese spider crab (*Macrocheira kaempferi*) has the greatest leg span of any known arthropod, reaching up to 3.7 m (12.1 ft) from claw to claw. They can live at depths between 50 and 600 m (160 and 1970 ft). In the aphotic zone, other unique larger invertebrates can grow to extreme sizes such as the giant squid *Mesonychoteuthis hamiltoni* (Famiy Cranchiidae), which can reach 9–10 m 495 kg (1091 lb). This may have been the squid that brought fear and inspired stories among the ancient mariners. Similarly, many species in the phylum Chordata can move, swim, and thus, are not sessile. Salps, larvaceans, doliolids, and pyrosomes can swim. Species in the subphylum Tunicata (urochordates) are an exception with about 3000 species that live either as solitary individuals or as colonies attached to the seafloor (Corbo et al., 2001; Zeng et al., 2006; Brown and Swalla, 2012). Bottrylids (colonial ascidians) exhibit allorecognition capabilities, which could be seen as forerunners to immunorecognition (Burnet, 1971; Saito et al., 2002). Perhaps the most notable chordates we can count among the benthos are the fishes. They represent the most speciose group of chordates with over 34,000 named taxa. However, most of the extant species are not demersal (benthic) and thus, mostly mobile with wide potential to disperse. Therefore, we will not consider all the possible fish taxa here because we only intend to focus on taxa with relatively low mobility.

Crustacea

Molluscs and crustaceans have some of the largest and most interesting invertebrate radiations of the benthos. Crustaceans number about 60,000 named species, and their evolution also dates back over 500 million years. There is little disagreement that the deep oceans are inaccessible to most people, including scientists, due simply to the high costs and the complex logistics of reaching an average depth of 4300 m in the dark. Upon browsing some of the remote videos produced by NOAA or EV *Nautilus*, we often find at least one or more crustacean species in every video. They are ubiquitous in multiple habitats and situations—scavenging, foraging, filter, and detritus feeding. Yet how much do we really know about these denizens of the deep?

The crustacean Class Malacostraca represents some of the most innovative crustacean radiations, which encompass over 40,065 species, and includes shrimp, crabs, lobsters, crayfish, krill, woodlice, and mantis shrimp. Within the class, order Isopoda has over 4500 marine species. They have radiated and inhabit diverse habitats, from the terrestrial wood lice (*Oniscus asellus*) to shallow Microcerberidea to the deep benthic Giant isopods (Bathymonas), which can live at greater than 2300 m depth.

Several crustaceans such as order Decapoda provide sustenance for millions of people worldwide. Prior to the 2010 Deepwater Horizon oil spill, Gulf of Mexico, landings of shrimp species such as Brown Shrimp (*Farfantepenaeus aztecus*), Pink shrimp (*Farfantepenaeus duorarum*), White shrimp (*Litopenaeus setiferus*), Royal red shrimp (*Hymenopenaeus robustus*) led the United States with 177.2 million pounds valued at $340 million dockside, accounting for about 82% of US total. Texas led all Gulf states with about 77.0 million pounds; Louisiana with 74.1 million pounds; Florida (west coast) with 11.8 Alabama with 10.0 million pounds; and Mississippi with 4.1 million pounds.

With over 9900 total species, 80% of which are marine, crustacean order Amphipoda species holds some ubiquitous species, which can fill important niches as mesograzers and scavengers in many marine habitats. However, they are not all meiofaunal, as the order has size ranges from 1 to 340 mm. Deep-sea isopods, along with turrid gastropods and bivalves helped map new latitudinal species diversity gradients (LSDGs) in the bathyal (500–4000 m) benthos of the North Atlantic. Similar to shallow seas, species richness declined towards the poles (Rex et al., 2000). In a study at the deep-sea Clarion-Clipperton Zone (CCZ) spanning about 1500 km, Bribiesca-Contreras et al. (2021) carried out a genetic connectivity study of scavenging amphipods in the superfamilies Alicelloidea (Lowry and De Broyer,

2008) and Lysianassoidea Dana 1849. These taxa are abundant in various abyssal ecosystems. A total of 17 different morphospecies of scavenging amphipods were identified, which include at least 30 genetic species delimited by a fragment of the cytochrome c oxidase subunit I (COI) barcode gene.

Overall geographic genetic differentiation was not found, suggesting that connectivity of the species over evolutionary timescales between multiple CCZ areas was concluded. We will return to the CCZ later in the book. In another study, Pardaliscidae amphipods have been found at hadal depths (Jamieson et al., 2012), and thus, this order could serve as an important indicator species for deep benthic ecosystem function and health.

Echinodermata

Over 7000 species of the bilaterian Echinodermata phylum (Greek for "spiny skin") have been named, with a wide distribution from shallow intertidal areas to abyssal depths. The phylum includes brittle stars (Ophiuroidea), sea cucumbers (Holothuroidea), crinoids (Crinoidea), sea urchins (Echinoidea), and perhaps the most conspicuous starfish (Asteroidea). Many echinoderm species have motility, but some crinoids such as sea lilies convert to a sessile lifestyle on benthic habitats. Echinoderms uniquely move with "tubular feet" or pedicellariae, which can be strong enough for gripping and providing synchronic traction. The movement is used to scour their habitat. Problematic starfish, such as the "crown of thorns" (family Acanthasteridae), has decimated some reefs during outbreaks (Hughes et al., 2014). Echinoderms do not compose fisheries as large as crustaceans, but nonetheless, 129,000 tons of sea urchins and sea cucumbers were harvested for food in 2019.

Feeding styles vary and help delineate echinoderm groups: urchins graze, starfish are active hunters, sea cucumbers feed on deposits while many crinoids are suspension feeders. Urchins can help sustain kelp forests by removing macroalgae (Harrold, 1987), while asteroids, ophiuroids, and holothuroids play ecological roles as planktivores, detritivores, and scavengers. With these activities, some echinoderms can fit into keystone roles (Jordán, 2001). For example, a large kelp forest runs along the northern Chilean coast for over 1000 km. In this habitat, asteroids *Heliaster helianthus* Lamarck, 1816, and *Meyenaster gelatinosus* Meyen, 1834, are considered *keystone* species through their predation of the mussel *Perumytilus purpuratus* in intertidal communities (Paine et al., 1985). Hermosillo-Núñez (2020) carried out simulations based on *keystone species complex* models with multiple resident species. The modeling confirmed that echinoderms such as *M. gelatinosus*, *Tetrapygus niger,* and *Loxechinus albus* were the most sensitive to perturbations in the kelp forest habitat.

Echinoderms' close interaction with benthic substrates through movements and foraging across their habitats may have propelled a need to develop effective immune defense systems against various microbes and potential pathogens that they may encounter. A first line of defense for the grazers appears as innate immunity at the molecular level, including Pattern Recognition Receptors (PRRs). These directly recognize molecules, which are conserved among a large group or class of microbes, called microbe-associated molecular patterns (MAMPs or PAMPS as a slight derivative of "pathogen-associated molecular patterns") (Palm and Medzhitov, 2009). Adaptive immunity arising later in evolution has allowed organisms to recognize "self" from nonself, foreign individuals (Janeway, 1989; Flajnik, 2002). For example, the genome of *Strongylocentrotus purparatus* urchin has been found to possess a large quiver of basic innate immune arrows such as "222 Toll-like receptors (TLR), a superfamily of more than 200 NACHT domain–leucine-rich repeat proteins (similar to nucleotide-binding and oligomerization domain (NOD) and NALP proteins of vertebrates), and a large family of scavenger receptor cysteine-rich proteins" (Hibino et al., 2006; Rast et al., 2006; Flajnik and Kasahara, 2010). More immunological insights stemmed from this genome data, where potential homologs to vertebrate immune receptors were identified such as RAG1/2 (Recombination Activating Genes 1 and 2) (Hibino et al., 2006). RAG proteins facilitate the rearrangement of immunoglobulin or T-cell receptors, which are central to MAMP. Solek (2012) studied transcription factors *SpGatac*, an ortholog of vertebrate *Gata-1/2/3* and *SpScl*, an ortholog of *Scl/Tal-2/Lyl-1* in sea urchin embryos and demonstrated immunocyte development. The ancestral trail of potential immunological molecules such as antibody and MHC genes ends at jawless fishes, gnathostomes, as no genes show homology to the adaptive major histocompatability complex or antibody-like PRR molecules in invertebrate lineages. Nonetheless, immune systems and forerunners to antibodies and more sophisticated immune recognition systems may be uncovered as more genome data accumulate. These would then be grouped under novel "services" that earlier metazoan ancestors developed and passed on to their descendants.

Moreover, critical assessment of these feeding habits underscores the importance of organismal functions, which can have impacts on habitat structure. For example, the surface deposit feeding of sea cucumbers will affect seabed biochemistries more differently than suspension feeders (Thrush and Lohrer, 2012). A biological traits analysis coupled with physical oceanographic measurements could provide better models for biodiversity trajectories (Bremner et al., 2006).

Cnidaria, Mollusca, and other phyla

This phylum of mostly sessile organisms arose early in metazoan radiation (Veron, 1995). Once named coelenterata, the cnidaria obtain their name from specialized cells called cnidocytes or nematocysts, which can deliver a sting and/or toxin after triggered by prey that appear in the vicinity of the cell (Holstein and Tardent, 1984). This early branching invertebrate phylum deserves to be placed on a special pedestal above some of the other metazoan, as cnidarians provide the basis of reefs, which host some of the highest levels of biodiversity on the planet, spanning a wide phylogenetic spectrum. The diverse cnidaria have been categorized into at least four major groups: Scyphozoa (jellyfish), Cubozoa (box jellies), and Hydrozoa, and the mostly sessile subphylum Anthozoa (sea anemones, corals, and sea pens). As we have a benthic focus and many of the hard, stony corals (Order Scleractinia) indelibly appear as various examples on relevant assessment and conservation topics in this book, we will turn attention more to the other sessile taxa and describe a few of the noncoral cnidarian taxa for now. For example, nonreef-building gorgonians (Octocorallia) often coincide with the scleractinia, but remain understudied. In shallow reef communities, Plexauridae and Gorgoniidae families comprise a large proportion and include sea fans, while sea anemones (Stichodactylidae: Actiniaria) form protective symbioses with pomacentrid fishes and perhaps help drive the radiation of the latter (Dunn, 1981). In a comparison of 242 gorgonian taxa, higher species diversity appeared on mesophotic reefs, suggesting a higher rate of species radiation compared with shallower area (Sánchez et al., 2021). This was especially evident in the family Ellisellidae. The faster diversification could also be coupled with lower rates of extinction in the deep-sea lineages.

I will always remember my first experience on a living coral reef in the Bahamas, during my dissertation work (on a very different nonmarine project) in graduate school. I was not yet marine biology-focused at the time. However, the reader will have to pardon my tendencies to fondly address reefs as living entities—each holds a great deal of personality. I took a tropical ecology course at George Mason University with Luther Brown, which included an experiential field component to

34 **Chapter 1** The seabed—Where life began and still evolves

the largest yet sparsely inhabited Andros Island, Bahamas. Here GMU had set up the Bahamas Environmental Research Center. As part of the course, the classes stayed at accommodations nearby with local residents in the Fresh Creek community, so that we could experience Bahamian culture as well as its natural habitats. Here after long but satisfying field days, we enjoyed home-cooked meals by Androsian Hortense Riley, who used land crabs (*Gecarcinus lateralis*) and beans to spice up her Bahamian rice dishes. The Fresh Creek settlement was also just a 20 min boat or dingy ride to the reefs and other natural habitats (Lopez et al., 2000).

On that snorkel, we saw up close, for the very first time, organisms that I had only encountered in books such Paul Humann's *Reef Identification* (Humman, 1992; Humann and Deloach, 2014)—e.g., soft and stony corals, brittle stars, and lively reef fish—blue-headed wrasse and thousands of grunts. Even more striking to me was the presence of large full-grown stands of *Acropora palmata*, which stretched north and south. They made the surface of the water and proximal horizon appear jagged, with the colonies poking above the water line at low tide. We swam toward and then among the conspicuous hard corals. We were in the shallow clear waters trying to avoid being cut by the sharp polyps. This was 1993, and only a few years later, these stands, and many like them up and down the Caribbean, would succumb to the first of several major bleaching events (Goreau et al., 2000; Manzello et al., 2007).

The east coast of Andros Island provides stark beauty at various depths and levels. The karst framework of the island the allows itself to be pockmarked with enticing blue holes and cenotes of various sizes, diameters, and depths. The class visited several and swam in blue holes and also snorkeled at several Andros Island reef sites. Andros Island has one of the largest barrier reefs in the Western Atlantic, also lying adjacent to the Tongue of the Ocean (TOTO), which is a deep underwater trench that plunges to a depth of 6600 ft., making it one of the deepest marine trenches in the world. Created by the collision of the North American plate and the Caribbean plate, causing the ocean floor to sink and form a trough, TOTO has a unique underwater environment characterized by steep walls, deep waters and strong currents, underwater ridges, and a flat floor. Not all corals are broadcast spawners and participate in the synchronous mass spawning events. For example, *Madracis* is cosmopolitan genus of scleractinian corals with many species that occur sympatrically across a range of 2 to >100 m in depth, but they reproduce by brooding larvae with limited dispersal (Brazeau et al., 2005).

Great attention has been paid to vibrant shallow tropical coral reefs for many good reasons. Reefs can be large three-dimensional habitats often visible from space and home to 600,000 to up to 9 million species (Reaka-Kudla et al., 1997). However, many corals occur in deep waters, below the photic zone. In fact, over 3350 species of corals occur at depths below 50 m (Cairns, 2007). Even at depth and in the absence of sunlight, the often small and solitary corals provide habitats, topography, and a basis for colonization to otherwise barren looking seafloor.

Moving to the largest marine phylum and second largest phylum of invertebrates, Mollusca exemplify the concept phenotypic diversity and ecological success across terrestrial, freshwater and shallow to the deep marine habitats. Over 85,000 mollusc species have been described (Brusca et al., 2022). This includes the highly intelligent octopus and cuttle fish, both of which could be considered partially benthic and pelagic depending on the species. Seven extant classes exist, which include Gastropoda, Bivalvia, Polyplacophora, Cephalopod, Scaphopoda, Aplacophora, and Monoplacophora. As the largest of these, gastropod species number up to 70,000 and spend most of their life either anchored or based on solid benthic or terrestrial habitats. Polyplacophora (chitons) also dwell on the seafloor.

For many years, Vermeij has studied various aspects of molluscan evolution, ecology, and systematics (Vermeij, 1977, 1994). The tautology between a permanently sessile lifestyle and motility that

Higher diversity of benthic macrofauna **35**

appears driven by successful predatory adaptations takes on multiple forms. Three hundred species of octopuses have been counted, and many embody the necessary skills of a successful, mixing both brain and braun. To support this feature, the dexterity of the octopuses tentacles is well known. Then, recent genome and transcriptome sequencing of *Octopus vulgaris* has revealed the expansion of genes needed for signaling, and cell communication, such as protocadherins and the expression of long interspersed element (LINE) L1, a non-LTR retrotransposon, which are associated with neuronal connectivity and behavioral plasticity important for learning (Albertin et al., 2015; Petrosino et al., 2022). There are 3000 molluscan nudibranch species alone. As one of the most conspicuously ornamented marine phenotypes, a gallery of photos to display the diverse phenotypes of species could be more appropriate than written descriptions. In this regard and if they were plants, they could be considered flowers of the sea.

At this juncture, I can point out how a fascinating intersection of natural products chemistry, gastropod evolution, and benthic ecology can lead to novel insights and discovery about benthic life. This is exemplified in the biology of the gastropod snails alone, where over 12,000 living species of predatory gastropods of the superfamily Conoidea exist. Within the Caenogastropoda, multiple families of shelled marine molluscs abound, which include the periwinkles, cowries, moon snails wentletraps, murexes, cone snails, and turrids and constitute about 60% of all living gastropods.

Over 700 species of predatory cone snails (genus *Conus*) exist and can generate 100–200 small venom peptides (colloquially known as conotoxins) (Olivera, 2006). Much of the early work was driven by pioneering biochemist and current US National Academy of Science member, Baldomero Olivero (also a codiscoverer of *E. coli* ligase), from the initial discovery and characterization of the wide mode of action these peptides have. In the early assays, the toxins affected behavior through their inhibition of N-type calcium channel (CaV2.2) or K-channel blocker. Indeed, the commercial nonopoid analgesic product, Prialt (generically called ziconitide), has been marketed and is identical to the natural peptide produced by the magician's cone snail, *Conus magus,* originally designated conotoxin MVIIA (Safavi-Hemami et al., 2019).

A closely related group to *Conus* snails is the turrids (Bouchet et al., 2009), which are much less well known. Seronay and others described innovative, low-tech methods, also known as *lumun lumun* to collect these snails, which display the highest abundances from 300 to 600 m depth in the Philippines (Seronay et al., 2010). In a single *lumun-lumun* community, the authors described up to 155 gastropod morphospecies. Venomous predators belonging to the superfamily Conoidea (36 morphospecies) appeared as the largest group, with the majority of morphospecies being micromolluscs in the family Turridae (sensu *lato*) (Bouchet et al., 2002), though other families (Drilliidae) can also produce venoms (Lu et al., 2020). Some deep-water turrid microsnails could also qualify as previously mentioned meiofauna.

From the first group indicated earlier, marine nematode worms represent the most abundant and cosmopolitan meiofaunal group in deep-sea sediments (Lambshead et al., 2001; Merckx et al., 2009). Because of their typical size of about 5–100 μm wide, and 0.1–2.5 mm length, nematodes qualify and are important members of meiofauna. Nematodes can represent from 50% to over 90% of the total meiofaunal abundance. Their geographical distribution, especially in coastal environments, is strongly influenced by the local environmental characteristics and vice versa.

Several studies have integrated multiple approaches to demonstrate how nematodes can help characterize the environmental quality status of diverse benthic habitats, usually close to the coast. For example, measuring sedimentary contents for total phytopigments, proteins, carbohydrates, lipids, and biopolymeric carbon has helped assess benthic trophic status in the northern Adriatic Sea.

36   **Chapter 1** The seabed—Where life began and still evolves

Individual nematode stoma morphology assists in delineating the functional traits of nematode assemblages. Bianchelli et al. (2018) found that the highest levels of nematode species statistically correlated with the highest sedimentary organic matter contents. For example, the Adriatic nematode genera of *Sabatieria, Paramonohystera, Metalinhomoeus, Theristus,* and *Odontophora* appeared as reliable indicators of organic enrichment.

We have yet to discuss members of the annelid phylum. Many taxa are not the most conspicuous of the invertebrates, but they can be very abundant across multiple benthic habitats. For example, more than 10,000 species of the paraphyletic Polychaeta class exist. Similar to the widespread molluscs, polychaete worms, such as bristle worms, have radiated to all of the ocean basins and various benthic habitats (Hutchings, 1998). This versatility in part stems from polychaetes' display of a wide range of temperature tolerance. The organisms have been found in 6°C water on deep abyssal plains as well as living near hydrothermal vents and their extreme high temperatures ranging from 60°C (140°F) up to as high as 464°C (867°F). Hartman (1966) recorded 457 species of polychaetes between the sub-Antarctic and the Antarctic Convergence. This wide range stems from a strong opportunistic ability to quickly recruit and respond to perturbations (Grassle and Grassle, 1974).

Polychaetes display diverse feeding methods ranging from surface deposit, suspension, mud swallowing, carnivory and herbivory (Carvalho et al., 2013). Very importantly, several worm species qualify as "ecosystem engineers." These are organisms that modify the benthic environment via their own physical structures (autogenic) or by transforming materials from one state to another (allogenic), with impacts on other species.

Then, we can continue to delve into some of the more obscure and cryptic meiofaunal groups. For example, the phylum Kinorhyncha (mud dragons) consists of small (100–1000 μm long), free living benthic animals inhabiting marine environments from the intertidal- to abyssal depth zones worldwide (Bang, 2020; Herranz et al., 2022). Osborn et al. (2012) have characterized the relatively understudied deep-sea acorn worms (Hemichordata, Enteropneusta) using ROV methods, while species of Loriciferans, such as *Spinoloricus turbatio* gen. et sp. nov., have been identified at the Galápagos Spreading Center (GSC) (Heiner and Neuhaus, 2007). In the wetlands of Kerala, India, three new species have been identified: amphipod *Victoriopisa cusatensis*, decapod *Psuedosesarma glabrum,* and bivalve *Indosphenia Kayalum* (Nandan et al., 2019).

Intriguing examples of genome evolution and novelties in animals including invertebrates appear plentiful, such as *Daphnia* gene duplications, mitochondrial genome rearrangements/expansions/contractions and nuclear mitochondrial (Numt) transpositions (Bensasson et al., 2001; Lopez et al., 1994; Rota-Stabelli et al., 2010; Erpenbeck et al., 2011). And what is the basis of the large genomes of amphipods, and why do crustaceans generally have a wide range in genome sizes (Gregory, 2005)? In another group, Bdelloid rotifers are extraordinarily desiccation-tolerant, resistant to doses of gamma irradiation capable of killing most other animals and have incorporated hundreds of nonmetazoan genes into their genomes. The published tetraploid rotifer genome supports the finding of an ameiotic lifestyle, abundant gene conversion, and genes protecting against oxidative stress (Flot et al., 2013). The genome of the marine annelid *Streblospio benedicti* promises to reveal the mechanisms underyling alternative developmental pathways such as poecilogony, polymorphic larval development (Zakas et al., 2022). The genome of another aquatic species, *Hydra magnipapillata*, appears to be shaped by frequent transposable elements, horizontal gene transfer and transsplicing, and simplification (Chapman et al., 2010). Hydras are immortal and, along with several other invertebrates, represent some of the longest-lived animals on the planet (Bosch, 2009; Boehm et al., 2012).

De novo assembly of the gorgonian octocoral sea fan, *Gorgonia ventalina,* transcriptome has revealed 210 genes coding involved in innate immunity pathways such as pattern recognition molecules, antimicrobial peptides, and reactive oxygen species (ROS) formation after exposure to parasites (Burge et al., 2013). Barshis et al. (2013) recently showed the upregulation of tumor necrosis factor receptor-associated factor (TRAF) homologs in corals responding to heat stress. These examples point to the upcoming trove of novel and valuable genetic sequences and metabolic pathways that will likely soon be revealed and continually uploaded for years to come from the benthos. More examples of this novelty will be explored further in Chapter 3.

Marine microbial diversity

As mentioned, biological diversity exists at many levels, and to gain a full understanding of biodiversity, it can help to understand all of these levels. Grappling and fully assessing biodiversity represent a major Herculean task. There are many authorities, whose opinions and approaches may starkly differ or conflict. This can involve counting or naming the taxa that compose diversity. Moreover, the field of biodiversity research has matured over the past three decades to include ecosystems functions, economic impacts (Vermeij, 1994).

At the penultimate level of diversity sits the cell, which can be considered one of the common denominators of organic life. Enormous metabolic activity occurs at the cellular, molecular, and biochemical levels in any living organism. The diversity of cell types, even within a single organism, can be fairly astounding. Cells carry out specialized functions for immunity (blood cells, macrophages), sensation (neurons), internal communication (neurons, glandular, hormone and cytokine secretors, granulocytes) structure (fibroblasts, dermal cells), and mobility (myocytes, collar cells). Similar importance could be attributed to nucleic acids, as the essential hereditary code for all life on earth. However, the cell is a different baseline, because cells and their components compose the basic building blocks of tissues and hence organisms and can be subjected to evolutionary selection because of their manifested phenotypes.

Within the cell, differential expression of gene, into a protein or new RNA, at the right time and place constitutes a phenotype, subject to selection as any bright coloration, enzyme shape, or adaptive behavior (Golding and Dean, 1998). The expressed genotypes in essence propose the development of a specific morphology, but their existence depends on whether that eventual morphology conferred an adaptive advantage to the individual or population. The cell is the first level of how the DNA code is read. If a protein does not function because of a single amino acid mutation, the cell may no longer function. The simple mutation could even create a cancer or runaway replication of cells in eukaryotes.

Even amidst the relative scarcity of the deep ocean, microbes will be found. In one of my early papers before next-generation sequencing was widely available, I described sponges as "oases of biodiversity" (for microbes) in the context of the apparent scarcity life of many abyssal plains (Lopez et al., 1999a,b). This can now be seen as a naïve assessment, as we now know that most habitats, including deep-sea sediments, have some of the most diverse communities of microorganisms (Whitman et al., 1998). Earlier I described how bacterial counts and biomass were estimated, which helped advance the impact of molecular microbiology studies in the 1990s. Also mentioned was how the characterization of microbial diversity gained momentum when Carl Woese championed the molecular taxonomic methods based on the ubiquitous rRNA genes (Woese et al., 1990).

38   **Chapter 1** The seabed—Where life began and still evolves

Microbes here again collectively represent bacteria, fungi, protozoans, and viruses—most of the living entities with DNA we cannot plainly see (Pedrós-Alió, 2006). However, I will not be able to fully cover most of these groups in the volume because (a) the groups are so large and (b) much of the research remains nascent. Very few studies, for example, have examined the distribution of protozoa or fungi at seafloor habitats, yet perhaps most are transported there as sinking particles (Orsi et al., 2013; Boeuf et al., 2019; Xu et al., 2018).

We can still speculate that many types of microorganisms will occur at depth and almost everywhere we look. Bacteria and other microbes inhabit almost every nook and cranny of the planet, from terrestrial subsurface to the abyssal and lower depths. More studies contend that because high hydrostatic pressure may represent the most common and constant environmental factor on the seafloor, this parameter may be one of the key drivers of microbial evolution (Zhao et al., 2020; Xiao et al., 2021). These deeper taxa should be eventually surveyed for specific adaptations when possible or affordable, to enable a complete benthic assessment, which integrates evolution and ecology. However, as I mentioned in a recent chapter on sponge and coral microcosms, our growing knowledge of what microbial taxa may be present in an environment may still not illuminate on their actual ecological functions, probability of redundancy or how they carry out niche partitioning within specific microbiomes (Lopez et al., 2019a,b).

We now have a deeper appreciation of microbial life after Carl Woese and colleagues ushered in the era of molecular microbiology in the late 1970s and 1980s. Woese was the first microbiologist to systematically propose using universal molecules such as ribosomal RNA (rRNA) as taxonomic markers for bacteria, as the majority of unicellular organisms could not be cultured in the laboratory. Ribosomes are organelles composed of both rRNA and proteins and carry out the essential process of translation of mRNA into proteins. Therefore, almost every living cell is expected to possess not one but multiple copies of active ribosomes. Woese and his colleagues reasoned correctly that rRNA genes would thus also be present in almost every cell—from bacteria to blue whales. This provided an opportunity to read and compare rRNA sequences across diverse species. After generating an ever-larger database of rRNA sequences, comparisons could be across more and more species. Soon universal PCR primers could be designed from alignments of multiple diverse rRNA sequences. This comparison indicated that the rRNA molecule was dynamic and had internal stretches of sequences, which were hyper-variable, ultra-conserved, or something in between. Eventually the rRNA sequence databases and approach allowed the construction of the first rRNA-based phylogenetic trees (Lane et al., 1985). At face value, a lay person would not expect to find the same molecule (albeit with variations) in different species, as they can be so different morphologically. Let us look at the differences between an eagle and manta ray or those two species and the various sharks. They are all elasmobranchs yet appear very different and have different lifestyles and natural histories. This group is even more different than pelagic teleosts, such as the tuna and marlin. The fishes differ from amphibia and reptiles and birds. Biodiversity spans many different levels from outward appearances to internal physiologies. Nonetheless, all of these organisms will have rRNA molecules in their cells.

Multiple studies supported the utility of rRNA. This exercise would end up causing a bona fide paradigm shift that affected the entire classification system. The three-"Domain" tree based on rRNA sequences that Woese et al. (1990) brilliantly drew indicated that Eubacteria, Archaea, and Eucarya composed the tree main branches. This represented a major paradigm shift for biological classification, a new structure of life, which was not immediately accepted. A subtle implication of the phylogenetic reconstruction was that the most common and longest branches of three-Domain tree were unicellular. The multicellular taxa—fungi, plants, and animals—that composed the bulk of traditional five

Kingdom classification (Margulis, 1974), were now relegated to just a few recently radiated tips on Eucarya branch of the three Domain tree. Soon, the "backbone" for the structuring of the second edition of Bergey's Manual of Systematic Bacteriology was based on 16S SSU rRNA (Ludwig and Klenk, 2001).

Further studies continued to propel microbiology. Cloning and large-scale sequencing of the microbial 16S small subunit (SSU) (and large subunit to a lesser extent) rRNA genes from more strains and various natural environmental samples without culturing confirmed that less than only ~1% of the microbial diversity has been cultured (Jannasch and Jones, 1959; Button et al., 1993; Giovannoni and Rappe, 2000).

The totality of the enhanced microbial systematics efforts eventually supported that up to 99% microbes are unculturable, depending on the habitat (Pace, 1997). The growing realization also hastened the establishment of the new field of *metagenomics*—which "beyond" a single genome. Coined by Jo Handelsman, at the University of Wisconsin-Madison, the term metagenomics was defined as the culture independent genomic analysis of an assemblage of microorganisms from environmental samples (Handelsman et al., 1998; Handelsman, 2004). This approach involved sequencing greater swaths of bacterial genomes, including protein coding genes besides the rRNA loci and grew alongside the rapidly developing high thoughput DNA sequencing technologies (HTS) (Mardis, 2011, 2017).

The metagenomics approach may have taken some of the air out of the culturing efforts as the majority microbes cannot be grown in the laboratory (Bull and Goodfellow, 2019). Nonetheless, some of my former colleagues at Harbor Branch Oceanographic Institution, such as Peter McCarthy and his laboratory, tried tirelessly to culture microbial symbionts from mostly benthic marine invertebrates. This eventually resulted in over 20,000 distinct marine cultures collected over 35 years of cruises. To profile the diversity, we conducted a molecular survey with universal 16S rRNA genes of about 10% of the HBOI microbial collection and found some cultures similar to previously isolated marine bacteria, while others appeared unique and cultured for the first time (Sfanos et al., 2005).

After the initial rRNA advances promoted by Woese, the methods were quickly applied to studying the ocean. Multiple investigators used the SSU rRNA gene to characterize both cultivable and uncultivable components of deep-sea marine microbial communities (DeLong, 1998, 2005). Giovannoni and Rappe' (2000) provided a timely review of molecular methods at the time. Then, Mitch Sogin at the Bay Paul Center of Marine Biological Laboratory (MBL) in Woods Hole Massachusetts also helped propel molecular evolution and the use of rRNA based taxonomies in the early 2000s. Sogin et al. (2006) published a pioneering paper that applied massively parallel sequencing of short sequence "tags" amplified from a segment of the rRNA gene from the microbial communities of diverse marine samples. Total DNA representing the community from meso- and bathypelagic waters, a submarine volcano on the Juan de Fuca Ridge in the northeast Pacific Ocean, and hydrothermal vent fluids collected from the 1998 eruption zone of Axial Seamount (Sogin et al., 2006) were sequenced, yielding about 118,000 PCR amplicons of the hypervariable 16S rRNA V6 region (no longer routinely used). Their efforts yielded reads ranging from 6505 to nearly 23,000 sequences, which was no small feat with the sequencing methods at the time (454 pyrosequencing) (Mardis, 2017). The most important insight from the study was the bulk of operational taxonomic units (OTU) (representing potentially unique microbial taxa) sequences appeared only once or a few times in the dataset, but the represented enormous phylogenetic breadth. This was not interpreted as sequence errors, although that was possible. Rather, most of the sequence diversity (e.g., richness) of microbial OTUs was spread out and uncommon. This pattern of finding "rare biospheres" in microbiomes was displayed in the diverse marine samples and also subsequently in many other habitats. In a few ways, this view echoes the adage

40 **Chapter 1** The seabed—Where life began and still evolves

"Everything is everywhere, but the environment selects" propounded by Dutch microbiologist Lourens Baas Becking (Baas-Becking, 1934). Whitman et al. (1998) were one of the first groups to quantitate the "silent majority" of bacterial biomass in the oceans. This seminal paper coincided with a greater appreciation for microbiology in general that began in the late 1980s. In the United States, the National Science Foundation (NSF) was led by its first female director, microbiologist Rita Colwell in 1998, who was instrumental in bringing microbiology into the foreground of American scientific priorities. Her success and journey also consequently raised the awareness of the challenges facing female researchers in the sciences (Colwell and McGrayne, 2021).

Around the same time, Sogin started a very influential and selective workshop/course on molecular evolution in 1988, which was one of the first of its kind by inviting expert and renowned faculty to provide their expertise on various molecular evolutionary principles: Joseph Felsenstein, Walter Fitch, Gary Olsen, etc. (At the time of the course's establishment, molecular evolution as a discipline was still fairly embryonic in many of its methods and principles—see Graur and Li, 2000). I believe I was fortunate to attend the workshop in the early 1990s, obtain extensive molecular phylogenetics training, sometimes taking time to work and dance with classmates with U2's *Achtung, Baby* playing in the background end to end. Later, many of these classmates would reach their own heights in research through the years (Nishiguchi and Nair, 2003). With respect to marine microbiology, MBL established another trend-setting workshop called "Strategies and Techniques for Analyzing Microbial Population Structures (STAMPS)." Currently run by Titus Brown and Amy Willis, this workshop provides students with practical experience running computer programs that will handle the large datasets of microbial genomes and communities that have been generated by the massively parallel sequencing (also known as "next-generation sequencing") in the mid-2000s.

Generation of biodiversity

We have now displayed a small cross-sectional sample of benthic biodiversity to whet the appetites of readers. The next question to follow is "how was this biodiversity generated?" The answer is complex, but relevant to assessment, which should be considered a dynamic enterprise. In addition, many decades of painstaking effort in genetics, ecology, and evolutionary biology have produced interesting and plausible answers to biodiversity generation.

One likely underlying basis of benthic biodiversity stems from the nearly endless *habitat heterogeneity* from the sublittoral to the abyssal that is available to well-equipped or adaptable taxa that could colonize potential niche spaces. Skimming over the mechanisms of how diverse seafloor structures were constructed geologically (Pérez-Díaz and Eagles, 2017), the existing spaces encompass the micro (micrometer) to the macro (meters to kilometer) scale. One recent well-documented study describes the biodiversity at bathyal (isobaths of 350 and 1000 m) "pockmarks" located along the Brazilian continental margin near the Santos Basin (SB) (Carrerette et al., 2022). A total of 182 taxa found have only 90 identified to genus or species level, underscoring the novelty of biodiversity that can be discovered by just looking. Annelida showed the highest relative abundances (60.6%) across all 27 sampling sites, followed by Echinodermata (23%), Arthropoda: Crustacea (7.1%), Mollusca (1.8%), Cnidaria (6.5%), Porifera (0.8%), and Bryozoa (0.2%). Although relatively low in number, the cnidaria point to the presence of deepwater coral reef communities in this area, which could serve as potential indicators sensitive to any future anthropogenic changes. Physical parameters such as water flows, presence of silt sediments, and nutrient concentrations likely affect these microbenthic communities. For example, many of pockmarks had higher concentration of nutrients such as nitrates, phosphates, and silicates, whereas surrounding carbonate mounds had a higher concentration of ammonia. Unfortunately, an assessment of meiofauna was not conducted.

This discussion leads to questions of species formation or population maintenance in benthic areas. Between shallow and deeper zones—which ones are sources or sinks for population rejuvenation, speciation or connectivity? (Pulliam, 1988; Assis et al., 2016). For example, do mesophotic reefs provide "refugia" to shallower species when environmental stresses occur? Similar questions can be asked of the deeper zones—e.g., what are the interactions between sublittoral-bathyal, bathyal-abyssal, abyssal-hadal fauna, etc., and can any of these provide effective sources and sinks (Rex et al., 2005; Hardy et al., 2015)? Understanding the mechanisms of *speciation* is not trivial and fraught with additional hurdles in the sea. However, determining species mechanisms has long been a goal among many marine biologists and directly relates to understanding biodiversity (Knowlton, 1993; Palumbi, 1992). Differences between shallow and deep speciation processes are likely (Wilson and Hessler, 1987). Speciation can sometimes a contentious topic, especially for organisms that can disperse long distances that is common in the ocean. At least the number of species (richness) remains a specific metric for comparing different regions and processes. For example, what boundaries do wide dispersal organisms encounter or that could prevent gene flow and hence speciation required for allopatric speciation? Many studies have addressed this process. Bowen et al. (2013), for example, have summarized many of the primary mechanisms that can lead to diversification and speciation in the ocean. For example, the coincident possibilities of allopatric and sympatric speciation in marine habitats are presented, along with exchange (export and import) of species between biodiversity centers and peripheries. Sympatric speciation appears possible with enhanced ecological partitioning, as exhibited in several examples: brooding corals (*Seriatopora histrix*), differential vocalization bygrunts (genus *Haemulon*), specific settlement cues in nudibranchs (genus *Phestilla*), and position of the intertidal zone staked out by Hawaiian limpets (*Cellana* spp.) (Faucci et al., 2007; Rocha et al., 2008; Bongaerts et al., 2010; Bird et al., 2011). Overall, multiple mechanisms likely exist for speciation, especially at diversity hot spots such as coral reefs.

As a molecular biologist, I have approached the topics of biodiversity from microscopic, molecular, or unseen points of view. This does not mean to take a fully reductionist viewpoint, since microbiologist Carl Woese once warned against self-imposed myopia (Woese, 2004). On the contrary, he believed that we could maintain the bigger picture and achieve our desired understanding of how biological systems function, while delving into the nuts and bolts of molecular mechanisms and microbes. Woese also approached diversity and function from an evolutionary point of view. There is this poignant passage from one of his last essays:

"A heavy price was paid for molecular biology's obsession with metaphysical reductionism. It stripped the organism from its environment; separated it from its history, from the evolutionary flow; and shredded it into parts to the extent that a sense of the whole—the whole cell, the whole multicellular organism, the biosphere—was effectively gone. Darwin saw biology as a 'tangled bank' (Darwin, 1859), with all of its aspects interconnected. Our task now is to resynthesize biology; put the organism back into its environment; connect it again to its evolutionary past; and let us feel that complex flow that is organism, evolution, and environment united. The time has come for biology to enter the nonlinear world." The references in this book try to follow this advice.

Marine symbioses—Getting to know each other better

Most organisms live with distinct partner(s), ranging from beneficial mutualisms to neutral commensals to detrimental parasites (Douglas, 2021). These intimate, often long-term associations define *symbiosis* and range from microbial communities to complex host multicellular interactions to systems of political economies and digital communications. Perhaps overshadowed by the concept of competition and selective processes in evolutionary theory, symbiosis sensu *latu* epitomizes a vital biological phenomenon. Developmental biology yielded a clever anecdotal adage in the 1980's attributed to South African-born British biologist Lews Wolpert, who formulated the French Flag Model of spatial development: "It is not birth, marriage, or death, but *gastrulation,* which is truly the most important time in your life" (Hopwood, 2022). This phrasing refers to the pivotal reorganization of the embryo via cell movements that begin transforming the early spherical ball-shaped blastula, into a multilayered organism with differentiated tissues. During gastrulation, many of the cells at or near the surface of metazoan embryos move to a new, more interior location. As discussed herein, the phenomenon of symbiosis is likewise just as important as gastrulation or perhaps even more influential due to the former's role in speciation, diversification, ecology, physiology, and evolution. If so, symbiosis requires its own adage: "Little is accomplished in biological isolation. Symbioses affects multiple levels of life, across a wide phylogenetic and physiological spectrum." This needs work and requires more brainstorming later.

Symbiosis as a biological concept dates back to the 1800s (de Bary, 1879) and is defined as the intimate, long-term associations among more than one symbiont and symbiotic hosts. Most organisms live in multipartner relationships ranging from beneficial mutualism (both partners benefit) from neutral commensalism (one benefits while another is unperturbed but present) to detrimental parasitism (degradation of one partner, usually the host, at the expense and benefit of the second symbiont). Examples of diverse types of symbiosis abound throughout nature (Douglas, 1994), and there is consensus that most living organisms live symbiotically with at least one other species (McFall-Ngai et al., 2013). Symbioses between macrosymbionts (hosts) and microsymbionts (microbial partners) display diverse features and modalities: shared metabolic pathways, chemautotrophic dependencies, assistance with digestion, mutual defense via mechanical structures or biologics. Some refer to the total organism as a "holobiont," compared with the individual member symbionts. I will not have the space to explore all of the possible examples of symbiosis in this book though I refer to many studies and reviews throughout. Recent relevant studies to benthic biodiversity will be introduced here and also peppered throughout the remainder of the volume. Moreover, because symbiosis is a nearly universal feature of life, it cannot be applied as a proxy or metric for benthic biodiversity per se, though each unique symbiosis can provide context for each specific habitat (e.g., how important is symbiosis in keystone species?)

A modern champion of cellular evolution, the importance of microorganisms and symbiosis was Lynn Margulis. She had asserted and proposed multiple theories covering various symbiotic scenarios, i.e., the common interdependence of dissimilar organisms. She was one of the first scientists to discuss supernumerary genomes within a single organism. The acceptance of the idea that the origin of symbiotic cells via the engulfment of bacterial prey, which eventually morphed to become mitochondria or chloroplasts, stemmed largely from her efforts (Margulis, 1971; Margulis and Fester, 1991; Margulis, 1993). This Serial Endosymbiosis theory cannot be overstated as it proposed the likely source of the first eukaryotic cells.

The concept of symbiosis has become very prevalent in the literature and even popular thinking and should approach the theory of evolution in terms of prominence in biology. However, because each symbiosis is most likely unique, we may not see a grand theory or unification of symbiosis in the near future. Moreover, symbiosis research remains a relatively understudied field—compared with cancer biology or genomics—yet more diverse and frequent symbiosis studies reflect a growing recognition of the importance and prevalence of symbiotic interactions in biology (Margulis, 1993; Nakabachi et al., 2005; Martin and Nehls, 2009; Lopez, 2019). Of course, as in most areas, we place ourselves, *Homo sapiens*, above all of the rest, and so the earliest and most extensive, well-funded initiatives in symbiosis research and microbial communities (also known as "microbiomes") research were of ourselves in the form of the Human Microbiome Project (HMP). Starting in the late 2000s, the US National Institutes of Health (NIH) embarked upon a decade-long project to characterize the microbial passengers, which live on and within our bodies with current molecular methods. HMP Phase I ran from 2008 to 2013 and aimed at characterizing the microbiomes at various anatomical sites from 300 healthy individuals and generated over 32 terabytes of data, which are all still publicly available through the HMP Data portal—https://portal.hmpdacc.org/ (Human Microbiome Project (HMP) Consortium, 2012). Deep sequencing characterized microbiomes in the gut, mouth, skin, nose, genitals, and 10 other sites. For example, multiple HMP studies confirmed that our GI tracts had the most variable microbiomes (beta diversity), dominated by unculturable Bacteroides. In addition, human skin microbiomes also displayed sufficiently high beta diversity to the extent it could be used for basic differentiation of different individuals if not unequivocal forensic identification (Hampton-Marcell et al., 2017).

The HMP project continues to generate a large amount of useful data and technical innovations that have provided interesting and unexpected impacts in the fields of medicine and medical microbiology and its subfields of immunology, infectious disease, epidemiology, and more distantly ecology and genomics (Knight et al., 2012; Caporaso et al., 2011; Lloyd-Price et al., 2017). Because of the reliance on culture-independent metagenomics methods, this field also blossomed with the various HMP studies.

In parallel, examples of obligate symbiosis in the sea appear boundless. Marine symbiosis research between microbial communities and their eukaryotic hosts has made significant progress (Haygood et al., 1999; Cavanaugh et al., 2006; Goffredi et al., 2007; Hentschel et al., 2003; Webster and Taylor, 2012; Voolstra et al., 2009; Thomas et al., 2016; Easson et al., 2020; Berger et al., 2021; McKenna et al., 2021; Medina et al., 2021). Each association is unique, such as whether symbiont transfer is vertical (through gametes) or horizontal (from the environment). This variation makes studying symbiosis much more interesting.

One of the best-characterized marine symbioses is the one between *Euprymna scolopes* (Hawaiian bobtail squid) and *Vibrio fischeri* (Koropatnick et al., 2004; McFall-Ngai et al., 2012; Nyholm and McFall-Ngai, 2004, 2021). This system involves specific molecular recognition, and migration to a light organ for bioluminescence by *Vibrio fischeri*, which is established at each host generation. The bioluminescence generated by the bacteria serves as an antipredatory defense. *Euprymna* releases chitin oligosaccharides to attract the Vibrio, while specific squid microbe-associated molecular patterns (MAMPs) (a slight derivative of "pathogen-associated molecular patterns" and previously mentioned PRRs) have been characterized that help mediate immune responses. MAMPs are known across the phylogenetic spectrum. Nyholm and McFall-Ngai (2021) discuss the rhythm of gene products such as peptidoglycan recognition protein 2 (PGRP2), alkaline phosphatase (AP), galaxin, cathepsin L, complement protein C3, and blood pigment haemocyanin, which has a C-terminal peptide with antimicrobial activity, and help to maintain the symbiosis into squid adulthood. The diversity of aquatic taxa and natural histories provides multiple opportunities to characterize common mechanisms for symbiont initiation and maintenance at the genetic level.

44 **Chapter 1** The seabed—Where life began and still evolves

Photosymbiotic traits tend to increase organismal biomass, provide another source of energy, and encompass a wide phylogenetic spectrum including corals, hydra, sponges, acoel worms, molluscs, protists, and tunicates. For example, kleptoplastic *Elysia* saccoglossans only sporadically incorporate chloroplasts into the host tissues after they consume algae. Multiple species of *Elysia* carry out kleptoplasty (Pierce et al., 2007; Rumpho et al., 2008; Middlebrooks et al., 2019). The algae consumed are exclusively coenocytic, a condition characterized by a lack of cell walls. This results in multinucleate unicellular macroalgae. Usually, this is an exclusive relationship between a specific species pair, such as between *Elysia chlorotica* and *Vaucheria litorea*, or *Elysia timida* and *Acetabularia acetabulum*. One study of the sea slug *Elysia timida* indicated that host-mediated behavioral and kleptoplast physiological photoprotection would be required to counter reactive oxygen species (ROS) from the stolen organelles (Cartaxana et al., 2019). *Elysia crispata* is a South Florida species, which retains functional chloroplasts up to 120 days after feeding, and its genome has been submitted to the ASG for whole genome sequencing. Ephemeral *Elysia* symbioses can be contrasted with multiple longer-term coral—*Symbiodinium* associations, which can be disrupted by environmental stresses (e.g., high sea surface temperatures). Medina et al. (2021) have been developing the upside-down jellyfish *Cassiopea xamachana* as a versatile model to address photosymbiosis, and recently identified dinoflagellate symbiont carotenoids which can promote the development of *C. xamachana* from the sessile polyp to the sexual medusa stage (Ohdera et al., 2022).

I have recently expounded on various features of sponge-microbial symbiosis systems (Lopez, 2019), so I will not retread too much of that here. Through a continuous flurry of activity, sponge symbioses research has advanced even though the number of potential bacterial symbionts per host species is superfluous (Thomas et al., 2016; Pita et al., 2018). When sponge-associated microbiome research confirmed that specific sponge species could harbor unique and taxon-dependent consortia of symbionts, wider biodiversity conservation implications could be immediately recognized (Lopez et al., 1999a,b; Hentschel et al., 2003, 2012; Taylor et al., 2007; Webster et al., 2010; Thomas et al., 2016; Moitinho-Silva et al., 2017; Cleary et al., 2019). This endeavor in itself could be endless as the microbiomes of all 8000+ Poriferan species have barely been characterized. Moreover, the "rare microbial biosphere" paradigm that most habitats, including unique symbiotic ones, will apply and predict that many species will contain novel microbial diversity.

One study has recently analyzed deep-sea sponge symbioses by surveying 52 locations, 1077 host individuals from 169 sponge species (including understudied glass sponges) (Busch et al., 2022). The results support previous deep-sea sponge analyses, by confirming the dominance of *Chloroflexi,* as well as Anck6, *Dadabacteria, Entotheonellaeota, Nitrospirota,* PAUC34f, and Spirochaetota in sponge mesohyl compared with sediment and seawater controls. A wide microbial diversity was characterized, including a total of 53,756 amplified sequence variants (ASVs) of 16S rRNA, which represented 201 bacterial classes, 379 orders, 463 families, and 747 genera. A comparison of the more recent deepwater sponge results was also performed in an earlier microbiome study by Thomas et al. (2016), which involved 81 species of mostly shallow sponges. OTU classifications also had to be converted to ASV equivalents. Although both are valid methods to characterize amplified 16S rRNA gene fragment sequences from a mostly uncultured community or sample, there are significant differences. To briefly explain, ASVs are generated by using high-throughput sequencing techniques to read unique DNA sequences that can differ by a single nucleotide. This potentially identifies unique amplicons generated from each bacterial species present in a sample. Alternatively, the older approach of OTU calling uses algorithms to cluster similar sequences based on a chosen similarity threshold, such as 97% similarity threshold to define groups of 16S rRNA sequences as belonging to the same species. Although the OTU method is simpler and computationally less intensive than ASV-based approaches, the former may underestimate overall microbial diversity. For example, different sequences that differ by only a few nucleotides may be classified as belonging to the same OTU. Additionally, ASV-based methods have been shown to provide more accurate and reproducible results than OTU-based methods, particularly for low-abundance and rare microbial taxa in complex microbial communities (Callahan et al., 2016). The comparison of both deep and shallow datasets agreed with the consensus that sponge host taxonomy remains a major driver of microbial community composition, and low microbial abundance (LMA) sponges had fewer host-specific ASVs and was affected more by environmental factors. Moreover, high microbial abundance (HMA) sponges had significantly higher taxonomic richness than LMA sponges. Again, one implication relevant to benthic biodiversity is that the loss of a single host species could translate to the concomitant loss of a unique microbial ecosystem, replete with its potential services, novel taxa and gene sequences, metabolic pathways, and cooccurrances (Lopez et al., 1999a; Paoli et al., 2022).

Corals provide the structure for many benthic ecosystems

As will be discussed more throughout this volume, one of the most well-known sources of benthic biodiversitry and marine symbioses are the reef building stony (scleractinian) coral hosts and their dependency on single-cell dinoflagellate algae (also called zooxanthellae) in the family Symbiodiniaceae, which helps produce sufficient energy and carbon to produce calcium carbonate skeletons (Muscatine and Porter, 1977; Trench, 1993). This partnership enables the construction of shallow coral reefs, which are one of the most biodiverse habitats on the planet (Knowlton et al., 2010; Fisher et al., 2015). Photosymbiosis is a specific subset of general symbioses and involves a photosynthetic partner, which can fix carbon dioxide from the atmosphere and convert it to glucose. This dependency has evolved several times throughout evolution (Goulet et al., 2019). For example, the versatility of *Symbiodinium* symbiosis and its variants can be observed through the wide spectrum hosts the algae has—corals, anemones, jellyfish, nudibranchs, Ciliophora, Foraminifera, zoanthids, sponges, and giant clams (Stat et al., 2006).

The nutritional interactions between corals and their algal symbionts create the existence of coral reefs in oligotrophic waters, one of the largest (and sometimes oldest) biogenic structures on the planet (Veron, 1995; Morris et al., 2019). Tight nutrient recycling within the holobiont provides the algal symbionts with respiratory CO_2 and nitrogenous waste products, and in exchange the coral host receives photosynthetically fixed carbon (Weiss, 2008; Davy et al., 2012). Still many questions remain regarding the establishment, maintenance, and causes for the loss of the symbiosis during stress (Rivera and Davies, 2021). Because of the relatively singular (1:1) association between different organisms, the scleractinian-dinoflagellate holobiont system has been elevated as a major symbiosis model. Upon greater examination, coral symbiosis can be just as complicated as other systems, e.g., multiple *Symbiodinium* strains can occur in the same coral colony or host (LaJeunesse, 2002; Kemp et al., 2008). For example, several million *Symbiodinium* cells can inhabit a square centimeter of coral tissue, which makes their presence significant in the metabolism and nourishment of the system (Muscatine and Porter, 1977).

46   Chapter 1 The seabed—Where life began and still evolves

Muscatine provided pioneering studies of dinoflagellates associated with corals (Muscatine, 1974; Muscatine and Porter, 1977; Hoegh-Guldberg et al., 2007). This was further classified into groups based on the specificity of their host associations through multiple innovative studies and then molecular characterizations. For example, while the author carried out a postdoctoral position at the Smithsonian Tropical Research Station (STRI) in Panama in the mid-1990s, it was quite exhilarating to be able to observe how Rob Rowan performed his molecular classifications of various Symbiodinium using rRNA amplicons. These were mainly late-night gels or electrophoresis by starlight. At that stage of development, the Symbiodinium clades were designated only by a simple lettering system. Rowan explored the mechanisms of thermotolerance between coral and their symbiotic partners, how they interact and exchange nutrients. Rowan has also investigated the potential for corals to adapt to changing environmental conditions through symbiotic associations with different types of dinoflagellates (Rowan et al., 1997; Rowan, 2004). In addition, contemporaneous at this time was working with Andrew Baker, who was carrying out similar in situ work on the corals for his dissertation, which also took place I was also in the Knowlton laboratory working as a postdoc on molecular fingerprinting of the *Montastraea* (later changed to *Orbicella*) *annularis* species complex (Lopez et al., 1999a,b). Baker went on to propose the adaptive potential of bleaching from the zooxanthellae perspective, which was met by some debate in the coral science community (Baker et al., 2004).

After years of honing the methods to refine the molecular taxonomy of *Symbiodinium*, such as with the nontrivial denaturing gradient gel electrophoresis (DGGE), LaJeunesse et al. (2018) clarified the systematics of the various *Symbiodinium* "clades" by asserting their equivalence to genera within the family Symbiodiniaceae. This effectively codified the taxonomy and provided official Latin binomials for the major groups, standardizing the nomenclature. The literature is replete with more revisions and characterizations of this important microalgal-coral symbiosis, and I will not go into these in detail here (Hoadley et al., 2021; Hume et al., 2016). However, as this symbiosis occurs in multiple species of stony corals and forms the crux of one of the most biodiverse ecosystems on the planet, its importance cannot be overstated.

What may be more impactful are the mechanisms that cause and maintain symbioses. The knowledge of interactions at the molecular level continues to accumulate. Levy et al. (2021) have summarized a dense atlas of newly sequenced coding RNA that affects coral symbiosis, calcification, and immunity. They have applied current cutting-edge methods such as single-cell RNAseq with fluorescence-activated cell sorting (FACS) to profile the tissue and cell-specific expression of genes in the reef-building Indo-Pacific stony coral *Stylophora pistillata*. For example, 353 host cell-specific genes with multiple metabolic functions were overexpressed in algal-associated cells compared with nonsymbiotic gastrodermal cells. This included lipid transporters, *ApoD* and *NPC1*, amino acid and peptide transporters in the folliculin complex, and key enzymes in the glutathione pathway, which can protect against ROS caused by photosynthesis. We also know several cnidarian taxa that do not host zooxanthellae, such as *Tubastraea*, and the group cannot accrete carbonate for skeleton structures.

As in other symbiotic microbiome studies, more results point to evidence that the most stable symbiotic bacterial microbiomes can protect their coral hosts from ROS arising from heat stress (Ziegler et al., 2017). Similar to the important role of bacterial symbionts for many sponge species, coral microbiome stability and composition have impact for the holobiont (Rohwer et al., 2002; Zaneveld et al., 2017). For example, taxa such as *Endozoicomonas,* Midichloriaceae, and Spirochaetaceae appear correlated to coral thermal tolerance (Maher et al., 2020; Palacio-Castro et al., 2022). More evidence points to microbial symbionts' effects on a host's methylome, an epigenomic dimension, sometimes in response to environmental stressors (Durante et al., 2019; Tong et al., 2020; Mohamed et al., 2023). At this time, all of the specific mechanisms of how the symbionts protect corals are not known. Common coral symbiont *Endozoicomonas* may protect against coral pathogens via competitive exclusion after evidence of bleaching (Neave et al., 2017). Interestingly, this common coral symbiont appears absent in deep-sea corals, perhaps due to lower pH and temperatures (Kellogg, 2019). Kawamura et al. (2021) have recently proposed using more intensive cell culture approaches to dissect the host and algal interactions at the molecular level. The researchers performed in vitro experiments with the coral *Acropora tenuis* and thedinoflagellate, *Breviolum minutum* to study the conditions required for symbiosis. The presence of the dinoflagellates caused changes in coral cell mobility and morphology, and overall the coral cells provided a suitable environment for the dinoflagellates to perform photosynthesis.

Switching to the deep oceans briefly, studies of the symbiosis of giant vestimentiferan tube worms (*Riftia pachyptila*) found at deep-sea vents have been iconic and show dependence on symbiotic bacteria. These carry out chemosynthetic metabolism based on the reducing power in hydrogen sulfide, gases, and minerals of the vent fluids to convert them into organic matter, which the tube worm (and other vent macrofauna) can then consume. In turn, the tube worms provide the bacteria with a structure, safe environment, and essential compounds for their survival. Thus, living in deep habitats requires symbiotic interactions, because there are limited sources of energy when there is no sunlight. The discovery of the tube worms in 1977 by Jack Corliss was a startling surprise on the expedition of the American bathyscaphe DSV *Alvin* to the Galapagos Rift. The worms lack a digestive tract, mouths, and anuses, which reflect their total dependence on their symbiotic bacteria (Smith and Reef, 1985). The mutualistic symbiosis was elegantly described by Cavanaugh et al. (1981) and others (Van Dover et al., 2001).

Other chemosymbiotic associations have been well documented at deep seafloor habitats such as hypersaline anoxic basins, sediments, seeps, and brine pools. For example, Danovaro et al. (2010) described three new species of Loriciferans (*Spinoloricus* nov. sp., *Rugiloricus* nov. sp., and *Pliciloricus* nov. sp.) in deep-sea sediments of the L'Atalante basin (Mediterranean Sea). This was the first evidence of survival in totally anoxic sediments supported by likely endosymbiotic anaerobic bacteria. In another example, cold seep communities have been characterized along the continental shelf of the Gulf of Mexico (GoM) since 1984 at depths ranging from 500 to 1000 m (Cordes et al., 2009). The release of dissolved methane, hydrogen sulfide, and hydrocarbons can have toxic effects, and the deep waters themselves allow this zone to be dubbed extreme. In these H_2S-rich environments, marine invertebrates have evolved several strategies, such as a mitochondrial sulfur detoxification pathway, which converts sulfide into sulfite to neutralize toxic sulfide effects (Tobler et al., 2016). In this context, at Brine Pool NR1, a brine-filled pockmark in the Gulf of Mexico, the mussel, *Bathymodiolus childressi* thrives with methanotrophic endosymbionts (Smith et al., 2000). Further analyses of feeding habits, temporal gradients, and comparison with other proximal sites indicate more flexibility and supplementation of symbioses with particulate suspension feeding (Riekenberg et al., 2016).

The diversity of aquatic taxa and natural histories provides multiple opportunities to explore and characterize common mechanisms for symbiont initiation and maintenance at the genetic level. Genomic and metatranscriptomic sequence data of holobionts have added novel gene discoveries and their relative expression levels, which in turn will provide important insights into the presence and role of molecular signals affecting metazoan-microbial symbioses. Initiated in 2019, the Aquatic Symbiosis Genome (ASG) Project has developed into a recent manifestation of whole genome consortia.

48 **Chapter 1** The seabed—Where life began and still evolves

Spearheaded by funding from the Gordon and Betty Moore Foundation (GBMF) in partnership with the Wellcome Sanger Institute (UK) (McKenna et al., 2021 and https://www.sanger.ac.uk/collaboration/aquatic-symbiosis-genomics-project/), ASG cooperates and dovetails with Sanger's Darwin Tree of Life Project, which aims to determine whole genome sequences to describe the biodiversity in the British Isles (DToL, 2022). The ASG plan calls for the chromosome level, whole genome sequencing up to 1000 aquatic organismal holobionts, consisting of at least 500 pairs of macro- (host) and their associated microsymbiotic partners, which together amount to 1000 new genomes.

With some of these factors in mind, and through an open competition, the GBMF has established the following 10 multicollaborator hubs for whole genome sequencing based on either taxonomic focus or question to determine whether genome sequences could answer longstanding marine symbioses questions:

- Sponges as symbiont communities
- Photosymbiosis in marine animals
- Coral symbiosis sensitivity to environmental change
- Evolution of new symbioses in single-celled eukaryotes
- Ciliates as models for symbiosis
- Genomic signatures behind the origin of multiple cephalopod symbiotic organs
- Bacterial symbiosis as an adaptation to extreme environments in annelid worms
- Evolution and ecology of fungal-algal symbioses across marine and freshwater environments
- Symbiosis as a driver for molluscan diversity
- Symbioses in 3D: diversity and dynamics in pelagic symbioses across the tree of life

The ASG project intends to generate massive genome datasets, about 100 species per hub, which will be released quickly to the public (e.g., to trainees) through portals and Wellcome Sanger GenomeNote publications. At the time of writing, the ASG project is still ongoing and has been extended through 2024, due to some delays caused by the COVID-19 pandemic. This enables the nascent theories revolving around diverse symbioses keep pace with growing sequence datasets. For example, González-Pech et al. (2023) emphasize how a "holobiont" approach will be required to fully grasp the mechanics of each unique symbiotic association and thus marine ecology in general.

Sequences alone probably cannot provide all of the answers, however. Therefore, new tools and approaches are now being earnestly applied to illuminating the mechanisms of symbiosis establishment and maintenance (Engelberts et al., 2021). This includes NanoSIMS and Raman spectroscopy to track isotopic incorporation of N and C in DOM and POM, and modified Fluorescence in situ hybridization (FISH) methods to identify the spatial location of specific symbionts (Lopez, 2019; Engelberts et al., 2021).

From analyses of these various types of symbioses, we can observe that several of the associations are fragile and can be easily broken. This leads to the question of why this dependency is so tenuous, or what are the evolutionary advantages to the holobiont? The coral bleaching events indicate that fitness is much reduced when the environment turns for the worse. It seems a more permanent relationship, such as chloroplast organelles, which are integral parts of plant cells, would be more adaptive than allowing symbiotic dinoflagellates to leave whenever the environment gets too hot (above 31°C).

We can propose that symbiosis is a major driver of biological evolution especially on the seabed, as the phenomenon provides almost endless new sources of adaptive innovation, metabolism, and genes for new recombinations in an energy starved habitat. Indeed many examples have now accumulated that both intracellular and extracellular horizontal gene transfers occur between symbiotic partners or organelles (Lopez et al., 1994; Rumpho et al., 2008; Degnan, 2014; Perreau and Moran, 2022). The role of symbiosis in benthic habitats cannot be overstated. Perhaps the sessile lifestyle lends itself to these unique interactions.

Our university once had a visit from an alumnus in the mid-2010s, who worked at NASA and presented the agency's future plans for Mars exploratory programs. Around the same time, ice was verified on the red planet (Schorghofer and Forget, 2012), and then the Hollywood movie "The Martian" was released. With all of this coincidental activity, I noted that none of events mentioned any aspects of microbiology, including when showing the plants fictitiously thriving on Martian soil, but were likely missing essential microbial plant symbionts. The rhizomes and mycorrhiza support various plant growths on earth by helping them obtain nitrogen in the atmosphere or minerals from the soil through the roots (Douglas, 2021). This deficiency led us to opine about the need for developing more experimental microbiological mesocosms to be used in space (Lopez et al., 2019a,b).

In this context, although many of our labs have been focused on marine habitats such as the deep sea and coral reefs, we can still view these benthic habitats as very "alien"-like, because people cannot easily or routinely visit these unique places. However, we could envision a future where visiting the ocean bed could be easier and more routine. In addition, the wide diversity of relatively unknown organisms that live under the waves, make the oceans a great place to explore, just like those who want to explore the solar system. The main difference is that the oceans are in our proverbial backyards, teeming with life but sometimes deteriorating too quickly. Thus, alien marine habitats need our help and protection, and drawing the contrast with space was meant to increase an overall appreciation for our planet's resources. A quote by Lynn Margulis I regularly convey to my undergraduate students in microbiology class is relevant to this topic: "We could wipe out all of the plants and the animals in the world, and life would eventually return. However, if we wiped out all of the microbes, our planet would become as barren as the moon." (Margulis, 1993).

Possible origins of biodiversity

Biological diversity (and its underlying genetic variation) is a cornerstone of evolution because current theory posits that forces such as natural selection or genetic drift act upon the existing variation in extant populations. Very rarely does a new variant appear out of thin air or from nothing, i.e., *creation ex nihilo* (Fig. 1.3). This can also invoke the childhood conundrum and riddle of "which came first—the chicken or the egg?" which can also apply to the concept of origins and speciation we have been discussing. Biologists know the answer should be the egg. The chicken had to already exist in order to lay the egg, and thus during the fertilization and development processes of the given organism, changes (such as mutations to its genetic code) could occur that ultimately affect the creature's life history and trajectory after it emerges. This applies to genetics as well, as new functional DNA code takes time to develop and often uses preexisting sequences. Now, data actually exist to address this riddle in molecular terms and in the context of de novo "gene birth": does a (i) basic gene "open reading frame (ORF)" gene structure (presence of regulatory promoter, start codon, etc.) or (ii) "transcription" first model appear more evident (Schlötterer, 2015)? In *Drosophila*, ORFs appear before they are transcribed into RNA (Zhao et al., 2014). We return to envisioning the possible birth of genes later in Chapter 3.

FIG. 1.3

Sculpture representing humanity in the context of "Creation ex nihilo" by Frederick Hart at Flagler College, St. Augustine.

The currently accepted theories of evolution are based on what was found in the solid geological and fossil records and promoted by Charles Lyell. His influence on Darwin's theory of natural selection was immense. However, these influences do not allude to the origin of cells or life itself. Complicating this perennial question is that microbes generally did not leave very good fossil records, and so we lack concrete evidence of their history and mechanisms of ascendance. David and Alm (2011) assert that genes can represent these fossils, like those appearing during the Great Oxidation Event around 2.5 billion years ago.

Another possibility to consider is that even at the very dawn of life, the ooze of benthic substrates could have been pivotal. Molecules sometimes need a platform to carry out their activities, and early molecules such as RNAzymes would not be an exception (Lazcano and Miller, 1996). We are interested in the type of molecular and cellular diversity that radiated from humble origins, possibly from marine sediments. Life likely began as a single-celled prokaryote, which then evolved slowly and gradually from there to form more complex multicellular aggregates, metabolism, and organisms. Several elaborate theories and discussions of how multicellular organisms could have evolved, have been proposed (Sepkoski, 1991; King, 2004; Mikhailov et al., 2009). Traditional origin ideas have pointed to a planktonic origin for metazoa. This is supported by free-swimming choanoflagellates, which potentially lead to sponges as the first steps in multicellularity and thus innovation. The fossil record supports the origin of metazoa from benthic habitats, as Cambrian benthic communities displayed relatively few suspension feeders (Schopf and Klein, 1992; Signor and Vermeij, 1994). As expected, the transition from plankton to an irreversibly attached juvenile sessile state may not have been trivial. The transformation requires cellular differentiation apparatuses and cells that can mediate and provide adhesive properties (Gleason and Hofmann, 2011). The whole development process of transformation and ability to respond to settlement cues was so appreciated that the term *competence* was coined by Waddington and Needham (1936). Thus, settlement to the seabed itself likely requires specific biochemical cues. Poriferan species were likely the first suspension feeders arising in the late Proterozoic (Ausich and Bottjer, 1985), and abundant bacterioplankton and DOM could have provided the sustenance for this lifestyle. Various authors have suggested seafloor and even deep-sea possibilities.

With respect to origins, hydrothermal vents and the various expanding mid ocean rifts provide sufficient energy to sustain complex communities (Van Dover et al., 2002). Several studies provide evidence that deep-water hydrothermal vents could sustain an origin story, as the habitats yield sufficient energy, nutrients to support the essential biochemical transformations for life. The available CO_2-reducing geochemistry is likely suitable for life formation (Russell and Martin, 2004; Martin et al., 2008; Russell et al., 2010). At hydrothermal vents, methanogens and acetogens satisfy biological carbon needs through the acetyl-coenzyme A pathway, an energy-releasing pathway of CO_2 fixation, when there is sufficient environmental H_2 and carbon in the water column. Vents, however, fall behind shallower photochemical origins of life in total energy flow, yet both have the proven capacity to produce most of the essential amino acid and nucleic acid building blocks (Rimmer and Shorttle, 2019). Nonetheless, vent biology is based on chemosynthesis as hydrothermal vents typically occur well below the photic zone. Various basal clades of animals exist at hydrothermal vents, such as deep-sea mussels, crabs, and carnivorous Cladorhiza sponges, which lie at the periphery of vent zones (Georgieva et al., 2020). These sponges appear to host chemoautotrophic Gammaproteobacteria belonging to the Thioglobaceae and Methylomonaceae lineages.

Hydrothermal vent habitats underscore how the environment factors can supersede the organism and its biology in particular deep-sea situations. The dominance of physical oceanographic factors (temperature, pressure, convection, lack of organic carbon) becomes even more apparent at greater depths and dictates the composition of this community. However, the versatility of bacterial metabolism and diversity helped create symbiotic innovations, which enabled the vent community existence in the first place. We have just begun to describe some of these benthic habitats and organisms, with their specific roles, dependencies, and requirements. Before we continue with the biodiversity assessments, we will first review philosophical and technical approaches for the assessments.

References

Acinas, S.G., Sánchez, P., Salazar, G., Cornejo-Castillo, F.M., Sebastián, M., Logares, R., Gasol, J.M., 2021. Deep ocean metagenomes provide insight into the metabolic architecture of bathypelagic microbial communities. Commun. Biol. 4 (1), 604.

52 Chapter 1 The seabed—Where life began and still evolves

Addadi, L., Weiner, S., 2014. Biomineralization: mineral formation by organisms. Phys. Scr. 89 (9), 098003.

Ahn, J.E., Ronan, A.D., 2020. Development of a model to assess coastal ecosystem health using oysters as the indicator species. Estuar. Coast. Shelf Sci. 233, 106528.

Aitken, F., Foulc, J.-N., 2019. From Deep Sea to Laboratory. 1: The First Explorations of the Deep Sea by H.M.S. Challenger (1872–1876). Iste-Wiley, London.

Albertin, C.B., Simakov, O., Mitros, T., Wang, Z.Y., Pungor, J.R., Edsinger-Gonzales, E., et al., 2015. The octopus genome and the evolution of cephalopod neural and morphological novelties. Nature 524, 220–224.

Alldredge, A.L., Gotschalk, C.C., 1990. The relative contribution of marine snow of different origins to biological processes in coastal waters. Cont. Shelf Res. 10 (1), 41–58.

Alldredge, A.L., Silver, M.W., 1988. Characteristics, dynamics and significance of marine snow. Prog. Oceanogr. 20 (1), 41–82.

Alldredge, A.L., Passow, U., Haddock, H.D., 1998. The characteristics and transparent exopolymer particle (TEP) content of marine snow formed from thecate dinoflagellates. J. Plankton Res. 20 (3), 393–406.

Alonso-Sáez, L., Waller, A.S., Mende, D.R., Bakker, K., Farnelid, H., Yager, P.L., Bertilsson, S., 2012. Role for urea in nitrification by polar marine archaea. Proc. Natl. Acad. Sci. U. S. A. 109 (44), 17989–17994.

Alves, A.S., Caetano, A., Costa, J.L., Costa, M.J., Marques, J.C., 2015. Estuarine intertidal meiofauna and nematode communities asindicator of ecosystem's recovery following mitigation measures. Ecol. Indic. 54, 184–196. https://doi.org/10.1016/j.ecolind.2015.02.013.

Amorim, J., Abreu, I., Rodrigues, P., Peixoto, D., Pinheiro, C., Saraiva, A., Oliva-Teles, L., 2019. Lymnaea stagnalis as a freshwater model invertebrate for ecotoxicological studies. Sci. Total Environ. 669, 11–28.

Andersson, D.I., Hughes, D., 1996. Muller's ratchet decreases fitness of a DNA-based microbe. Proc. Natl. Acad. Sci. U. S. A. 93 (2), 906–907.

Andradi-Brown, D., Appeldoorn, R.S., Baker, E., Ballantine, D., 2016. Mesophotic coral ecosystems: a lifeboat for coral reefs? In: UN Environment. GRID-Arendal.

Anthony, M.A., Bender, S.F., van der Heijden, M.G., 2023. Enumerating soil biodiversity. Proc. Natl. Acad. Sci. U. S. A 120 (33), e2304663120.

Appeltans, W., Ahyong, S.T., Anderson, G., Angel, M.V., Artois, T., Bailly, N., Błażewicz-Paszkowycz, M., 2012. The magnitude of global marine species diversity. Curr. Biol. 22 (23), 2189–2202.

Araújo, M.B., Anderson, R.P., Márcia Barbosa, A., Beale, C.M., Dormann, C.F., Early, R., Rahbek, C., 2019. Standards for distribution models in biodiversity assessments. Sci. Adv. 5 (1), eaat4858.

Assis, J., Coelho, N.C., Lamy, T., Valero, M., Alberto, F., Serrão, E.Á., 2016. Deep reefs are climatic refugia for genetic diversity of marine forests. J. Biogeogr. 43 (4), 833–844.

Ausich, W.I., Bottjer, D.J., 1985. Phanerozoic tiering in suspension-feeding communities on soft substrata: Implications for diversity. In: Valentine, J.W. (Ed.), Phanerozoic Diversity Patterns: Profiles in Macroevolution. Princeton University Press, pp. 255–274.

Avila, C., Angulo-Preckler, C., 2021. A minireview on biodiscovery in Antarctic marine benthic invertebrates. Front. Mar. Sci. 8, 686477.

Baas-Becking, L.G.M., 1934. Geobiologie of inleiding tot de milieukunde. W.P. Van Stockum & Zoon, The Hague, the Netherlands.

Baird, D., Ulanowicz, R.E., 1989. The seasonal dynamics of the Chesapeake Bay ecosystem. Ecol. Monogr. 59 (4), 329–364.

Baker, A.C., Starger, C.J., McClanahan, T.R., Glynn, P.W., 2004. Corals' adaptive response to climate change. Nature 430 (7001), 741.

Baker, C., Potter, A., Tran, M., 2008. Sedimentology and geomorphology of the north west region: a spatial analysis. Geoscience Australia Record 2008/07. Geoscience Australia, Canberra, p. 220.

Bang, A.E., 2020. The Biodiversity of Mud Dragons (Kinorhyncha) in the Fjords of Møre Og Romsdal, Norway. Unversity of Oslo (Master's Thesis).

Barnes, D.K., Griffiths, H.J., 2008. Biodiversity and biogeography of southern temperate and polar bryozoans. Glob. Ecol. Biogeogr. 17 (1), 84–99.

Barnett, R., Yamaguchi, N., Barnes, I., Cooper, A., 2006. The origin, current diversity and future conservation of the modern lion (Panthera leo). Proc. R. Soc. B Biol. Sci. 273 (1598), 2119–2125.

Barshis, D.J., Ladner, J.T., Oliver, T.A., Seneca, F.O., Traylor-Knowles, N., Palumbi, S.R., 2013. Genomic basis for coral resilience to climate change. Proc. Natl. Acad. Sci. U. S. A. 110 (4), 1387–1392. https://doi.org/10.1073/pnas.1210224110 (Epub 2013 Jan 7).

Beal, L.M., Hummon, J.M., Williams, E., Brown, O.B., Baringer, W., Kearns, E.J., 2008. Five years of Florida current structure and transport from the Royal Caribbean Cruise Ship explorer of the seas. J. Geophys. Res. Oceans 113 (C6).

Bean, T.P., Khatir, Z., Lyons, B.P., van Aerle, R., Minardi, D., Bignell, J.P., Leitão, A., 2020. De novo transcriptome assembly of the Qatari pearl oyster Pinctada imbricata radiata. Mar. Genomics 51, 100734.

Bensasson, D., Zhang, D.X., Hartl, D.L., Hewitt, G.M., 2001. Mitochondrial pseudogenes: evolution's misplaced witnesses. Trends Ecol. Evol. 16, 314–321.

Berger, L., Blackwelder, P., Frank, T., Sutton, T., Pruzinsky, N., Slayden, N., Lopez, J.V., 2021. Microscopic and genetic characterization of bacterial bioluminescent symbionts of the Gulf of Mexico Pyrosome, *Pyrosoma atlanticum*. Front. Mar. Sci. 8, 606818. https://www.frontiersin.org/articles/10.3389/fmars.2021.606818/full.

Bergquist, P., 1978. Sponges. Hutchinson, London. 268 pp.

Bianchelli, S., Buschi, E., Danovaro, R., Pusceddu, A., 2018. Nematode biodiversity and benthic trophic state are simple tools for the assessment of the environmental quality in coastal marine ecosystems. Ecol. Indic. 95, 270–287.

Billett, D.S.M., Lampitt, R.S., Rice, A.L., Mantoura, R.F.C., 1983. Seasonal sedimentation of phytoplankton to the deep-sea benthos. Nature (London) 302, 520–522.

Bird, C.E., Holland, B.S., Bowen, B.W., Toonen, R.J., 2011. Diversification of sympatric broadcast-spawning limpets (Cellana spp.) within the Hawaiian archipelago. Mol. Ecol. 20 (10), 2128–2141.

Boehm, A.M., Khalturin, K., Anton-Erxleben, F., Hemmrich, G., Klostermeier, U.C., Lopez-Quintero, J.A., Oberg, H.H., Puchert, M., Rosenstiel, P., Wittlieb, J., et al., 2012. FoxO is a critical regulator of stem cell maintenance in immortal hydra. Proc. Natl. Acad. Sci. U. S. A. 109, 19697–19702.

Boeuf, D., Edwards, B.R., Eppley, J.M., Hu, S.K., Poff, K.E., Romano, A.E., DeLong, E.F., 2019. Biological composition and microbial dynamics of sinking particulate organic matter at abyssal depths in the oligotrophic open ocean. Proc. Natl. Acad. Sci. U. S. A. 116 (24), 11824–11832.

Bonfil, R., Meÿer, M., Scholl, M.C., Johnson, R., O'Brien, S., Oosthuizen, H., Paterson, M., 2005. Transoceanic migration, spatial dynamics, and population linkages of white sharks. Science 310 (5745), 100–103.

Bongaerts, P., Riginos, C., Ridgway, T., Sampayo, E.M., van Oppen, M.J., Englebert, N., Hoegh-Guldberg, O., 2010. Genetic divergence across habitats in the widespread coral Seriatopora hystrix and its associated Symbiodinium. PLoS One 5 (5), e10871.

Bosch, T.C., 2009. Hydra and the evolution of stem cells. BioEssays 31 (4), 478–486.

Bouchet, P., Lozouet, P., Maestrati, P., Heros, V., 2002. Assessing the magnitude of species richness in tropical marine environments: high numbers of molluscs at a New Caledonia site. Biol. J. Linn. Soc. 75, 421–436.

Bouchet, P., Lozouet, P., Sysoev, A., 2009. An inordinate fondness for turrids. Deep-Sea Res. II Top. Stud. Oceanogr. 56 (19-20), 1724–1731.

Bowen, B.W., Rocha, L.A., Toonen, R.J., Karl, S.A., 2013. The origins of tropical marine biodiversity. Trends Ecol. Evol. 28 (6), 359–366.

Bowen, B.W., Forsman, Z.H., Whitney, J.L., Faucci, A., Hoban, M., Canfield, S.J., Toonen, R.J., 2020. Species radiations in the sea: what the flock? J. Hered. 111 (1), 70–83.

Bower, A.S., Lozier, M.S., Gary, S.F., Böning, C.W., 2009. Interior pathways of the North Atlantic meridional overturning circulation. Nature 459, 243.

Boyd, P.W., Claustre, H., Levy, M., Siegel, D.A., Weber, T., 2019. Multi-faceted particle pumps drive carbon sequestration in the ocean. Nature 568, 327–335. https://doi.org/10.1038/s41586-019-1098-2.

Brazeau, D.A., Sammarco, P.W., Gleason, D.F., 2005. A multi-locus genetic assignment technique to assess sources of *Agaricia agaricites* larvae on coral reefs. Mar. Biol. 147 (5), 1141–1148. https://doi.org/10.1007/s00227-005-0022-5.

Bremner, J., Rogers, S.I., Frid, C.L.J., 2006. Matching biological traits to environmental conditions in marine benthic ecosystems. J. Mar. Syst. 60 (3-4), 302–316.

Bribiesca-Contreras, G., Dahlgren, T.G., Horton, T., Drazen, J.C., Drennan, R., Jones, D.O., Glover, A.G., 2021. Biogeography and connectivity across habitat types and geographical scales in Pacific abyssal scavenging amphipods. Front. Mar. Sci. 8, 705237.

Broecker, W.S., 1987. The biggest chill. Nat. Hist. 10, 74.

Brown, F.D., Swalla, B.J., 2012. Evolution and development of budding by stem cells: ascidian coloniality as a case study. Dev. Biol. 369 (2), 151–162.

Browne, N., Braoun, C., McIlwain, J., Nagarajan, R., Zinke, J., 2019. Borneo coral reefs subject to high sediment loads show evidence of resilience to various environmental stressors. PeerJ 7, e7382.

Brum, J.R., Ignacio-Espinoza, J.C., Roux, S., Doulcier, G., Acinas, S.G., Alberti, A., Sullivan, M.B., 2015. Patterns and ecological drivers of ocean viral communities. Science 348 (6237), 1261498.

Brusca, R., Giribet, G., Moore, W., 2022. Invertebrates, fourth ed. Sinauer Associates.

Buesseler, K.O., Lampitt, R.S., 2008. Introduction to "understanding the ocean's biological pump: results from VERTIGO". Deep-Sea Res. II Top. Stud. Oceanogr. 55 (14-15), 1519–1521.

Bull, A.T., Goodfellow, M., 2019. Dark, rare and inspirational microbial matter in the extremobiosphere: 16 000 m of bioprospecting campaigns. Microbiology 165 (12), 1252–1264.

Burge, C.A., Mouchka, M.E., Harvell, C.D., Roberts, S., 2013. Immune response of the Caribbean sea fan, Gorgonia ventalina, exposed to an Aplanochytrium parasite as revealed by transcriptome sequencing. Front. Physiol. 4, 180.

Burnet, F.M., 1971. Self-recognition in colonial marine forms and flowering plants in relation to the evolution of immunity. Nature 232, 230–235.

Busch, K., Slaby, B.M., Bach, W., Boetius, A., Clefsen, I., Colaço, A., Hentschel, U., 2022. Biodiversity, environmental drivers, and sustainability of the global deep-sea sponge microbiome. Nat. Commun. 13 (1), 5160.

Button, D.K., Schut, F., Quang, P., Martin, R., Robertson, B.R., 1993. Viability and isolation of marine bacteria by dilution culture: theory, procedures, and initial results. Appl. Environ. Microbiol. 59 (3), 881–891.

Cairns, S.D., 2004. The azooxanthellate Scleractinia (Coelenterata: Anthozoa) of Australia. Rec. Aust. Mus. 56, 259–329.

Cairns, S.D., 2007. Deep-water corals: an overview with special reference to diversity and distribution of deep-water scleractinian corals. Bull. Mar. Sci. 81 (3), 311–322.

Cairns, S.D., 2011. Global diversity of the stylasteridae (Cnidaria: Hydrozoa: Athecatae). PLoS One 6 (7), e21670.

Callahan, B.J., McMurdie, P.J., Rosen, M.J., Han, A.W., Johnson, A.J.A., Holmes, S.P., 2016. DADA2: high-resolution sample inference from Illumina amplicon data. Nat. Methods 13 (7), 581–583. https://doi.org/10.1038/n.

Caporaso, J.G., Lauber, C.L., Walters, W.A., Berg-Lyons, D., Lozupone, C.A., Turnbaugh, P.J., et al., 2011. Global patterns of 16S rRNA diversity at a depth of millions of sequences per sample. Proc. Natl. Acad. Sci. U. S. A. 108, 4516–4522.

Carballo, J.L., Bautista, E., Nava, H., Cruz-Barraza, J.A., Chávez, J.A., 2013. Boring sponges, an increasing threat for coral reefs affected by bleaching events. Ecol. Evol. 3 (4), 872–886.

Caron, J.-B., Morris, S.C., Cameron, C.B., 2013. Tubicolous enteropneusts from the Cambrian period. Nature 495, 503–506. https://doi.org/10.1038/nature12017.

Carranza, A., Defeo, O., Beck, M., 2009. Diversity, conservation status and threats to native oysters (Ostreidae) around the Atlantic and Caribbean coasts of South America. Aquat. Conserv. Mar. Freshwat. Ecosyst. 19, 344–353.

Carrerette, O., Güth, A.Z., Bergamo, G., Souza, B.H., Banha, T.N., Nagata, P.D., Sumida, P.Y., 2022. Macrobenthic Assemblages across Deep-Sea Pockmarks and Carbonate Mounds at Santos Basin, SW Atlantic. Ocean and Coastal Research, p. 70.

Cartaxana, P., Morelli, L., Jesus, B., Calado, G., Calado, R., Cruz, S., 2019. The photon menace: kleptoplast protection in the photosynthetic sea slug Elysia timida. J. Exp. Biol. 222 (12), jeb202580.

Carvalho, R., Wei, C.L., Rowe, G., Schulze, A., 2013. Complex depth-related patterns in taxonomic and functional diversity of polychaetes in the Gulf of Mexico. Deep-Sea Res. I Oceanogr. Res. Pap. 80, 66–77.

Carvalho, M.d.S., Lopes, D.A., Cosme, B., Hajdu, E., 2016. Seven new species of sponges (Porifera) from deep-sea coral mounds at Campos Basin (SW Atlantic). Helgol. Mar. Res. 70 (1), 1–33.

Cavalier-Smith, T., Chao, E.E.Y., 2006. Phylogeny and megasystematics of phagotrophic heterokonts (kingdom Chromista). J. Mol. Evol. 62, 388–420.

Cavanaugh, C.M., Gardiner, S.L., Jones, M.L., Jannasch, H.W., Waterbury, J.B., 1981. Prokaryotic cells in the hydrothermal vent tube worm Riftia pachyptila Jones: possible chemoautotrophic symbionts. Science 213 (4505), 340–342.

Cavanaugh, C.M., McKiness, Z.P., Newton, I.L., Stewart, F.J., 2006. Marine chemosynthetic symbioses. The Prokaryotes 1, 475–507.

Ceballos, G., Ehrlich, P.R., Barnosky, A.D., García, A., Pringle, R.M., Palmer, T.M., 2015. Accelerated modern human–induced species losses: entering the sixth mass extinction. Sci. Adv. 1 (5), e1400253.

Cerca, J., Meyer, C., Purschke, G., Struck, T.H., 2020. Delimitation of cryptic species drastically reduces the geographical ranges of marine interstitial ghost-worms (Stygocapitella; Annelida, Sedentaria). Mol. Phylogenet. Evol. 143, 106663.

Chapman, J.A., Kirkness, E.F., Simakov, O., Hampson, S.E., Mitros, T., Weinmaier, T., Steele, R.E., 2010. The dynamic genome of hydra. Nature 464 (7288), 592–596.

Chaves-Fonnegra, A., Feldhei, K.A., Secord, J., Lopez, J.V., 2015. Population structure and dispersal of the coral excavating sponge Cliona delitrix. Mol. Ecol. 24, 1447–1466. https://doi.org/10.1111/mec.13134.

Cleary, D.F., Swierts, T., Coelho, F.J., Polónia, A.R., Huang, Y.M., Ferreira, M.R., de Voogd, N.J., 2019. The sponge microbiome within the greater coral reef microbial metacommunity. Nat. Commun. 10 (1), 1644.

Coles, V.J., Brooks, M.T., Hopkins, J., Stukel, M.R., Yager, P.L., Hood, R.R., 2013. The pathways and properties of the Amazon River plume in the tropical North Atlantic Ocean. J. Geophys. Res. Oceans 118 (12), 6894–6913.

Colwell, R.R., McGrayne, S.B., 2021. A Lab of One's Own: One Woman's Personal Journey Through Sexism in Science. Simon and Schuster, New York.

Conley, D.J., Paerl, H.W., Howarth, R.W., Boesch, D.F., Seitzinger, S.P., Havens, K.E., Likens, G.E., 2009. Controlling eutrophication: nitrogen and phosphorus. Science 323 (5917), 1014–1015.

Cook, A.B., Bernard, A.M., Boswell, K.M., Bracken-Grissom, H., D'Elia, M., DeRada, S., Sutton, T.T., 2020. A multidisciplinary approach to investigate deep-pelagic ecosystem dynamics in the Gulf of Mexico following deepwater horizon. Front. Mar. Sci. 7, 548880.

Cooley, S.R., Yager, P.L., 2006. Physical and biological contributions to the western tropical North Atlantic Ocean carbon sink formed by the Amazon River plume. J. Geophys. Res. Oceans 111 (C8).

Cooley, S.R., Coles, V.J., Subramaniam, A., Yager, P.L., 2007. Seasonal variations in the Amazon plume-related atmospheric carbon sink. Glob. Biogeochem. Cycles 21 (3).

Corbo, J.C., Di Gregorio, A., Levine, M., 2001. The ascidian as a model organism in developmental and evolutionary biology. Cell 106 (5), 535–538.

56 Chapter 1 The seabed—Where life began and still evolves

Cordes, E.E., Bergquist, D.C., Fisher, C.R., 2009. Macro-ecology of Gulf of Mexico cold seeps. Annu. Rev. Mar. Sci. 1, 143–168.

Craig, M.T., Eble, J.A., Bowen, B.W., Robertson, D.R., 2007. High genetic connectivity across the Indian and Pacific Oceans in the reef fish Myripristis berndti (Holocentridae). Mar. Ecol. Prog. Ser. 334, 245–254.

Crisp, M.D., Trewick, S.A., Cook, L.G., 2011. Hypothesis testing in biogeography. Trends Ecol. Evol. 26 (2), 66–72.

Czekanski-Moir, J.E., Rundell, R.J., 2020. Endless forms most stupid, icky, and small: the preponderance of non-charismatic invertebrates as integral to a biologically sound view of life. Ecol. Evol., 12638–12649.

Dall'Olmo, G., Dingle, J., Polimene, L., Brewin, R.J., Claustre, H., 2016. Substantial energy input to the meso-pelagic ecosystem from the seasonal mixed-layer pump. Nat. Geosci. 9 (11), 820–823.

Dambach, J., Rödder, D., 2011. Applications and future challenges in marine species distribution modeling. Aquat. Conserv. Mar. Freshwat. Ecosyst. 21 (1), 92–100.

Danovaro, R., Gambi, C., Dell'Anno, A., Corinaldesi, C., Fraschetti, S., Vanreusel, A., Gooday, A.J., 2008. Exponential decline of deep-sea ecosystem functioning linked to benthic biodiversity loss. Curr. Biol. 18 (1), 1–8.

Danovaro, R., Dell'Anno, A., Pusceddu, A., Gambi, C., Heiner, I., Kristensen, R.M., 2010. The first metazoa living in permanently anoxic conditions. BMC Biol. 8 (1), 30.

Darling, J.A., Carlton, J.T., 2018. A framework for understanding marine cosmopolitanism in the Anthropocene. Front. Mar. Sci. 5, 293.

Darling, E.S., McClanahan, T.R., Maina, J., Gurney, G.G., Graham, N.A., Januchowski-Hartley, F., Mouillot, D., 2019. Social–environmental drivers inform strategic management of coral reefs in the Anthropocene. Nat. Ecol. Evol. 3 (9), 1341–1350.

Darwin, C., 1859. On the Origin of Species. Reprint, Modern Library, New York, NY, 1998.

Darwin Tree of Life Project Consortium, 2022. Sequence locally, think globally: the Darwin tree of life project. Proc. Natl. Acad. Sci. U. S. A. 119 (4), e2115642118.

David, L.A., Alm, E.J., 2011. Rapid evolutionary innovation during an Archaean genetic expansion. Nature 469 (7328), 93–96.

Davy, S.K., Allemand, D., Weis, V.M., 2012. Cell biology of cnidarian-dinoflagellate symbiosis. Microbiol. Mol. Biol. Rev. 76 (2), 229–261.

de Bary, H.A., 1879. Die Erscheinung der Symbiose. Trübner, Strassburg.

de Goeij, J.M., van Oevelen, D., Vermeij, M.J., Osinga, R., Middelburg, J.J., de Goeij, A.F.P.M., et al., 2013. Surviving in a marine desert: the sponge loop retains resources within coral reefs. Science 342, 108–110.

de Goeij, J.M., Lesser, M.P., Pawlik, J.R., 2017. Nutrient fluxes and ecological functions of coral reef sponges in a changing ocean. In: Carballo, J. (Ed.), Climate Change, Ocean Acidification and Sponges: Impacts Across Multiple Levels of Organization. Springer, pp. 373–410.

De Vargas, C., Audic, S., Henry, N., Decelle, J., Mahé, F., Logares, R., Velayoudon, D., 2015. Eukaryotic plankton diversity in the sunlit ocean. Science 348 (6237), 1261605.

Degnan, S.M., 2014. Think laterally: horizontal gene transfer from symbiotic microbes may extend the phenotype of marine sessile hosts. Front. Microbiol. 5, 638.

Delandmeter, P., Lewis, S.E., Lambrechts, J., Deleersnijder, E., Legat, V., Wolanski, E., 2015. The transport and fate of riverine fine sediment exported to a semi-open system. Estuar. Coast. Shelf Sci. 167, 336–346.

DeLong, E.F., 1998. Molecular phylogenetics: New perspective on the ecology, evolution and biodiversity of marine organisms. In: Cooksey, K.E. (Ed.), Molecular Approaches to the Study of the Ocean. Chapman-Hall, London, pp. 1–28.

Delong, E.F., 2005. Microbial community genomics in the ocean. Nat. Rev. Microbiol. 3, 459–469.

DeLong, E.F., Franks, D.G., Alldredge, A.L., 1993. Phylogenetic diversity of aggregate-attached vs. free-living marine bacterial assemblages. Limnol. Oceanogr. 38 (5), 924–934.

Dengler, M., et al., 2004. Break-up of the Atlantic deep western boundary current into eddies at 8° S. Nature 432, 1018.

Dentzien-Dias, P.C., Poinar Jr., G., de Figueiredo, A.E.Q., Pacheco, A.C.L., Horn, B.L.D., Schultz, C.L., 2013. Tapeworm eggs in a 270 million-year-old shark coprolite. PLoS One 8, e55007. https://doi.org/10.1371/journal.pone.0055007.

DeVries, T., 2022. The ocean carbon cycle. Annu. Rev. Environ. Resour. 47, 317–341.

D'hondt, S., Inagaki, F., Zarikian, C.A., Abrams, L.J., Dubois, N., et al., 2015. Presence of oxygen and aerobic communities from sea floor to basement in deep-sea sediments. Nat. Geosci. 8 (4), 299–304.

Diamond, J.M., 1975. The island dilemma: lessons of modern biogeographic studies for the design of natural reserves. Biol. Conserv. 7 (2), 129–146.

Diaz, C.M., Rutzler, K., 2001. Sponges: essential component of Caribbean coral reefs. Bull. Mar. Sci. 69 (2), 532–546.

Dolan, M.F., Grehan, A.J., Guinan, J.C., Brown, C., 2008. Modelling the local distribution of cold-water corals in relation to bathymetric variables: adding spatial context to deep-sea video data. Deep-Sea Res. I Oceanogr. Res. Pap. 55 (11), 1564–1579.

Donovan, M.K., Friedlander, A.M., Lecky, J., Jouffray, J., et al., 2018. Combining fish and benthic communities into multiple regimes reveals complex reef dynamics. Sci. Rep. 8 (1), 1–11.

Douglas, A.E., 1994. Symbiotic Interactions (No. 577.85 D733s). Oxford University Press, Oxon, GB.

Douglas, A.E., 2021. The Symbiotic Habit. Princeton University Press.

Duffy, J.E., Cardinale, B.J., France, K.E., McIntyre, P.B., Thébault, E., Loreau, M., 2007. The functional role of biodiversity in ecosystems: incorporating trophic complexity. Ecol. Lett. 10 (6), 522–538.

Duineveld, G.C.A., Lavaleye, M.S.S., Berghuis, E.M., 2004. Particle flux and food supply to a seamount cold-water coral community (Galicia Bank, NW Spain). Mar. Ecol. Prog. Ser. 277, 13–23.

Dunn, D.F., 1981. The clownfish sea anemones: Stichodactylidae (Coelenterata: Actiniaria) and other sea anemones symbiotic with pomacentrid fishes. Trans. Am. Philos. Soc. 71 (1), 3–115.

Dunn, C.W., Hejnol, A., Matus, D.Q., Pang, K., Browne, W.E., Smith, S.A., Giribet, G., 2008. Broad phylogenomic sampling improves resolution of the animal tree of life. Nature 452 (7188), 745–749.

Durante, M.K., Baums, I.B., Williams, D.E., Vohsen, S., Kemp, D.W., 2019. What drives phenotypic divergence among coral clonemates of Acropora palmata? Mol. Ecol. 28 (13), 3208–3224.

Durden, J.M., Bett, B.J., Huffard, C.L., Ruhl, H.A., Smith, K.L., 2019. Abyssal Deposit-Feeding Rates Consistent with the Metabolic Theory of Ecology.

Easson, C.G., Chaves-Fonnegra, A., Thacker, R.W., Lopez, J.V., 2020. Host population genetics and biogeography structure the microbiome of the sponge Cliona delitrix. Ecol. Evol. 10 (4), 2007–2020.

Eble, J.A., Rocha, L.A., Craig, M.T., Bowen, B.W., 2011. Not all larvae stay close to home: insights into marine population connectivity with a focus on the brown surgeonfish (*Acanthurus nigrofuscus*). J. Mar. Biol. 2011.

Edgecombe, G.D., Giribet, G., Dunn, C.W., Hejnol, A., Kristensen, R.M., 2011. Higher-level metazoan relationships: recent progress and remaining questions. Org. Divers. Evol. 11 (2), 151–172.

Egbeocha, C.O., Malek, S., Emenike, C.U., Milow, P., 2018. Feasting on microplastics: ingestion by and effects on marine organisms. Aquat. Biol. 27, 93–106.

Eklund, R.L., Knapp, L.C., Sandifer, P.A., Colwell, R.C., 2019. Oil spills and human health: contributions of the Gulf of Mexico research initiative. GeoHealth 3 (12), 391–406.

Engelberts, J.P., Robbins, S.J., Damjanovic, K., Webster, N.S., 2021. Integrating novel tools to elucidate the metabolic basis of microbial symbiosis in reef holobionts. Mar. Biol. 168, 1–19.

Erpenbeck, D., Voigt, O., Adamski, M., Woodcroft, B.J., Hooper, J.N., Wörheide, G., Degnan, B.M., 2011. NUMTs in the sponge genome reveal conserved transposition mechanisms in metazoans. Mol. Biol. Evol. 28 (1), 1–5. https://doi.org/10.1093/molbev/msq217.

Etter, R.J., Caswell, H., 1994. The advantages of dispersal in a patchy environment: Effects of disturbance in a cellular automaton model. In: Young, C.M., Eckelbarger, K.J. (Eds.), Reproduction, Larval Biology, and Recruitment of the Deep-Sea Benthos. Columbia University Press, New York, pp. 284–305.

Etter, R.J., Grassle, J.F., 1992. Patterns of species diversity in the deep sea as a function of sediment particle size diversity. Nature 360 (6404), 576–578.

58 **Chapter 1** The seabed—Where life began and still evolves

Eyre, B.D., Glud, R.N., Patten, N., 2008. Mass coral spawning: A natural large-scale nutrient addition experiment. Limnol. Oceanogr. 53 (3), 997–1013.

Faucci, A., Toonen, R.J., Hadfield, M.G., 2007. Host shift and speciation in a coral-feeding nudibranch. Proc. R. Soc. B Biol. Sci. 274 (1606), 111–119.

Fedor, P., Zvaríková, M., 2019. Biodiversity indices. Encycl. Ecol. 2, 337–346.

Fisher, R., O'Leary, R.A., Low-Choy, S., Mengersen, K., Knowlton, N., Brainard, R.E., Caley, M.J., 2015. Species richness on coral reefs and the pursuit of convergent global estimates. Curr. Biol. 25 (4), 500–505.

Flajnik, M.F., 2002. Comparative analyses of immunoglobulin genes: surprises and portents. Nat. Rev. Immunol. 2 (9), 688–698.

Flajnik, M.F., Kasahara, M., 2010. Origin and evolution of the adaptive immune system: genetic events and selective pressures. Nat. Rev. Genet. 11 (1), 47–59.

Floeter, S.R., Rocha, L.A., Robertson, D.R., Joyeux, J.C., Smith-Vaniz, W.F., Wirtz, P., Bernardi, G., 2008. Atlantic reef fish biogeography and evolution. J. Biogeogr. 35 (1), 22–47.

Flot, J.F., Hespeels, B., Li, X., Noel, B., et al., 2013. Genomic evidence for ameiotic evolution in the bdelloid rotifer Adineta vaga. Nature 500 (7643), 453–457.

Fogarty, N.D., Marhaver, K.L., 2019. Coral spawning, unsynchronized. Science 365 (6457), 987–988.

Foley, C.J., Feiner, Z.S., Malinich, T.D., Höök, T.O., 2018. A meta-analysis of the effects of exposure to micro-plastics on fish and aquatic invertebrates. Sci. Total Environ. 631, 550–559.

Freeman III, A.M., Herriges, J.A., Kling, C.L., 2014. The Measurement of Environmental and Resource Values: Theory and Methods. Routledge.

Freeman, C.J., Easson, C.G., Fiore, C.L., Thacker, R.W., 2021. Sponge–microbe interactions on coral reefs: multiple evolutionary solutions to a complex environment. Front. Mar. Sci. 8, 705053.

Fujiwara, Y., Jimi, N., Sumida, P.Y., Kawato, M., Kitazato, H., 2019. New species of bone-eating worm Osedax from the abyssal South Atlantic Ocean (Annelida, Siboglinidae). ZooKeys 814, 53.

Fuller, M., 1991. Summer on the Lakes in 1843. University of Illinois Press.

Futuyma, D., 2017. Evolution, fourth ed. Sinauer Associates.

Gage, J.D., 1997. High benthic species diversity in deep-sea sediments: the importance of hydrodynamics. In: Marine Biodiversity: Patterns and Processes. Cambridge University Press, Cambridge, pp. 148–177.

Garwood, R.J., Edgecombe, G.D., Charbonnier, S., Chabard, D., Sotty, D., Giribet, G., 2016. Carboniferous Onychophora from Montceau-les-mines, France, and onychophoran terrestrialization. Invertebr. Biol. 135, 179–190. https://doi.org/10.1111/ivb.12130.

Georgieva, M.N., Taboada, S., Riesgo, A., Díez-Vives, C., De Leo, F.C., Jeffreys, R.M., Glover, A.G., 2020. Evidence of vent-adaptation in sponges living at the periphery of hydrothermal vent environments: ecological and evolutionary implications. Front. Microbiol. 11, 1636.

Gibson, D.G., Glass, J.I., Lartigue, C., Noskov, V.N., Chuang, R.Y., Algire, M.A., Venter, J.C., 2010. Creation of a bacterial cell controlled by a chemically synthesized genome. Science 329 (5987), 52–56.

GIGA (Community of Scientists), 2014. The Global Invertebrate Genome Alliance (GIGA): developing community resources to study diverse invertebrates. J. Hered. 105, 1–18. https://doi.org/10.1093/jhered/est084.

Giovannoni, S.J., Rappe, M.S., 2000. Evolution, diversity, and molecular ecology of marine prokaryotes. In: Kirchman, D.L. (Ed.), Microbial Ecology of the Oceans. Wiley-Liss Inc., New York, NY, pp. 47–84.

Giribet, G., Edgecombe, G.D., 2020. The Invertebrate Tree of Life. Princeton University Press.

Gleason, D.F., 1998. Sedimentation and distributions of green and brown morphs of the Caribbean coral Porites astreoides Lamarck. J. Exp. Mar. Biol. Ecol. 230 (1), 73–89.

Gleason, D.F., Hofmann, D.K., 2011. Coral larvae: from gametes to recruits. J. Exp. Mar. Biol. Ecol. 408 (1-2), 42–57.

Goffredi, S.K., Orphan, V.J., Rouse, G.W., Jahnke, L., Embaye, T., et al., 2005. Evolutionary innovation: a bone-eating marine symbiosis. Environ. Microbiol. 7, 1369–1378.

Glud, R.N., Eyre, B.D., Patten, N., 2008. Biogeochemical responses to mass coral spawning at the Great Barrier Reef: effects on respiration and primary production. Limnol. Oceanogr. 53 (3), 1014–1024.

Goffredi, S.K., Johnson, S.B., Vrijenhoek, R.C., 2007. Genetic diversity and potential function of microbial symbionts associated with newly-discovered species of *Osedax* polychaete worms. Appl. Environ. Microbiol. 73, 2314–2323.

Goffredi, S.K., Tilic, E., Mullin, S.W., Dawson, K.S., Keller, A., Lee, R.W., Orphan, V.J., 2020. Methanotrophic bacterial symbionts fuel dense populations of deep-sea feather duster worms (Sabellida, Annelida) and extend the spatial influence of methane seepage. Sci. Adv. 6 (14), eaay8562.

Golding, G.B., Dean, A.M., 1998. The structural basis of molecular adaptation. Mol. Biol. Evol. 15 (4), 355–369.

González-Pech, R.A., Li, V.Y., Garcia, V., Boville, E., Mammone, M., Kitano, H., Ritchie, K.B., Medina, M., 2023. The evolution, assembly, and dynamics of marine holobionts. Annu. Rev. Mar. Sci. 16, 11.1–11.24. https://doi.org/10.1146/annurev-marine-022123-104345.

Gooday, A.J., Da Silva, A.A., Pawlowski, J., 2011. Xenophyophores (Rhizaria, Foraminifera) from the Nazaré Canyon (Portuguese margin, NE Atlantic). Deep-Sea Res. II Top. Stud. Oceanogr. 58 (23–24), 2401–2419.

Goreau, T., McClanahan, T., Hayes, R., Strong, A.L., 2000. Conservation of coral reefs after the 1998 global bleaching event. Conserv. Biol. 14 (1), 5–15.

Goulet, T.L., Lucas, M.Q., Schizas, N.V., 2019. Symbiodiniaceae genetic diversity and symbioses with hosts from shallow to mesophotic coral ecosystems. In: Mesophotic Coral Ecosystems. Springer, pp. 537–551.

Grabowski, J.H., Brumbaugh, R.D., Conrad, R.F., Keeler, A.G., Opaluch, J.J., et al., 2012. Economic valuation of ecosystem services provided by oyster reefs. Bioscience 62 (10), 900–909.

Graeber, D., Wengrow, D., 2021. The Dawn of Everything: A New History of Humanity. Penguin UK.

Graf, G., Rosenberg, R., 1997. Bioresuspension and biodeposition: a review. J. Mar. Syst. 11 (3-4), 269–278.

Grant, R.E., 1826. Observations and experiments on the structure and functions of the Sponge. Edinbu. New Phil. J. 14, 336–341.

Grassle, J.F., Grassle, J.P., 1974. Oportunistic life histories and genetic systems in marine benthic polychaetes. J. Mar. Res. 32, 253–284.

Graur, D., Li, W.-S., 2000. Fundamentals of Molecular Evolution. Oxford University Press.

Grebmeier, J.M., McRoy, C.P., Feder, H.M., 1988. Pelagic-benthic coupling on the shelf of the northern Bering and Chukchi seas. 1. Food supply source and benthic biomass. Mar. Ecol. Prog. Ser. 48 (1), 57–67.

Gregory, T.R., 2005. Genome size evolution in animals. In: Gregory (Ed.), The Evolution of the Genome. Academic Press, pp. 3–87.

Gribble, K.E., Mark Welch, D.B., 2017. Genome-wide transcriptomics of aging in the rotifer Brachionus manjavacas, an emerging model system. BMC Genomics 18, 1–14.

Griffiths, J.R., Kadin, M., Nascimento, F.J., Tamelander, T., Törnroos, A., Bonaglia, S., Winder, M., 2017. The importance of benthic–pelagic coupling for marine ecosystem functioning in a changing world. Glob. Chang. Biol. 23 (6), 2179–2196.

Guy, G., Metaxas, A., 2022. Recruitment of deep-water corals and sponges in the Northwest Atlantic Ocean: implications for habitat distribution and population connectivity. Mar. Biol. 169 (8), 107.

Hampton-Marcell, J.T., Lopez, J.V., Gilbert, J., 2017. The human microbiome: an emerging tool in forensics. Microb. Biotechnol. 10 (2), 228–230. https://doi.org/10.1111/1751-7915.12699.

Handelsman, J., 2004. Metagenomics: application of genomics to uncultured microorganisms. Microbiol. Mol. Biol. Rev. 68 (4), 669–685.

Handelsman, J., Rondon, M.R., Brady, S.F., Clardy, J., Goodman, R.M., 1998. Molecular biological access to the chemistry of unknown soil microbes: a new frontier for natural products. Chem. Biol. 5 (10), R245–R249.

Hardy, S.M., Smith, C.R., Thurnherr, A.M., 2015. Can the source–sink hypothesis explain macrofaunal abundance patterns in the abyss? A modelling test. Proc. R. Soc. B Biol. Sci. 282 (1808), 20150193.

Harrold, C., 1987. The ecological role of echinoderms in kelp forests. Echinoderm Stud. 2, 137–233.

60 **Chapter 1** The seabed—Where life began and still evolves

Hartman, O., 1966. Polychaeta Myzostomidae and Sedentaria of Antarctica. Antarctic Res. Ser. 7, 1–158.

Hawkes, N., et al., 2019. Glass sponge grounds on the Scotian Shelf and their associated biodiversity. Mar. Ecol. Prog. Ser. 614, 91–109.

Haygood, M.G., Schmidt, E.W., Davidson, S.K., Faulkner, D.J., 1999. Microbial symbionts of marine invertebrates: opportunities for microbial biotechnology. J. Mol. Microbiol. Biotechnol. 1, 33–43.

Hector, A., Bagchi, R., 2007. Biodiversity and ecosystem multifunctionality. Nature 448 (7150), 188–190.

Heiner, I., Neuhaus, B., 2007. Loricifera from the deep sea at the Galápagos Spreading Center, with a description of Spinoloricus turbatio gen. et sp. nov.(Nanaloricidae). Helgol. Mar. Res. 61 (3), 167–182.

Hellberg, M.E., 2009. Gene flow and isolation among populations of marine animals. Annu. Rev. Ecol. Evol. Syst. 40, 291–310.

Hentschel, U., Fieseler, L., Wehrl, M., Gernert, C., Steinert, M., Hacker, J., Horn, M., 2003. Microbial diversity of marine sponges. In: Müller, W.E.G. (Ed.), Molecular Marine Biology of Sponges. Springer Verlag, Heidelberg, pp. 60–88.

Hentschel, U., Piel, J., Degnan, S.M., Taylor, M.W., 2012. Genomic insights into the marine sponge microbiome. Nat. Rev. Microbiol. 10 (9), 641–654.

Hermosillo-Núñez, B.B., 2020. Contribution of echinoderms to keystone species complexes and macroscopic properties in kelp forest ecosystems (northern Chile). Hydrobiologia 847 (3), 739–756.

Herranz, M., Stiller, J., Worsaae, K., Sørensen, M.V., 2022. Phylogenomic analyses of mud dragons (Kinorhyncha). Mol. Phylogenet. Evol. 168, 107375.

Herrera, S., Shank, T.M., Sánchez, J.A., 2012. Spatial and temporal patterns of genetic variation in the widespread antitropical deep-sea coral P aragorgia arborea. Mol. Ecol. 21 (24), 6053–6067.

Hibino, T., Loza-Coll, M., Messier, C., Majeske, A.J., Cohen, A.H., Terwilliger, D.P., Rast, J.P., 2006. The immune gene repertoire encoded in the purple sea urchin genome. Dev. Biol. 300 (1), 349–365.

Hoadley, K.D., Pettay, D.T., Lewis, A., Wham, D., Grasso, C., Smith, R., Warner, M.E., 2021. Different functional traits among closely related algal symbionts dictate stress endurance for vital Indo-Pacific reef-building corals. Glob. Chang. Biol. 27 (20), 5295–5309.

Hoegh-Guldberg, O., Muller-Parker, G., Cook, C.B., Gates, R.D., Gladfelter, E., Trench, R.K., Weis, V.M., 2007. Len Muscatine (1932–2007) and his contributions to the understanding of algal-invertebrate endosymbiosis. Coral Reefs 26, 731–739.

Hoegh-Guldberg, O., Hughes, L., McIntyre, S., Lindenmayer, D.B., Parmesan, C., Possingham, H.P., Thomas, C. D., 2008. Assisted colonization and rapid climate change. Science 321 (5887), 345–346.

Hoelzel, A.R., 1998. Genetic structure of cetacean populations in sympatry, parapatry, and mixed assemblages: implications for conservation policy. J. Hered. 89 (5), 451–458.

Holstein, T., Tardent, P., 1984. An ultrahigh-speed analysis of exocytosis: nematocyst discharge. Science 223 (4638), 830–833.

Hooper, J.N.A., Van Soest, R.W.M. (Eds.), 2002. Systema Porifera: A Guide to the Classification of Sponges. 2 volumes. Kluwer Academic/Plenum Publishers, New York, Boston, Dordrecht, London, Moscow, https://doi.org/10.1007/978-1-4615-0747-5_1. 1708 + xlviii pp. (printed version).

Hooper, J.N.A., Van Soest, R.W.M., Debrenne, F., 2002. Phylum Porifera Grant, 1826. In: Hooper, J.N.A., van Soest, R.W.M. (Eds.), Systema Porifera: A Guide to the Classification of Sponges. Vol. 1. Kluwer Academic/Plenum Publishers, New York, Boston, Dordrecht, London, Moscow, pp. 9–14.

Hopwood, N., 2022. 'Not birth, marriage or death, but gastrulation': the life of a quotation in biology. Br. J. Hist. Sci. 55 (1), 1–26.

Horne, J.B., van Herwerden, L., Choat, J.H., Robertson, D.R., 2008. High population connectivity across the Indo-Pacific: congruent lack of phylogeographic structure in three reef fish congeners. Mol. Phylogenet. Evol. 49 (2), 629–638.

Howell, K.L., Holt, R., Endrino, I.P., Stewart, H., 2011. When the species is also a habitat: comparing the predictively modelled distributions of Lophelia pertusa and the reef habitat it forms. Biol. Conserv. 144 (11), 2656–2665.

Hubbell, S.P., 1979. Tree dispersion, abundance, and diversity in a tropical dry forest: that tropical trees are clumped, not spaced, alters conceptions of the organization and dynamics. Science 203 (4387), 1299–1309.

Hucke-Gaete, R., Bedrinana-Romano, L., Viddi, F.A., Ruiz, J.E., Torres-Florez, J.P., Zerbini, A.N., 2018. From Chilean Patagonia to Galapagos, Ecuador: novel insights on blue whale migratory pathways along the Eastern South Pacific. PeerJ 6, e4695.

Hughes, R.N., Hughes, D.J., Smith, I.P., 2014. Limits to understanding and managing outbreaks of crown-of-thorns starfish (Acanthaster spp.). Oceanogr. Mar. Biol. Annu. Rev. 52, 133–200.

Human Microbiome Project (HMP) Consortium, et al., 2012. Structure, function and diversity of the healthy human microbiome. Nature 486, 207–214.

Humann, P., Deloach, N., 2014. Reef fish identification: Florida, Caribbean, Bahamas. New World Publications.

Hume, B.C.C., Voolstra, C.R., Arif, C., D'Angelo, C., Burt, J.A., Eyal, G., Wiedenmann, J., 2016. Ancestral genetic diversity associated with the rapid spread of stress-tolerant coral symbionts in response to Holocene climate change. Proc. Natl. Acad. Sci. U. S. A. 113 (16), 4416–4421.

Humman, P., 1992. Reef Creature Identification. New World Publications.

Hutchings, P., 1998. Biodiversity and functioning of polychaetes in benthic sediments. Biodivers. Conserv. 7 (9), 1133–1145.

Hutchings, P., Kupriyanova, E., 2018. Cosmopolitan polychaetes–fact or fiction? Personal and historical perspectives. Invertebr. Syst. 32 (1), 1–9.

Hutchison III, C.A., Chuang, R.Y., Noskov, V.N., Assad-Garcia, N., Deerinck, T.J., Ellisman, M.H., Venter, J.C., 2016. Design and synthesis of a minimal bacterial genome. Science 351 (6280), aad 6253.

Jackson, J.B., Kirby, M.X., Berger, W.H., Bjorndal, K.A., Botsford, L.W., Bourque, B.J., Warner, R.R., 2001. Historical overfishing and the recent collapse of coastal ecosystems. Science 293 (5530), 629–637.

Jamieson, A.J., 2015. The Hadal Zone: Life in the Deepest Oceans. Cambridge Univ. Press.

Jamieson, A.J., 2018. A contemporary perspective on hadal science. Deep-Sea Res. II Top. Stud. Oceanogr. 155, 4–10.

Jamieson, A.J., Lörz, A.N., Fujii, T., Priede, I.G., 2012. In situ observations of trophic behaviour and locomotion of *Princaxelia amphipods* (Crustacea: Pardaliscidae) at hadal depths in four West Pacific trenches. J. Mar. Biol. Assoc. U. K. 92 (1), 143–150.

Janeway, C.A., 1989. Approaching the asymptote? Evolution and revolution in immunology. In: Cold Spring Harbor Symposia on Quantitative Biology. Vol. 54. Cold Spring Harbor Laboratory Press, pp. 1–13.

Jannasch, H.W., Jones, G.E., 1959. Bacterial populations in sea water as determined by different methods of enumeration. Limnol. Oceanogr. 4 (2), 128–139.

Jenkins, C.J., 1984. Erosion and deposition at abyssal depths in the Tasman Sea. A seismic stratigraphic study of the bottom-current patterns. In: University of Sydney Ocean Sciences Institute Report. 4. University of Sydney, p. 53.

Jing, H., Xiao, X., Zhang, Y., Li, Z., Jian, H., Luo, Y., Han, Z., 2022. Composition and ecological roles of the core microbiome along the Abyssal-Hadal transition zone sediments of the Mariana Trench. Microbiol. Spectr. 10 (3). e01988-21.

Johnston, M.W., Milligan, R.J., Easson, C.G., DeRada, S., English, D.C., Penta, B., Sutton, T.T., 2019. An empirically validated method for characterizing pelagic habitats in the Gulf of Mexico using ocean model data. Limnol. Oceanogr. Meth. 17 (6), 362–375.

Jones, R., Fisher, R., Bessell-Browne, P., 2019. Sediment deposition and coral smothering. PLoS One 14 (6), e0216248.

Jordán, F., 2001. Trophic fields. Community Ecol. 2, 181–185.

Jørgensen, B.B., Boetius, A., 2007. Feast and famine—microbial life in the deep-sea bed. Nat. Rev. Microbiol. 5 (10), 770–781'.

Juravel, K., Porras, L., Höhna, S., Pisani, D., Wörheide, G., 2023. Exploring genome gene content and morphological analysis to test recalcitrant nodes in the animal phylogeny. PLoS One 18 (3), e0282444.

62 **Chapter 1** The seabed—Where life began and still evolves

Kawamura, K., Sekida, S., Nishitsuji, K., Shoguchi, E., Hisata, K., Fujiwara, S., Satoh, N., 2021. In vitro symbiosis of reef-building coral cells with photosynthetic dinoflagellates. Front. Mar. Sci. 900.

Keene, J., Baker, C., Tran, M., Potter, A., 2008. Sedimentology and Geomorphology of the East Marine Region of Australia. 10 Geoscience Australia, Record.

Keller, N.B., 1976. The deep-sea madreporarian corals of the genus Fungiacyathus from the Kurile-Kamchatka, Aleutian Trenches and other regions of the world oceans [in Russian, English translation deposited at Smithsonian]. Trudy Instit. Okean. 99, 31–44.

Kellogg, C.A., 2019. Microbiomes of stony and soft deep-sea corals share rare core bacteria. Microbiome 7, 1–13.

Keltner, D., 2023. Awe: The New Science of Everyday Wonder and How It Can Transform Your Life. Penguin Press.

Kemp, D.W., Fitt, W.K., Schmidt, G.W., 2008. A microsampling method for genotyping coral symbionts. Coral Reefs 27, 289–293.

Kennedy, V.S., 2018. Shifting Baselines in the Chesapeake Bay: An Environmental History. Johns Hopkins University Press.

King, N., 2004. The unicellular ancestry of animal development. Dev. Cell 7 (3), 313–325.

Kirby, M.X., 2004. Fishing down the coast: historical expansion and collapse of oyster fisheries along continental margins. Proc. Natl. Acad. Sci. U. S. A. 101, 13096.

Knight, R., Jansson, J., Field, D., Fierer, N., Desai, N., Fuhrman, J.A., et al., 2012. Unlocking the potential of metagenomics through replicated experimental design. Nat. Biotechnol. 30, 513–520.

Knowlton, N., 1993. Sibling species in the sea. Annu. Rev. Ecol. Syst. 24 (1), 189–216.

Knowlton, N., Brainard, R.E., Fisher, R., Moews, M., Plaisance, L., Caley, M.J., 2010. Coral reef biodiversity. In: Life in the world's oceans: diversity distribution and abundance. Wiley and Sons, pp. 65–74.

Kocot, K.M., Todt, C., Mikkelsen, N.T., Halanych, K.M., 2019. Phylogenomics of Aplacophora (Mollusca, Aculifera) and a solenogaster without a foot. Proc. R. Soc. B 286 (1902), 20190115.

Könneke, M., Bernhard, A.E., de La Torre, J.R., Walker, C.B., Waterbury, J.B., Stahl, D.A., 2005. Isolation of an autotrophic ammonia-oxidizing marine archaeon. Nature 437 (7058), 543–546.

Koropatnick, T.A., Engle, J.T., Apicella, M.A., Stabb, E.V., Goldman, W.E., McFall-Ngai, M.J., 2004. Microbial factor-mediated development in a host-bacterial mutualism. Science 306 (5699), 1186–1188.

Körtzinger, A., 2003. A significant CO_2 sink in the tropical Atlantic Ocean associated with the Amazon River plume. Geophys. Res. Lett. 30 (24), 2287. https://doi.org/10.1029/2003GL018841.

Kough, A.S., Paris, C.B., Butler IV, M.J., 2013. Larval connectivity and the international management of fisheries. PLoS One 8 (6), e64970.

Krausfeldt, L., Lopez, J.V., Bilodeau, C., Lee, H.W., Casali, S., 2023. Change and stasis of distinct sediment microbiomes across Port Everglades inlet (PEI) and the adjacent coral reefs. PeerJ 11, e14288. https://doi.org/10.7717/peerj.14288.

Kritzer, J.P., Sale, P.F., 2004. Metapopulation ecology in the sea: from Levins' model to marine ecology and fisheries science. Fish Fish. 5 (2), 131–140.

LaJeunesse, T.J.M.B., 2002. Diversity and community structure of symbiotic dinoflagellates from Caribbean coral reefs. Mar. Biol. 141, 387–400.

LaJeunesse, T.C., Parkinson, J.E., Gabrielson, P.W., Jeong, H.J., Reimer, J.D., Voolstra, C.R., Santos, S.R., 2018. Systematic revision of Symbiodiniaceae highlights the antiquity and diversity of coral endosymbionts. Curr. Biol. 28 (16), 2570–2580.

Lambshead, P.J.D., Tietjen, J., Glover, A., Ferrero, T., et al., 2001. Impact of large-scale natural physical disturbance on the diversity of deep-sea North Atlantic nematodes. Mar. Ecol. Prog. Ser. 214, 121–126.

Lane, D.J., Pace, B., Olsen, G.J., Stahl, D.A., Sogin, M.L., Pace, N.R., 1985. Rapid determination of 16S ribosomal RNA sequences for phylogenetic analyses. Proc. Natl. Acad. Sci. U. S. A. 82 (20), 6955–6959.

Lartillot, N., Philippe, H., 2004. A Bayesian mixture model for across-site heterogeneities in the amino-acid replacement process. Mol. Biol. Evol. 21 (6), 1095–1109.

Lartillot, N., Brinkmann, H., Philippe, H., 2007. Suppression of long-branch attraction artefacts in the animal phylogeny using a site-heterogeneous model. BMC Evol. Biol. 7 (1), 1–14.

Laumer, C.E., Bekkouche, N., Kerbl, A., Goetz, F., Neves, R.C., Sørensen, M.V., Worsaae, K., 2015. Spiralian phylogeny informs the evolution of microscopic lineages. Curr. Biol. 25 (15), 2000–2006.

Lazcano, A., Miller, S.L., 1996. The origin and early evolution of life: prebiotic chemistry, the pre-RNA world, and time. Cell 85 (6), 793–798.

Lefèvre, N., Moore, G., Aiken, J., Watson, A., Cooper, D., Ling, R., 1998. Variability of pCO2 in the tropical Atlantic in 1995. J. Geophys. Res. Oceans 103 (C3), 5623–5634.

Lester, S.E., Rassweiler, A., McCoy, S.J., Dubel, A.K., et al., 2020. Caribbean reefs of the Anthropocene: variance in ecosystem metrics indicates bright spots on coral depauperate reefs. Glob. Chang. Biol. 26 (9), 4785–4799.

Levin, L.A., Gage, J.D., 1998. Relationships between oxygen, organic matter and the diversity of bathyal macrofauna. Deep Sea Res. Part II Top. Stud. Oceanogr. 45, 129–163.

Levin, L.A., Plaia, G.R., Huggett, C.L., 1994. The influence of natural organic enhancement on life histories and community structure of bathyal polychaetes. In: Young, C.M., Eckelbarger, K.J. (Eds.), Reproduction, Larval Biology, and Recruitment of the Deep-Sea Benthos. Columbia University Press, New York, pp. 261–283.

Levinton, J.S., 2017. Marine Biology: Function, Biodiversity and Ecology, fifth ed. Oxford University Press.

Levitan, D.R., Fogarty, N.D., Jara, J., Lotterhos, K.E., Knowlton, N., 2011. Genetic, spatial, and temporal components of precise spawning synchrony in reef building corals of the Montastraea annularis species complex. Evolution 65 (5), 1254–1270.

Levy, S., Elek, A., Grau-Bové, X., Menéndez-Bravo, S., Iglesias, M., Tanay, A., Sebé-Pedrós, A., 2021. A stony coral cell atlas illuminates the molecular and cellular basis of coral symbiosis, calcification, and immunity. Cell 184 (11), 2973–2987.

Li, Y., Shen, X.X., Evans, B., Dunn, C.W., Rokas, A., 2021. Rooting the animal tree of life. Mol. Biol. Evol. 38 (10), 4322–4333.

Lignell, R., Heiskanen, A.S., Kuosa, H., Gundersen, K., Kuuppo-Leinikki, P., Pajuniemi, R., Uitto, A., 1993. Fate of a phytoplankton spring bloom: sedimentation and carbon flow in the planktonic food web in the northern Baltic. Mar. Ecol. Prog. Ser. 94, 239–252.

Lim, X., 2021. Microplastics are everywhere—but are they harmful. Nature 593 (7857), 22–25.

Lima-Mendez, G., Faust, K., Henry, N., Decelle, J., Colin, S., Carcillo, F., Raes, J., 2015. Determinants of community structure in the global plankton interactome. Science 348 (6237), 1262073.

Linnaeus, C., 1758. Systema Naturae. 1 Laurentii Salvii, Stockholm, pp. 1–824.

Lloyd-Price, J., Mahurkar, A., Rahnavard, G., Crabtree, J., Orvis, J., Hall, A.B., Huttenhower, C., 2017. Strains, functions and dynamics in the expanded human microbiome project. Nature 550 (7674), 61–66.

Lopez, J.V., 2019. After the taxonomic identification phase: addressing the functions of symbiotic communities within marine invertebrates. In: Li, Z. (Ed.), Symbionts of Marine Sponges and Corals. Springer-Verlag, Dordrecht, The Netherlands, pp. 105–144, https://doi.org/10.1007/978-94-024-1612-1_8.

Lopez, J.V., McCarthy, P.J., Janda, K.E., Willoughby, R., Pomponi, S.A., 1999a. Molecular techniques reveal wide phyletic diversity of heterotrophic microbes associated with the sponge genus *Discodermia* (Porifera: Demospongiae). Proceedings of the 5th International Sponge Symposium. Mem. Queensl. Mus. 44, 329–341. Brisbane ISSN 0079-8835.

Lopez, J.V., Kersanach, R., Rehner, S.A., Knowlton, N., 1999b. Molecular determination of species boundaries in corals: genetic analysis of the *Montastraea annularis* complex using amplified fragment length polymorphisms and a microsatellite marker. Biol. Bull. 196, 80–93.

Lopez, J.V., Peterson, C.L., Morales, F., Brown, L., 2000. Molecular studies of biodiversity in the field: Andros Island Flora and Fauna in the new millennium. Bahamas J. Sci. 8, 32–41.

Lopez, J.V., Yuhki, N., Modi, W., Masuda, R., O'Brien, S.J., 1994. *Numt*, a recent transfer and tandem amplification of mitochondrial DNA in the nuclear genome of the domestic cat. J. Mol. Evol. 39, 171–190.

64 Chapter 1 The seabed—Where life began and still evolves

Lopez, J.V., Kamel, B., Medina, M., Collins, T., Baums, I.B., 2019a. Multiple facets of marine invertebrate conservation genomics. Annu. Rev. Anim. Biosci. 7, 473–497. http://www.annualreviews.org/eprint/bSr3PRCJCfbmPz39sH5Q/full/10.1146/annurev-animal-020518-115034.

Lopez, J.V., Peixoto, R., Rosado, A.S., 2019b. Inevitable future: space colonization beyond earth with microbes first. FEMS Microbiol. Ecol. 95 (10), fiz127. https://doi.org/10.1093/femsec/fiz127. https://academic.oup.com/femsec/article/95/10/fiz127/5553461.

Lovvorn, J.R., Cooper, L.W., Brooks, M.L., De Ruyck, C.C., Bump, J.K., Grebmeier, J.M., 2005. Organic matter pathways to zooplankton and benthos under pack ice in late winter and open water in late summer in the north-central Bering Sea. Mar. Ecol. Prog. Ser. 291, 135–150.

Lowry, J.K., De Broyer, C., 2008. Alicellidae and Valettiopsidae, two new callynophorate families (Crustacea: Amphipoda). Zootaxa 1843 (1), 57–66.

Lozier, M.S., 2010. Deconstructing the conveyor belt. Science 328 (5985), 1507–1511.

Lu, A., Watkins, M., Li, Q., Robinson, S.D., Concepcion, G.P., Yandell, M., Fedosov, A.E., 2020. Transcriptomic profiling reveals extraordinary diversity of venom peptides in unexplored predatory gastropods of the genus Clavus. Genome Biol. Evol. 12 (5), 684–700.

Ludwig, W., Klenk, H., 2001. Overview: a phylogenetic backbone and taxonomic framework for procaryotic systematics. In: Garrity, G.M. (Ed.), Bergey's Manual of Systematic Bacteriology, second ed. Springer-Verlag, New York, NY, pp. 49–65.

Lundsten, L., Schlining, K.L., Frasier, K., Johnson, S.B., Kuhnz, L.A., Harvey, J.B., Vrijenhoek, R.C., 2010. Time-series analysis of six whale-fall communities in Monterey Canyon, California, USA. Deep-Sea Res. I Oceanogr. Res. Pap. 57 (12), 1573–1584.

Maher, R.L., Schmeltzer, E.R., Meiling, S., McMinds, R., Ezzat, L., Shantz, A.A., Vega Thurber, R., 2020. Coral microbiomes demonstrate flexibility and resilience through a reduction in community diversity following a thermal stress event. Front. Ecol. Evol. 8, 555698.

Maldonado, M., Beazley, L., López-Acosta, M., Kenchington, E., Casault, B., Hanz, U., Mienis, F., 2021. Massive silicon utilization facilitated by a benthic-pelagic coupled feedback sustains deep-sea sponge aggregations. Limnol. Oceanogr. 66 (2), 366–391.

Maloof, A.C., Porter, S.M., Moore, J.L., Dudas, F.O., Bowring, S.A., Higgins, J.A., Fike, D.A., Eddy, M.P., 2010. The earliest Cambrian record of animals and ocean geochemical change. Geol. Soc. Am. Bull. 122 (11−12), 1731–1774. bibcode: 2010GSAB..122.1731M https://doi.org/10.1130/B30346.1.

Manzello, D.P., Berkelmans, R., Hendee, J.C., 2007. Coral bleaching indices and thresholds for the Florida reef tract, Bahamas, and St. Croix, US Virgin Islands. Mar. Pollut. Bull. 54 (12), 1923–1931.

Mardis, E.R., 2011. A decade's perspective on DNA sequencing technology. Nature 470 (7333), 198–203.

Mardis, E.R., 2017. DNA sequencing technologies: 2006–2016. Nat. Protoc. 12 (2), 213–218.

Margulis, L., 1971. Symbiosis and evolution. Sci. Am. 225 (2), 48–61.

Margulis, L., 1974. Five-Kingdom Classification and the Origin and Evolution of Cells. Springer US, pp. 45–78.

Margulis, L., 1993. Symbiosis in Cell Evolution. WH Freeman.

Margulis, L., Fester, R. (Eds.), 1991. Symbiosis as a Source of Evolutionary Innovation: Speciation and Morphogenesis. MIT Press.

Marris, E., 2004. There are worse places to do research, but for the Caribbean cruise ship explorer of the seas, home to two. Nature 431, 394–395.

Marshall, C.E., Glegg, G.A., Howell, K.L., 2014. Species distribution modelling to support marine conservation planning: the next steps. Mar. Policy 45, 330–332.

Martin, W., Baross, J., Kelley, D., Russell, M.J., 2008. Hydrothermal vents and the origin of life. Nat. Rev. Microbiol. 6, 805.

Martin, F., Nehls, U., 2009. Harnessing ectomycorrhizal genomics for ecological insights. Curr. Opin. Plant Biol. 12 (4), 508–515.

Martini, S., Kuhnz, L., Mallefet, J., Haddock, S.H., 2019. Distribution and quantification of bioluminescence as an ecological trait in the deep sea benthos. Sci. Rep. 9 (1), 14654.

Martinson, J., 1999. Cruise line fined $18m for dumping waste at sea. The Guardian.

Mayr, E., 1970. Populations, Species, and Evolution: An Abridgment of Animal Species and Evolution. Vol. 19 Harvard University Press.

McDonald, J.H., Seed, R., Koehn, R.K., 1991. Allozymes and morphometric characters of three species of Mytilus in the northern and southern hemispheres. Mar. Biol. 111, 323–333.

McFall-Ngai, M., Heath-Heckman, E.A., Gillette, A.A., Peyer, S.M., Harvie, E.A., 2012. The secret languages of coevolved symbioses: Insights from the *Euprymna scolopes–Vibrio fischeri* symbiosis. In: Seminars in Immunology. Vol. 24, No. 1. Academic Press, pp. 3–8.

McFall-Ngai, M., Hadfield, M.G., Bosch, T.C., Carey, H.V., Domazet-Lošo, T., Douglas, A.E., Wernegreen, J.J., 2013. Animals in a bacterial world, a new imperative for the life sciences. Proc. Natl. Acad. Sci. U. S. A. 110 (9), 3229–3236.

McKenna, V., Archibald, J.M., Beinart, R., Dawson, M.N., Hentschel, U., Keeling, P.J., Lopez, J.V., Martín-Durán, J.M., Petersen, J.M., Sigwart, J.D., Simakov, O., Sutherland, K.R., Sweet, M., Talbot, N., Thompson, A.W., Bender, S., Harrison, P.W., Rajan, J., Cochrane, G., Berriman, M., Lawniczak, M., Blaxter, M., 2021. The Aquatic Symbiosis Genomics Project: probing the evolution of symbiosis across the tree of life. [version 1; peer review: awaiting peer review]. Wellcome Open Res. 6, 254. https://wellcomeopenresearch.org/articles/6-254/v1.

McMurray, S.E., Blum, J.E., Pawlik, J.R., 2008. Redwood of the reef: growth and age of the giant barrel sponge Xestospongia muta in the Florida Keys. Mar. Biol. 155, 159–171.

McMurray, S.E., Henkel, T.P., Pawlik, J.R., 2010. Demographics of increasing populations of the giant barrel sponge Xestospongia muta in the Florida Keys. Ecology 91 (2), 560–570.

Medina, M., Sharp, V., Ohdera, A., Bellantuono, A., Dalrymple, J., Gamero-Mora, E., Fitt, W.K., 2021. The upside-down jellyfish Cassiopea xamachana as an emerging model system to study cnidarian–algal symbiosis. In: Handbook of Marine Model Organisms in Experimental Biology. CRC Press, pp. 149–171.

Melo-Merino, S.M., Reyes-Bonilla, H., Lira-Noriega, A., 2020. Ecological niche models and species distribution models in marine environments: a literature review and spatial analysis of evidence. Ecol. Model. 415, 108837.

Menard, H.W., Smith, S.M., 1966. Hypsometry of ocean basin provinces. J. Geophys. Res. 71, 4305–4325.

Merckx, B., Goethals, P., Steyaert, M., Vanreusel, A., Vincx, M., Vanaverbeke, J., 2009. Predictability of marine nematode biodiversity. Ecol. Model. 220, 1449–1458. https://doi.org/10.1016/j.ecolmodel.2009.03.016.

Middlebrooks, M.L., Curtis, N.E., Pierce, S.K., 2019. Algal sources of sequestered chloroplasts in the Sacoglossan Sea slug *Elysia crispata* vary by location and ecotype. Biol. Bull. 236 (2), 88–96.

Mikhailov, K.V., Konstantinova, A.V., Nikitin, M.A., Troshin, P.V., et al., 2009. The origin of Metazoa: a transition from temporal to spatial cell differentiation. Bioessays 31, 758–768.

Milligan, R.J., Bernard, A.M., Boswell, K.M., Bracken-Grissom, H.D., D'Elia, M.A., DeRada, S., Easson, C.G., English, D., Eytan, R.I., Hu, C., Lembke, C., Lopez, J.V., Penta, B., Richards, T., Romero, I.C., Shivji, M., Timm, L., Warren, J.D., Weber, M., Wells, R.J.D., Sutton, T.T., 2019. The application of novel research technologies by the deep pelagic nekton dynamics of the Gulf of Mexico (DEEPEND) consortium. Mar. Technol. 52 (6), 81–86.

Mohamed, A.R., Ochsenkühn, M.A., Kazlak, A.M., Moustafa, A., Amin, S.A., 2023. The coral microbiome: towards an understanding of the molecular mechanisms of coral–microbiota interactions. FEMS Microbiol. Rev. 47 (2), fuad005.

Moitinho-Silva, L., Nielsen, S., Amir, A., Gonzalez, A., Ackermann, G.L., Cerrano, C., Thomas, T., 2017. The sponge microbiome project. Gigascience 6 (10), gix077.

Monaco, C.J., Porporato, E.M., Lathlean, J.A., Tagliarolo, M., Sarà, G., McQuaid, C.D., 2019. Predicting the performance of cosmopolitan species: dynamic energy budget model skill drops across large spatial scales. Mar. Biol. 166, 1–13.

Monniot, C., Monniot, F., 1994. Additions to the inventory of eastern tropical Atlantic ascidians; arrival of cosmopolitan species. Bull. Mar. Sci. 54 (1), 71–93.

Mora, C., Sale, P.F., 2002. Are populations of coral reef fish open or closed? Trends Ecol. Evol. 17 (9), 422–428.

Moroz, L.L., Kocot, K.M., Citarella, M.R., Dosung, S., Norekian, T.P., Povolotskaya, I.S., Kohn, A.B., 2014. The ctenophore genome and the evolutionary origins of neural systems. Nature 510 (7503), 109–114.

Morris, L.A., Voolstra, C.R., Quigley, K.M., Bourne, D.G., Bay, L.K., 2019. Nutrient availability and metabolism affect the stability of coral–Symbiodiniaceae symbioses. Trends Microbiol. 27 (8), 678–689.

Morrison, C.L., Ross, S.W., Nizinski, M.S., Brooke, S., Järnegren, J., Waller, R.G., King, T.L., 2011. Genetic discontinuity among regional populations of Lophelia pertusa in the North Atlantic Ocean. Conserv. Genet. 12, 713–729.

Murgarella, M., Puiu, D., Novoa, B., Figueras, A., Posada, D., Canchaya, C., 2016. A first insight into the genome of the filter-feeder mussel *Mytilus galloprovincialis*. PLoS One 11 (3), e0151561.

Muscatine, L., 1974. Endosymbiosis of cnidarians and algae. In: Muscatine, L., Lenhoff, H. (Eds.), Coelenterate Biology: Reviews and New Perspectives. Academic Press, New York, pp. 359–395.

Muscatine, L., Porter, J.W., 1977. Reef corals: mutualistic symbioses adapted to nutrient-poor environments. Bioscience 27 (7), 454–460.

Nakabachi, A., Shigenobu, S., Sakazume, N., Shiraki, T., et al., 2005. Transcriptome analysis of the aphid bacteriocyte, the symbiotic host cell that harbors an endocellular mutualistic bacterium, Buchnera. Proc. Natl. Acad. Sci. U.S.A. 102 (15), 5477–5482.

Nandan, B.S., Joseph, P., Jayachandran, R.P., 2019. The recent benthic observations from coastal wetlands of central Kerala, India. In: Compendium on Advances in Benthic Studies. Cochin University of Science and Technology Kochi, p. 272. https://www.researchgate.net/profile/P-R-Jayachandran/publication/331635223_The_ recent_benthic_observations_from_coastal_wetlands_of_central_Kerala_India/links/ 5c88679f299bf14e7e783114/The-recent-benthic-observations-from-coastal-wetlands-of-central-Kerala-India.pdf.

National Academies of Sciences, Engineering, and Medicine (NASEM), 2021. A Research Strategy for Ocean-Based Carbon Dioxide Removal and Sequestration. National Academies Press, Washington DC.

Neave, M.J., Michell, C.T., Apprill, A., Voolstra, C.R., 2017. *Endozoicomonas* genomes reveal functional adaptation and plasticity in bacterial strains symbiotically associated with diverse marine hosts. Sci. Rep. 7, 40579.

Nei, M., Kumar, S., 2000. Molecular Evolution and Phylogenetics. Oxford University Press.

Neumann, B., Vafeidis, A.T., Zimmermann, J., Nicholls, R.J., 2015. Future coastal population growth and exposure to sea-level rise and coastal flooding-a global assessment. PLoS One 10 (3), e0118571.

Nishiguchi, M.K., Nair, V.S., 2003. Evolution of symbiosis in the Vibrionaceae: a combined approach using molecules and physiology. Int. J. Syst. Evol. Microbiol. 53 (6), 2019–2026.

Norkko, J., Gammal, J., Hewitt, J.E., Josefson, A.B., Carstensen, J., Norkko, A., 2015. Seafloor ecosystem function relationships: in situ patterns of change across gradients of increasing hypoxic stress. Ecosystems 18, 1424–1439.

Nyholm, S.V., McFall-Ngai, M., 2004. The winnowing: establishing the squid–vibrio symbiosis. Nat. Rev. Microbiol. 2 (8), 632–642.

Nyholm, S.V., McFall-Ngai, M.J., 2021. A lasting symbiosis: how the Hawaiian bobtail squid finds and keeps its bioluminescent bacterial partner. Nat. Rev. Microbiol. 19 (10), 666–679.

Ohdera, A.H., Avila-Magaña, V., Sharp, V., Watson, K., et al., 2022. Symbiosis-driven development in an early branching metazoan. bioRxiv, 2022–2027.

O'Leary, B.C., Roberts, C.M., 2018. Ecological connectivity across ocean depths: implications for protected area design. Glob. Ecol. Conserv. 15, e00431.

Olivera, B., 2006. Conus peptides: biodiversity-based discovery and exogenomics*. J. Biol. Chem. 281 (42), 31173–31177.

Olsen, A., Triñanes, J.A., Wanninkhof, R., 2004. Sea–air flux of CO2 in the Caribbean Sea estimated using in situ and remote sensing data. Remote Sens. Environ. 89 (3), 309–325.

Orsi, W., Biddle, J.F., Edgcomb, V., 2013. Deep sequencing of subseafloor eukaryotic rRNA reveals active fungi across marine subsurface provinces. PLoS One 8 (2), e56335.

Osborn, K.J., Kuhnz, L.A., Priede, I.G., Urata, M., Gebruk, A.V., Holland, N.D., 2012. Diversification of acorn worms (Hemichordata, Enteropneusta) revealed in the deep sea. Proc. R. Soc. B Biol. Sci. 279 (1733), 1646–1654.

Osman, E.O., Weinnig, A.M., 2022. Microbiomes and obligate symbiosis of deep-sea animals. Annu. Rev. Anim. Biosci. 10, 151–176.

Pace, N.R., 1997. A molecular view of microbial diversity and the biosphere. Science 276 (5313), 734–740.

Pachiadaki, M.G., Sintes, E., Bergauer, K., Brown, J.M., Record, N.R., Swan, B.K., Stepanauskas, R., 2017. Major role of nitrite-oxidizing bacteria in dark ocean carbon fixation. Science 358 (6366), 1046–1051.

Paine, R.T., Castillo, J.C., Cancino, J., 1985. Perturbation and recovery patterns of starfish-dominated intertidal assemblages in Chile, New Zealand, and Washington state. Am. Nat. 125 (5), 679–691.

Palacio-Castro, A.M., Rosales, S.M., Dennison, C.E., Baker, A.C., 2022. Microbiome signatures in *Acropora cervicornis* are associated with genotypic resistance to elevated nutrients and heat stress. Coral Reefs 41 (5), 1389–1403.

Palm, N.W., Medzhitov, R., 2009. Pattern recognition receptors and control of adaptive immunity. Immunol. Rev. 227 (1), 221–233.

Palumbi, S.R., 1992. Marine speciation on a small planet. Trends Ecol. Evol. 7 (4), 114–118.

Palumbi, S.R., 2009. Speciation and the evolution of gamete recognition genes: pattern and process. Heredity 102 (1), 66–76.

Paoli, L., Ruscheweyh, H.J., Forneris, C.C., Hubrich, F., Kautsar, S., Bhushan, A., Sunagawa, S., 2022. Biosynthetic potential of the global ocean microbiome. Nature 607 (7917), 111–118.

Parfrey, L.W., Lahr, D.J., Knoll, A.H., Katz, L.A., 2011. Estimating the timing of early eukaryotic diversification with multigene molecular clocks. Proc. Natl. Acad. Sci. U. S. A. 108 (33), 13624–13629.

Pascal, N., Allenbach, M., Brathwaite, A., Burke, L., Le Port, G., Clua, E., 2016. Economic valuation of coral reef ecosystem service of coastal protection: a pragmatic approach. Ecosyst. Serv. 21, 72–80.

Paul, V.J., Ritson-Williams, R., 2008. Marine chemical ecology. Nat. Prod. Rep. 25 (4), 662–695.

Paul, V.J., Freeman, C.J., Agarwal, V., 2019. Chemical ecology of marine sponges: new opportunities through "-omics". Integr. Comp. Biol. 59 (4), 765–776.

Pawlik, J.R., McMurray, S.E., 2020. The emerging ecological and biogeochemical importance of sponges on coral reefs. Annu. Rev. Mar. Sci. 12, 315–337.

Pawlik, J.R., McMurray, S.E., Erwin, P., Zea, S., 2015. A review of evidence for food limitation of sponges on Caribbean reefs. Mar. Ecol. Prog. Ser. 519, 265–283.

Pawlik, J.R., Burkepile, D.E., Thurber, R.V., 2016. A vicious circle? Altered carbon and nutrient cycling may explain the low resilience of Caribbean coral reefs. Bioscience 66 (6), 470–476.

Payandeh, A.R., Justic, D., Mariotti, G., Huang, H., Valentine, K., Walker, N.D., 2021. Suspended sediment dynamics in a deltaic estuary controlled by subtidal motion and offshore river plumes. Estuar. Coast. Shelf Sci. 250, 107137.

Pedrós-Alió, C., 2006. Marine microbial diversity: can it be determined? Trends Microbiol. 14 (6), 257–263.

Pérez-Díaz, L., Eagles, G., 2017. South Atlantic paleobathymetry since early cretaceous. Sci. Rep. 7 (1), 11819.

Perreau, J., Moran, N.A., 2022. Genetic innovations in animal–microbe symbioses. Nat. Rev. Genet. 23 (1), 23–39.

Pesant, S., Not, F., Picheral, M., Kandels-Lewis, S., Le Bescot, N., Gorsky, G., Searson, S., 2015. Open science resources for the discovery and analysis of Tara Oceans data. Sci. Data 2 (1), 1–16.

Petrosino, G., Ponte, G., Volpe, M., Zarrella, I., Ansaloni, F., Langella, C., Sanges, R., 2022. Identification of LINE retrotransposons and long non-coding RNAs expressed in the octopus brain. BMC Biol. 20 (1), 1–22.

Philippe, H., Brinkmann, H., Lavrov, D.V., Littlewood, D.T.J., Manuel, M., Wörheide, G., Baurain, D., 2011. Resolving difficult phylogenetic questions: why more sequences are not enough. PLoS Biol. 9 (3), e1000602.

Pierce, S.K., Curtis, N.E., Hanten, J.J., Boerner, S.L., Schwartz, J.A., 2007. Transfer, integration and expression of functional nuclear genes between multicellular species. Symbiosis 43, 57–64.

68 Chapter 1 The seabed—Where life began and still evolves

Pineda, J., Caswell, H., 1998. Bathymetric species-diversity patterns and boundary constraints on vertical range distribution. Deep Sea Res. Part II Top. Stud. Oceanogr. 45, 83–101.

Pisani, D., Pett, W., Dohrmann, M., Feuda, R., Rota-Stabelli, O., Philippe, H., Wörheide, G., 2015. Genomic data do not support comb jellies as the sister group to all other animals. Proc. Natl. Acad. Sci. U. S. A. 112 (50), 15402–15407.

Pita, L., Hoeppner, M.P., Ribes, M., Hentschel, U., 2018. Differential expression of immune receptors in two marine sponges upon exposure to microbial-associated molecular patterns. Sci. Rep. 8 (1), 1–15.

Poinar Jr., G., Kerp, H., Hass, H., 2008. *Palaeonema phyticum* gen. n., sp n. (Nematoda: Palaeonematidae fam. n.), a Devonian nematode associated with early land plants. Nematology 10, 9–14.

Puglisi, M.P., Sneed, J.M., Ritson-Williams, R., Young, R., 2019. Marine chemical ecology in benthic environments. Nat. Prod. Rep. 36 (3), 410–429.

Pulliam, H.R., 1988. Sources, sinks, and population regulation. Am. Nat. 132 (5), 652–661.

Raffaelli, D., Bell, E., Weithoff, G., Matsumoto, A., Cruz-Motta, J.J., Kershaw, P., Jones, M., 2003. The ups and downs of benthic ecology: considerations of scale, heterogeneity and surveillance for benthic–pelagic coupling. J. Exp. Mar. Biol. Ecol. 285, 191–203.

Rainbow, P.S., Phillips, D.J., 1993. Cosmopolitan biomonitors of trace metals. Mar. Pollut. Bull. 26 (11), 593–601.

Ramirez-Llodra, E., Brandt, A., Danovaro, R., De Mol, B., Escobar, E., German, C.R., Levin, L.A., Martinez Arbizu, P., Menot, L., Buhl-Mortensen, P., Narayanaswamy, B.E., Smith, C.R., Tittensor, D.P., Tyler, P. A., Vanreusel, A., Vecchione, M., 2010. Deep, diverse and definitely different: unique attributes of the world's largest ecosystem. Biogeosciences 7 (9), 2851–2899.

Ramirez-Llodra, E., Tyler, P.A., Baker, M.C., Bergstad, O.A., Clark, M.R., Escobar, E., Van Dover, C.L., 2011. Man and the last great wilderness: human impact on the deep sea. PLoS One 6 (8), e22588.

Randin, C.F., Ashcroft, M.B., Bolliger, J., Cavender-Bares, J., Coops, N.C., Dullinger, S., Payne, D., 2020. Monitoring biodiversity in the Anthropocene using remote sensing in species distribution models. Remote Sens. Environ. 239, 111626.

Rast, J.P., Smith, L.C., Loza-Coll, M., Hibino, T., Litman, G.W., 2006. Genomic insights into the immune system of the sea urchin. Science 314 (5801), 952–956.

Reaka-Kudla, M.L., Wilson, D.E., Wilson, E.O. (Eds.), 1997. Biodiversity II: Understanding and Protecting Our Biological Resources. Joseph Henry Press.

Redmond, A.K., McLysaght, A., 2021. Evidence for sponges as sister to all other animals from partitioned phylogenomics with mixture models and recoding. Nat. Commun. 12 (1), 1783.

Redmond, N.E., Morrow, C.C., Thacker, R.W., Diaz, M.C., Boury-Esnault, N., Cárdenas, P., 2013. Phylogeny and systematics of Demospongiae in light of new small-subunit ribosomal DNA (18S) sequences. Integr. Comp. Biol. 53 (3), 388–415.

Reece, J.S., et al., 2011. Long larval duration in moray eels (Muraenidae) ensures ocean-wide connectivity despite differences in adult niche breadth. Mar. Ecol. Prog. Ser. 437, 269–277.

Reed, J.K., 2002. Deep-water Oculina coral reefs of Florida: biology, impacts, and management. Hydrobiologia 471 (1-3), 43–55.

Reeder-Myers, L., Braje, T.J., Hofman, C.A., Elliott Smith, E.A., Garland, C.J., Grone, M., Rick, T.C., 2022. Indigenous oyster fisheries persisted for millennia and should inform future management. Nat. Commun. 13 (1), 2383.

Reiswig, H.M., 2002. Class Hexactinellida Schmidt, 1870. In: Systema Porifera. Springer, Boston MA, pp. 1201–1210.

Renegar, D.A., Turner, N.R., 2021. Species sensitivity assessment of five Atlantic scleractinian coral species to 1-methylnaphthalene. Sci. Rep. 11 (1), 529. https://doi.org/10.1038/s41598-020-80055-0.

Rex, M.A., 1981. Community structure in the deep-sea benthos. Annu. Rev. Ecol. Syst. 12, 331–353.

Rex, M.A., Stuart, C.T., Coyne, G., 2000. Latitudinal gradients of species richness in the deep-sea benthos of the North Atlantic. Proc. Natl. Acad. Sci. U. S. A. 97 (8), 4082–4085.

Rex, M.A., McClain, C.R., Johnson, N.A., Etter, R.J., Allen, J.A., Bouchet, P., Warén, A., 2005. A source-sink hypothesis for abyssal biodiversity. Am. Nat. 165 (2), 163–178.

Riekenberg, P.M., Carney, R.S., Fry, B., 2016. Trophic plasticity of the methanotrophic mussel Bathymodiolus childressi in the Gulf of Mexico. Mar. Ecol. Prog. Ser. 547, 91–106.

Rimmer, P.B., Shorttle, O., 2019. Origin of life's building blocks in carbon- and nitrogen-rich surface hydrothermal vents. Life 9. https://doi.org/10.3390/life9010012.

Rivera, H.E., Davies, S.W., 2021. Symbiosis maintenance in the facultative coral, Oculina arbuscula, relies on nitrogen cycling, cell cycle modulation, and immunity. Sci. Rep. 11 (1), 21226.

Rix, L., de Goeij, J.M., van Oevelen, D., Struck, U., Al-Horani, F.A., Wild, C., Naumann, M.S., 2018. Reef sponges facilitate the transfer of coral-derived organic matter to their associated fauna via the sponge loop. Mar. Ecol. Prog. Ser. 589, 85–96.

Roberts, C.M., 1997. Connectivity and management of Caribbean coral reefs. Science 278 (5342), 1454–1457.

Roberts, J.M., Wheeler, A.J., Freiwald, A., 2006. Reefs of the deep: the biology and geology of cold-water coral ecosystems. Science 312 (5773), 543–547.

Rocha, L.A., Robertson, D.R., Rocha, C.R., Van Tassell, J.L., Craig, M.T., Bowen, B.W., 2005. Recent invasion of the tropical Atlantic by an Indo-Pacific coral reef fish. Mol. Ecol. 14 (13), 3921–3928.

Rocha, L.A., Lindeman, K.C., Rocha, C.R., Lessios, H.A., 2008. Historical biogeography and speciation in the reef fish genus Haemulon (Teleostei: Haemulidae). Mol. Phylogenet. Evol. 48 (3), 918–928.

Rodil, I.F., Lucena-Moya, P., Tamelander, T., Norkko, J., Norkko, A., 2020. Seasonal variability in benthic–pelagic coupling: quantifying organic matter inputs to the seafloor and benthic macrofauna using a multi-marker approach. Front. Mar. Sci. 7, 404.

Rogers, A.D., 1999. The biology of Lophelia pertusa (Linnaeus 1758) and other deep-water reef-forming corals and impacts from human sctivities. Int. Rev. Hydrobiol. 84 (4), 315–406.

Rohwer, F., Seguritan, V., Azam, F., Knowlton, N., 2002. Diversity and distribution of coral-associated bacteria. Mar. Ecol. Prog. Ser. 243, 1–10.

Rota-Stabelli, O., Kayal, E., Gleeson, D., Daub, J., Boore, J.L., Telford, M.J., Pisani, D., Blaxter, M., Lavrov, D.V., 2010. Ecdysozoan mitogenomics: evidence for a common origin of the legged invertebrates, the Panarthropoda. Genome Biol. Evol. 2, 425–440.

Rouse, G.W., Goffredi, S.K., Vrijenhoek, R.C., 2004. Osedax: bone-eating marine worms with dwarf males. Science 305 (5684), 668–671.

Rowan, R., 2004. Thermal adaptation in reef coral symbionts. Nature 430 (7001), 742.

Rowan, R., Knowlton, N., Baker, A., Jara, J., 1997. Landscape ecology of algal symbionts creates variation in episodes of coral bleaching. Nature 388 (6639), 265–269.

Rumpho, M.E., Worful, J.M., Lee, J., Kannan, K., Tyler, M.S., Bhattacharya, D., Manhart, J.R., 2008. Horizontal gene transfer of the algal nuclear gene psbO to the photosynthetic sea slug Elysia chlorotica. Proc. Natl. Acad. Sci. U. S. A. 105 (46), 17867–17871.

Rusch, D.B., Halpern, A.L., Sutton, G., Heidelberg, K.B., Williamson, S., Yooseph, S., Venter, J.C., 2007. The sorcerer II global ocean sampling expedition: northwest Atlantic through eastern tropical Pacific. PLoS Biol. 5 (3), e77.

Rushmore, M.E., Ross, C., Fogarty, N.D., 2021. Physiological responses to short-term sediment exposure in adults of the Caribbean coral Montastraea cavernosa and adults and recruits of Porites astreoides. Coral Reefs 40, 1579–1591.

Russell, M.J., Martin, W., 2004. The rocky roots of the acetyl-CoA pathway. Trends Biochem. Sci. 29, 358–363.

Russell, M., Hall, A., Martin, W., 2010. Serpentinization as a source of energy at the origin of life. Geobiology 8, 355–371.

Ryan, J.F., Pang, K., NISC Comparative Sequencing Program, Mullikin, J.C., Martindale, M.Q., Baxevanis, A.D., 2010. The homeodomain complement of the ctenophore Mnemiopsis leidyi suggests that Ctenophora and Porifera diverged prior to the ParaHoxozoa. EvoDevo 1, 1–18.

Ryan, J.F., Pang, K., Schnitzler, C.E., Nguyen, A.D., Nguyen, A.D., Moreland, R.T., et al., 2013. The genome of the ctenophore *Mnemiopsis leidyi* and its implications for cell type evolution. Science 342, 6164.

Safavi-Hemami, H., Brogan, S.E., Olivera, B.M., 2019. Pain therapeutics from cone snail venoms: from Ziconotide to novel non-opioid pathways. J. Proteome 190, 12–20.

Saito, Y., Hirose, E., Watanabe, H., 2002. Allorecognition in compound ascidians. Int. J. Dev. Biol. 38 (2), 237–247.

Samadi, S., Bottan, L., Macpherson, E., De Forges, B.R., Bosselier, M.C., 2006. Seamount endemism questioned by the geographic distribution and population genetic structure of marine invertebrates. Mar. Biol. 149 (6), 1463–1475.

Sánchez, J.A., González-Zapata, F.L., Prada, C., Dueñas, L.F., 2021. Mesophotic gorgonian corals evolved multiple times and faster than deep and shallow lineages. Diversity 13 (12), 650.

Sanders, H.L., 1968. Marine benthic diversity: a comparative study. Am. Nat. 102 (925), 243–282.

Sanders, H.L., Hessler, R.R., Hampson, G.R., 1965. An introduction to the study of deep-sea benthic faunal assemblages along the gay head-Bermuda transect. Deep-Sea Res. 12, 845–867.

Santagata, S., Ade, V., Mahon, A.R., Wisocki, P.A., Halanych, K.M., 2018. Compositional differences in the habitat-forming bryozoan communities of the Antarctic shelf. Front. Ecol. Evol. 6, 116.

Schlötterer, C., 2015. Genes from scratch–the evolutionary fate of de novo genes. Trends in Genet. 31 (4), 215–219.

SchoÈnberg, C.H., 2002. Substrate effects on the bioeroding demosponge Cliona orientalis. 1. Bioerosion rates. Mar. Ecol. 23 (4), 313–326.

Schopf, J.W., Klein, C. (Eds.), 1992. The Proterozoic Biosphere: A Multidisciplinary Study. Cambridge University Press.

Schorghofer, N., Forget, F., 2012. History and anatomy of subsurface ice on Mars. Icarus 220 (2), 1112–1120.

Schram, F.R., 1973. Pseudocoelomates and a nemertine from the Illinois Pennsylvanian. J. Paleontol. 47, 985–989.

Schultz, D.T., Haddock, S.H., Bredeson, J.V., Green, R.E., Simakov, O., Rokhsar, D.S., 2023. Ancient gene linkages support ctenophores as sister to other animals. Nature 618, 110–117.

Sepkoski Jr., J.J., 1991. Diversity in the Phanerozoic oceans: a partisan review. In: Dudley, E.C. (Ed.), The Unity of Evolutionary Biology. Proceedings of the Fourth International Congress on Systematic. Dioscorides Press, pp. 210–236.

Seronay, R.A., Fedosov, A.E., Astilla, M.A.Q., Watkins, M., Saguil, N., Heralde III, F.M., Olivera, B.M., 2010. Accessing novel conoidean venoms: biodiverse lumun-lumun marine communities, an untapped biological and toxinological resource. Toxicon 56 (7), 1257–1266.

Sevigny, J., Leasi, F., Simpson, S., Di Domenico, M., Jörger, K.M., Norenburg, J.L., Thomas, W.K., 2021. Target enrichment of metazoan mitochondrial DNA with hybridization capture probes. Ecol. Indic. 121, 106973.

Sfanos, K.A.S., Harmody, D.K., McCarthy, P.J., Dang, P., Pomponi, S.A., Lopez, J.V., 2005. A molecular systematic survey of cultured microbial associates of deep water marine invertebrates. Syst. Appl. Microbiol. 28 (3), 242–264.

Shen, M., Ye, S., Zeng, G., Zhang, Y., et al., 2020. Can microplastics pose a threat to ocean carbon sequestration? Mar. Pollut. Bull. 150, 110712.

Shlesinger, T., Loya, Y., 2019. Breakdown in spawning synchrony: a silent threat to coral persistence. Science 365 (6457), 1002–1007.

Siegel, D.A., Buesseler, K.O., Behrenfeld, M.J., Benitez-Nelson, C.R., Boss, E., Brzezinski, M.A., et al., 2016. Prediction of the export and fate of global ocean net primary production: the EXPORTS science plan. Front. Mar. Sci. 3, 1. https://doi.org/10.3389/fmars.2016.00022.

Signor, P.W., Vermeij, G.J., 1994. The plankton and the benthos: origins and early history of an evolving relationship. Paleobiology 20 (3), 297–319.

Simion, P., Philippe, H., Baurain, D., Jager, M., Richter, D.J., Di Franco, A., Lapebie, P., 2017. A large and consistent phylogenomic dataset supports sponges as the sister group to all other animals. Curr. Biol. 27 (7), 958–967.

Sinniger, F., Pawlowski, J., Harii, S., Gooday, A.J., Yamamoto, H., Chevaldonné, P., Creer, S., 2016. Worldwide analysis of sedimentary DNA reveals major gaps in taxonomic knowledge of deep-sea benthos. Front. Mar. Sci. 3, 92.

Smetacek, V.S., 1985. Role of sinking in diatom life-history cycles: ecological, evolutionary and geological significance. Mar. Biol. 84, 239–251.

Smidt, J., 2018. The first water in the universe. In: SC 18. supercomputing.org, pp. 1–2.

Smith, C.R., Reef, J.B., 1985. Symbiosis in marine animals. Annu. Rev. Ecol. Syst. 16 (1985), 145–179.

Smith, E.B., Scott, K.M., Nix, E.R., Korte, C., Fisher, C.R., 2000. Growth and condition of seep mussels (Bathymodiolus childressi) at a Gulf of Mexico brine pool. Ecology 81 (9), 2392–2403.

Smith, C.R., De Leo, F.C., Bernardino, A.F., Sweetman, A.K., Arbizu, P.M., 2008. Abyssal food limitation, ecosystem structure and climate change. Trends Ecol. Evol. 23 (9), 518–528.

Smith, C.R., Glover, A.G., Treude, T., Higgs, N.D., Amon, D.J., 2015. Whale-fall ecosystems: recent insights into ecology, paleoecology, and evolution. Annu. Rev. Mar. Sci. 7, 571–596.

Snelgrove, P.V., Soetaert, K., Solan, M., Thrush, S., Wei, C.L., Danovaro, R., Volkenborn, N., 2018. Global carbon cycling on a heterogeneous seafloor. Trends Ecol. Evol. 33 (2), 96–105.

Sogin, M.L., Morrison, H.G., Huber, J.A., Welch, D.M., Huse, S.M., Neal, P.R., Herndl, G.J., 2006. Microbial diversity in the deep sea and the underexplored "rare biosphere". Proc. Natl. Acad. Sci. U. S. A. 103 (32), 12115–12120.

Solek, C.M., 2012. Dissertation a Gene Regulatory Network for the Specification of Immunocytes in an Invertebrate Model System. University of Toronto, Canada.

Stasko, A.D., Bluhm, B.A., Michel, C., Archambault, P., Majewski, A., et al., 2018. Benthic-pelagic trophic coupling in an Arctic marine food web along vertical water mass and organic matter gradients. Mar. Ecol. Prog. Ser. 594, 1–19.

Stat, M., Carter, D., Hoegh-Guldberg, O., 2006. The evolutionary history of *Symbiodinium* and scleractinian hosts—symbiosis, diversity, and the effect of climate change. Perspect. Plant Ecol. Evol. Syst. 8 (1), 23–43.

Stoecker, D.K., Capuzzo, J.M., 1990. Predation on protozoa: its importance to zooplankton. J. Plankton Res. 12 (5), 891–908.

Stommel, H., 1958. The abyssal circulation. Deep-Sea Res. 5, 80.

Stratmann, T., van Oevelen, D., Martinez Arbizu, P., Wei, C.L., Liao, J.X., Cusson, M., Soetaert, K., 2020. The BenBioDen database, a global database for meio-, macro-and megabenthic biomass and densities. Sci. Data 7 (1), 206.

Sully, S., van Woesik, R., 2020. Turbid reefs moderate coral bleaching under climate-related temperature stress. Glob. Chang. Biol. 26 (3), 1367–1373.

Sunagawa, S., Coelho, L.P., Chaffron, S., Kultima, J.R., et al., 2015. Ocean plankton. Structure and function of the global ocean microbiome. Science 348, 1261359.

Sutton, T.T., Frank, T., Judkins, H., Romero, I.C., 2020. As gulf oil extraction goes deeper, who is at risk? Community structure, distribution, and connectivity of the deep-pelagic fauna. In: Murawski, S. (Ed.), Scenarios and Responses to Future Deep Oil Spills: Fighting the Next War. Springer, pp. 403–418.

Tamelander, T., Reigstad, M., Hop, H., Carroll, M.L., Wassmann, P., 2008. Pelagic and sympagic contribution of organic matter to zooplankton and vertical export in the Barents Sea marginal ice zone. Deep-Sea Res. II Top. Stud. Oceanogr. 55, 2330–2339. https://doi.org/10.1016/j.dsr2.2008.05.019.

72   Chapter 1 The seabed—Where life began and still evolves

Tanet, L., Martini, S., Casalot, L., Tamburini, C., 2020. Reviews and syntheses: bacterial bioluminescence–ecology and impact in the biological carbon pump. Biogeosciences 17 (14), 3757–3778.

Taylor, M.W., Radax, R., Steger, D., Wagner, M., 2007. Sponge-associated microorganisms: evolution, ecology, and biotechnological potential. Microbiol. Mol. Biol. Rev. 7 (2), 295–347.

Telford, M.J., Budd, G.E., Philippe, H., 2015. Phylogenomic insights into animal evolution. Curr. Biol. 25 (19), R876–R887.

Tenggardjaja, K.A., Bowen, B.W., Bernardi, G., 2014. Vertical and horizontal genetic connectivity in *Chromis verater*, an endemic damselfish found on shallow and mesophotic reefs in the Hawaiian archipelago and adjacent Johnston Atoll. PLoS One 9 (12), e115493.

Thiel, H., Schriever, G., 1989. Cruise Report DISCOL 1, SONNEFcruise 61, Balboa/PanamaFCallao/Peru with Contribution by C. Borowski, C. Bussau, D. Hansen, J. Melles, J. Post, K. Steinkamp, K. Watson. Vol. 3 Berichte aus dem Zentrum fur Meeres- und Klimaforschungder Universitat Hamburg. 75pp.

Thomas, T., Moitinho-Silva, L., Lurgi, M., Easson, C.G., Björk, J., Astudillo, C., et al., 2016. The global sponge microbiome: symbiosis insights derived from a basal metazoan phylum. Nat. Commun. 7 (11870), 1–12. https://doi.org/10.1038/ncomms11870.

Thomson, C.W., 1878. The Voyage of the " Challenger." The Atlantic; a Preliminary Account of the General Results of the Exploring Voyage of HMS "Challenger" During the Year 1873 and the Early Part of the Year 1876. vol. 1 Harper & Brothers. https://tinyurl.com/2cmm3dnb.

Thoram, S., Sager, W.W., Reed, W., Nakanishi, M., Zhang, J., 2022. Improved high-resolution bathymetry map of Tamu massif and southern Shatsky rise and its geologic implications. J. Geophys. Res. Solid Earth 127 (11). e2022JB024304.

Thrush, S.F., Lohrer, A.M., 2012. Why bother going outside: the role of observational studies in understanding biodiversity-ecosystem function relationships. In: Biodiversity, Ecosystem Functioning, & Human Wellbeing an Ecological and Econimic Perspective. Oxford University Press, Oxford, pp. 200–214.

Tobler, M., Passow, C.N., Greenway, R., Kelley, J.L., Shaw, J.H., 2016. The evolutionary ecology of animals inhabiting hydrogen sulfide–rich environments. Annu. Rev. Ecol. Evol. Syst. 47, 239–262.

Tong, H., Cai, L., Zhou, G., Zhang, W., Huang, H., Qian, P.Y., 2020. Correlations between prokaryotic microbes and stress-resistant algae in different corals subjected to environmental stress in Hong Kong. Front. Microbiol. 11, 686.

Trench, R.K., 1993. Microalgal-invertebrate symbiosis, a review. Endocytobiosis Cell Res. 9, 135–175.

Trestrail, C., Nugegoda, D., Shimeta, J., 2020. Invertebrate responses to microplastic ingestion: reviewing the role of the antioxidant system. Sci. Total Environ. 734, 138559.

Turner, N.R., Parkerton, T.F., Renegar, D.A., 2021. Toxicity of two representative petroleum hydrocarbons, toluene and phenanthrene, to five Atlantic coral species. Mar. Pollut. Bull. 169, 112560.

United Nations, 2017. https://www.un.org/sustainabledevelopment/wp-content/uploads/2017/05/Ocean-fact-sheet-package.pdf.

United Nations Environmental Program, 2000. Biological Diversity Assessment.

Van der Kaaden, A.S., Mohn, C., Gerkema, T., et al., 2021. Feedbacks between hydrodynamics and cold-water coral mound development. Deep Sea Res. Part I Oceanogr. Res. Pap. 178, 103641.

Van Dover, C.L., Humphris, S.E., Fornari, D., Cavanaugh, C.M., et al., 2001. Biogeography and ecological setting of Indian Ocean hydrothermal vents. Science 294 (5543), 818–823.

Van Dover, C.L., German, C.R., Speer, K.G., Parson, L.M., Vrijenhoek, R.C., 2002. Evolution and biogeography of deep-sea vent and seep invertebrates. Science 295 (5558), 1253–1257.

Van Oppen, M.J., Bongaerts, P.I.M., Underwood, J.N., Peplow, L.M., Cooper, T.F., 2011. The role of deep reefs in shallow reef recovery: an assessment of vertical connectivity in a brooding coral from west and east Australia. Mol. Ecol. 20 (8), 1647–1660.

Van Soest, R.W., Boury-Esnault, N., Vacelet, J., Dohrmann, M., Erpenbeck, D., De Voogd, N.J., Hooper, J.N., 2012. Global diversity of sponges (Porifera). PLoS One 7 (4), e35105.

Venter, J.C., 2014. Life at the Speed of Light: From the Double Helix to the Dawn of Digital Life. Penguin.

Venter, J.C., Adams, M.D., Myers, E.W., Li, P.W., Mural, R.J., Sutton, G.G., Kalush, F., 2001. The sequence of the human genome. Science 291 (5507), 1304–1351.

Venter, J.C., Remington, K., Heidelberg, J.F., Halpern, A.L., Rusch, D., Eisen, J.A., Smith, H.O., 2004. Environmental genome shotgun sequencing of the Sargasso Sea. Science 304 (5667), 66–74.

Vermeij, G.J., 1977. The Mesozoic marine revolution: evidence from snails, predators and grazers. Paleobiology 3 (3), 245–258.

Vermeij, G.J., 1994. The evolutionary interaction among species: selection, escalation, and coevolution. Annu. Rev. Ecol. Syst. 25 (1), 219–236.

Veron, J.E.N., 1995. Corals in Space and Time: The Biogeography and Evolution of the Scleractinia. Cornell University Press.

Vetter, E.W., Dayton, P.K., 1998. Macrofaunal communities within and adjacent to a detritus-rich submarine canyon system. Deep-Sea Res. II Top. Stud. Oceanogr. 45 (1-3), 25–54.

Via, R.K., Thomas, D.J., 2006. Evolution of Atlantic thermohaline circulation: Early Oligocene onset of deepwater production in the North Atlantic. Geology 34 (6), 441–444.

Vieira, H., Leal, M.C., Calado, R., 2020. Fifty shades of blue: How blue biotechnology is shaping the bioeconomy. Trends Biotechnol. 38 (9), 940–943.

Viles, H., Spencer, T., 2014. Coastal Problems: Geomorphology, Ecology and Society at the Coast. Routledge.

Vogel, S., 1977. Current-induced flow through living sponges in nature. Proc. Natl. Acad. Sci. U. S. A. 74 (5), 2069–2071.

Voolstra, C.R., Schwarz, J.A., Schnetzer, J., Sunagawa, S., Desalvo, M.K., Szmant, A.M., Coffroth, M.A., Medina, M., 2009. The host transcriptome remains unaltered during the establishment of coral-algal symbioses. Mol. Ecol. 8 (9), 1823–1833 (Epub 2009 Mar 20).

Waddington, C.H., Needham, J., 1936. Evocation, individuation and competence in amphibian organizer action. Proc. Kon. Akad. Wetensch. Amsterdam 39, 887–890.

Warren, B., Wunsch, C., 1981. Evolution of Physical Oceanography. MIT Press, Cambridge, MA.

Wassman, P., 1984. Sedimentation and benthic mineralization of organic detritus in a Norwegian fjord. Mar. Biol. 83, 83–94.

Webster, N.S., Taylor, M.W., 2012. Marine sponges and their microbial symbionts: love and other relationships. Environ. Microbiol. 14 (2), 335–346.

Webster, N.S., Taylor, M.W., Benham, F., Lücker, S., 2010. Deep sequencing reveals exceptional diversity and modes of transmission for bacterial sponge symbionts. Environ. Microbiol. 12 (8), 2070–2082.

Weiss, V.M., 2008. Cellular mechanisms of Cnidarian bleaching: stress causes the collapse of symbiosis. J. Exp. Biol. 211 (19), 3059–3066.

Whelan, N.V., Kocot, K.M., Moroz, L.L., Halanych, K.M., 2015. Error, signal, and the placement of Ctenophora sister to all other animals. Proc. Natl. Acad. Sci. U. S. A. 112 (18), 5773–5778.

Whitman, W.B., Coleman, D.C., Wiebe, W.J., 1998. Prokaryotes: the unseen majority. Proc. Natl. Acad. Sci. U. S. A. 95 (12), 6578–6583.

Williams, E., Prager, E., Wilson, D., 2002. Research combines with public outreach on a cruise ship. Eos 83 (50), 590–596.

Wilson, E.O., 1987. The little things that run the world (the importance and conservation of invertebrates). Conserv. Biol. 1, 344–346.

Wilson, E.O., 1992. The Diversity of Life. Belknap Press, Cambridge, MA. 242 pp.

Wilson, G.D., Hessler, R.R., 1987. Speciation in the deep sea. Annu. Rev. Ecol. Syst. 18 (1), 185–207.

Woese, C.R., 2004. A new biology for a new century. Microbiol. Mol. Biol. Rev. 68 (2), 173–186.

Woese, C.R., Kandler, O., Wheelis, M.L., 1990. Towards a natural system of organisms: proposal for the domains archaea, bacteria, and eucarya. Proc. Natl. Acad. Sci. U. S. A. 87 (12), 4576–4579.

Wolff, T., 1961. Animal life from a single abyssal trawling. Galathea Rep. 5, 129–162.

Wolff, T., 1970. The concept of the hadal or ultra-abyssal fauna. In: Deep Sea Research and Oceanographic Abstracts. Vol. 17, No. 6. Elsevier, pp. 983–1003.

Wuchter, C., Abbas, B., Coolen, M.J., Herfort, L., van Bleijswijk, J., Timmers, P., Sinninghe Damsté, J.S., 2006. Archaeal nitrification in the ocean. Proc. Natl. Acad. Sci. U. S. A. 103 (33), 12317–12322.

Xiao, X., Zhang, Y., Wang, F., 2021. Hydrostatic pressure is the universal key driver of microbial evolution in the deep ocean and beyond. Environ. Microbiol. Rep. 13 (1), 68–72.

Xu, Z., Wang, M., Wu, W., Li, Y., Liu, Q., Han, Y., Liu, H., 2018. Vertical distribution of microbial eukaryotes from surface to the hadal zone of the Mariana Trench. Front. Microbiol. 9, 2023.

Zakas, C., Harry, N.D., Scholl, E.H., Rockman, M.V., 2022. The genome of the poecilogonous annelid Streblospio benedicti. Genome Biol. Evol. 14 (2), evac008.

Zaneveld, J.R., McMinds, R., Vega Thurber, R., 2017. Stress and stability: applying the Anna Karenina principle to animal microbiomes. Nat. Microbiol. 2 (9), 1–8.

Zeng, L., Jacobs, M.W., Swalla, B.J., 2006. Coloniality has evolved once in stolidobranch ascidians. Integr. Comp. Biol. 46 (3), 255–268.

Zeng, Q., Huang, D., Lin, R., Wang, J., 2018. Deep-sea metazoan meiofauna from a polymetallic nodule area in the Central Indian Ocean Basin. Mar. Biodivers. 48 (2018), 395–405. https://doi.org/10.1007/s12526-017-0778-0.

Zeppilli, D., Vanreusel, A., Danovaro, R., 2011. Cosmopolitanism and biogeography of the genus Manganonema (Nematoda: Monhysterida) in the deep sea. Animals 1 (3), 291–305.

Zhao, L., Saelao, P., Jones, C.D., Begun, D.J., 2014. Origin and spread of de novo genes in Drosophila melanogaster populations. Science 343 (6172), 769–772.

Zhang, Z.-Q., 2011. Animal biodiversity: an outline of higher-level classificationand survey of taxonomic richness. Zootaxa 3148, 237.

Zhao, W., Ma, X., Liu, X., Jian, H., Zhang, Y., Xiao, X., 2020. Cross-stress adaptation in a piezophilic and hyperthermophilic archaeon from deep sea hydrothermal vent. Front. Microbiol. 11, 2081. https://doi.org/10.3389/fmicb.2020.02081.

Zibrowius, H., 1980. Les Scléractiniaires de la Méditerranée et de l'Atlantique nord-oriental. Mémoires de l'Institut océanographique, Monaco.

Ziegler, M., Seneca, F.O., Yum, L.K., Palumbi, S.R., Voolstra, C.R., 2017. Bacterial community dynamics are linked to patterns of coral heat tolerance. Nat. Commun. 8 (1), 14213.

CHAPTER

Multiple approaches to understanding the benthos

2

Living in the era of big science

With so many cross-currents of history apparently converging and crossing paths, creating noticeable turbulence in the Anthropocene epoch mentioned in Chapter 1, this is a good time to reflect on our human species. From an evolutionary perspective, we should realize how recently *Homo sapiens* emerged within the span of Earth's history (last 300,000 years versus over 4 billion years for the Earth) and also how well the performance has allowed us to reach our current status. For example, we can marvel that at the time of writing, only 107 years has passed since the Wright brothers successfully took to the air at Kitty Hawk. The term "scientist" has only existed since the middle 1800s, after being coined by William Whewell, while the scientific method of hypothesis testing and refutation is even younger (Popper, 1959). In addition, as mentioned earlier, **biodiversity** in peril as a viable concept that could pose potential mass extinction risks, and that should concern all citizens, only emerged in the lay consciousness about a generation ago.

Some scientific and technical advances over the past few years can be perceived to occur at a dizzying pace and perhaps with self-imposed deadlines to finish what is necessary (Tollefson and Gibney, 2022; Callaway et al., 2022). Although specific scientific disciplines or national programs may have well-organized networks for coordination, overall, only a few entities can organize and enforce large-scale scientific or economic policies, in spite of the best intentions. As societies healed after the twentieth century's great World Wars, concerned citizens have formed well-intentioned global organizations such as the United Nations (and its many focus groups and committees), the Committee on Space Research (COSPAR) established by the International Council of Scientific Unions to formulate policy on contamination of space; the International Atomic Energy Agency (IAEA) works to promote peaceful use of nuclear energy, or the UN's International Seabed Authority (ISA) which regulates and controls all mineral-related activities in the international seabed area for the benefit of mankind as a whole. Relevant to the topics in this volume, the ISA was established in 1996 based on Part XI of the UN's Law of the Sea (1982, and its Implementation Agreement of 1994) (https://www.sciencedirect.com/topics/earth-and-planetary-sciences/law-of-the-sea), with the dual purpose of a) organizing and controlling activities in the Area (Article 157) to regulate the extraction of mineral resources in areas beyond national jurisdiction and b) protection and conserve of the natural resources (Bräger et al., 2020). In addition, actions of the UN conference of parties (COP) together with the Secretariat of the Convention on Biological Diversity, the United Nations Environment Program, the International Union for Conservation of Nature (IUCN) all implemented the strategic Plan for Biodiversity

Assessments and Conservation of Biological Diversity From Coral Reefs to the Deep Sea. https://doi.org/10.1016/B978-0-12-824112-7.00004-2
Copyright © 2024 Elsevier Inc. All rights reserved.

75

76 **Chapter 2** Multiple approaches to understanding the benthos

2011–2020 (Secretariat of the Convention on Biological Diversity, 2011; https://www.cbd.int/decision/cop/?id=12268). These and many other agencies address large-scale issues that cross international boundaries and challenge human society. Large-scale international cooperation in the sciences and policy may find daylight through the intermingling of commercial with public funding. The International Space Station and other astronomical endeavors represent some tangible results of the initial idea of broad cooperation. However, pressing proprietary interests from certain profitable scientific developments (e.g., biotechnology) and lack of enforcement can sometimes create obstacles to more global resolutions (e.g., biodiversity treaties).

Because of the wide expanses of the oceans, by default, we scale our thinking to encompass the vastness along the way. This macro view may be facilitated by current events and culture, suggesting that we live in an era of "Big" projects. This paradigm is supported by humanity's activities, which still generate large physical structures (sprawling cities and taller skyscrapers, bigger cruise ships, extensive pipelines, and cables, etc.). We also collectively produce enormous digital files and more data, which can be stored in ever more efficient computers, solid-state drives, and "the cloud." These activities include astronomy, videography (YouTube), texting (Twitter), and genomics (Stephens et al., 2015). The large human-generated data sets have produced a reliance driven by computers and technological companies, which can act as gatekeepers (Google, Microsoft) (Malde et al., 2020). The statistics quoted here will quickly be out of date by the time this book gets published, which underscores the pace of computational and informational advances (Harari et al., 2019). For example, novel avenues for video expression such as Tik Tok may already be usurping the popular YouTube. From 1956 to 2015, computing power increased one trillion-fold. By 2025, the world is projected to produce more than 175 zettabytes (ZB or 1021) of various data per year, equivalent to one billion terabytes (TB) (Rydning et al., 2018).

And yet, to successfully accomplish complex ecosystem-level questions and address large scales, researchers often have to practice reductionism, breaking larger wholes into more manageable constituent components, delegate, and prioritize before reaching the ultimate conclusion. These are realistic necessities, and the same ideas of "divide and conquer" can and have been successfully applied to studying the benthos. We also have to agree on what composes a truly healthy ecosystem (Costanza and Mageau, 1999; Da Ros et al., 2019; Duarte et al., 2020), which is not an easy task when multiple disciplines are involved. The liquid oceans and the solid seafloor they overlay fall into this bin of thinking, qualifying precisely because each habitat is extremely vast. I now propose one practical way to view its assessment and eventual conservation.

Just as the ocean and benthos are spatially large, many human-conceived scientific frontiers and projects can also aspire to great heights and ambitions, to the point that programs are coined as "Big Science" projects. The Apollo moon program, developing safe energy from nuclear fusion and trying to cure the many faces of cancer, can certainly qualify as a big, expansive science project (Mukherjee, 2011). For the latter, Sporn (1996) recounts that "Carcinoma is not a disease of an individual cell. Carcinoma is ultimately a more complex failure in homoeostasis, a chronic, maladaptive tissue and organismic response to injury." These projects are aptly considered "big" because many people and experts are required for success, and typically large amounts of money and time must be invested. Handling the intrinsic difficulty of conducting big science, we can classify grand scientific endeavors into three approaches and concepts: (a) *The Big Experiment*, (b) *An Informative Map*, or (c) *the Leading Wedge* (Eddy, 2013). I borrow this classification because they provide a useful triptych for approaching the assessment of benthic habitats:

> "A big *experiment* is driven by a single question or hypothesis to test but requires a large-scale community investment. Like any experiment, it would generally include an experimental design, positive and negative controls, and validation experiments that test conclusions from multiple angles. A failure mode in a big experiment is the difficulty in planning experiments properly in a committee process. A *map* is a data resource—comprehensive, complete, closed ended—to be used by multiple groups, over a long time, for multiple purposes. The decision to build a map is a cost/benefit calculation, weighed against individual labs who are already making piecemeal maps in an ill coordinated fashion, especially when small groups lack technical expertise to make the map well. A failure mode with a map is to miscalculate the cost/benefit analysis and make a map that too few individual labs will use. A leading *wedge* is a massed technology development effort, in an area where we need radically better methods. There is a driving goal that is visionary, but perhaps arbitrary. The actual long-term goal is to develop and democratize a breakthrough technology, making it cheap and effective for routine use in individual labs, for individual well-designed experiments. A failure mode in a leading wedge is the inertia of the big science phase, failing to transition as quickly as it should to democratization to small labs."
>
> **(Eddy, 2013)**

Big experiments on the seafloor are difficult

In the of Big Science problems, scale does not necessarily pertain to physical size or broad spatial biogeographic regions. As quoted above, the big experiment is based on a testable hypothesis and large enough in scale to require significant community support. I also posit that the big experiment could encompass the intended or unintended major impacts of the results on science and society, such as Frederick Griffith's elegant laboratory experiment with virulent and avirulent strains of *Streptococcus pneumoniae* in 1928. After inoculating mice with different combinations of live and heat-killed strains, this experiment discovered a previously unknown "transforming" factor, which converted avirulent strains to virulence (Griffith, 1928). The transformant turned out to be DNA. This was not a large experiment in scale, but it was the first major step in confirming DNA as the hereditary material, which was confirmed by Oswald Avery less than 26 years later (Hotchkiss, 1965). Determining the genetic code of DNA based on the four nucleotides arranged in three base codons required more "big impact" experiments, albeit in a laboratory (Watson et al., 2013). To step back a little further, designing proper, effective experiments and formulating cleary hypotheses to test, are not trivial tasks in any science. The design is based on the *hypothetico-deductive* approach and the falsifiability of the hypotheses (Popper, 1959; Medawar, 1967). The scale of running in situ studies in oceanic settings may compel some to adopt the older *inductive* methods of empiricism, and this could be viewed as reasonable as well in some circles. I will avoid deep digressions on inherent tautologies and philosophical debates of carrying out science, interesting and important as they are. Refutation of incorrect hypotheses has been a strong cornerstone and served science for several decades. Thus, I reiterate that carrying out effective experiments requires as much forethought and planning, besides the physical implementation of hardware in order to refute the *original* idea, if one is to reach robust and valid conclusions.

With respect to ocean studies, each of these approaches has its own pros and cons. For example, executing big experiments can often be too difficult to implement on large geographic scales or

78   Chapter 2 Multiple approaches to understanding the benthos

extensive depths. I qualify this by specifying in situ experiments—Can big experiments be performed in the unpredictable ocean? Smaller-scale experiments are possible, and many have been conducted across various marine habitats. In one respect, most of the coral restoration, rescue efforts, and nurseries, which have propagated at many sites in an effort to reproduce and maintain a standing stock for reseeding deteriorating reefs, are some of the heroic, large-scale experiments (Vaughan, 2021). However, in general, experiments can work best in the laboratory, or smaller-scale mesocosms, where most variables can be controlled. I will mention a few provocative experiments on benthic taxa throughout this book. For example, Williams et al. (2017) carried out in situ biodiversity restoration experiments in Coral Triangle seagrass habitats with positive results. In another example, a focused set of ecotoxicology experiments have been carried out at the Nova Southeastern University Marine Toxicology laboratory at the Guy Harvey Oceanographic Center, as well as various aquaculture facilities. This laboratory has developed a comprehensive toxicity assay for scleractinian corals to establish the independent and interactive effects of environmental contaminants on the coral host and symbiont. Test facilities include three custom exposure systems (flow-through, recirculating, and static), which can be utilized to test the effects of a wide variety of current and emerging chemicals of concern, such as petroleum hydrocarbons, metals, herbicides, pesticides, UV filters, and other organic and inorganic contaminants (Renegar et al., 2017; Turner et al., 2021). The author's laboratory has similarly been attempting to establish the ubiquitous marine sponge *Cinanchyrella* as a model organism, alongside rotifers, nematodes, and mollusks (Li et al., 2014; Rodríguez-Romero et al., 2014; Desplat et al. (2023)).

Of course, scientific experiments are carried out deliberately to address specific questions and test hypotheses. However, in situ field experiments can require extensive logistics and extra costs for personnel, boats, and sometimes expensive equipment that can withstand the wear and tear of transport and corrosive saltwater. There have been many excellent ecological field experiments carried out on coral reefs, seagrasses and rocky intertidals, and other benthic habitats over the years. For example Dayton's experiments on rocky intertidal species interactions in the Pacific Northwest San Juan islands represent a monumental effort in field ecology (Dayton, 1971). There are too many to fully review, but I can describe a few that stand out, especially regarding the specific questions and special logistical issues. For example, to test the influence of top-down control on coral habitats by large and small herbivorous fishes, Hughes et al. (2007) carried out a long-term, large-scale ($300\,m^2$, 30 months) study in the Great Barrier reef. Their findings indicated a successfully generated phase shift to macroalgal dominance when predation by fishes was limited. To follow coral spawning from release to fertilization, Don Levitan, Nancy Knowlton, Javier Jara, and colleagues used a team of divers and chemically lit drogues in the field experiments at Lee Stocking Island in the Bahamas and San Blas, Panama (Levitan et al., 2004). Synchronous laboratory experiments were also conducted in parallel to test for fertilization compatibilities. Their results indicated precise timing of gamete bundle releases by each member of the *Montastraea* (now called *Orbicella*) sibling species complex. Genetically, closer *Oxyopes annularis* and *O. franksi* species showed temporally separated spawning times, which lessened the chances of gamete interactions at the surface (Knowlton, 1993; Knowlton et al., 1997; Fukami and Knowlton, 2005).

In another example, a long-term (10 month) controlled experiment was conducted to test the impact of ocean acidification on corals at the Hawaii Institute of Marine Biology (Jokiel et al., 2008). Acidification effects on coral spawning, growth, and larval recruitment of two corals *Pocillopora damicornis* and *Montipora capitata* were measured and compared. Indeed, the effects of lower pH showed decreases in larval recruitment linked to significant (86%) decrease in crustose coralline algae (CCA) coverage, while calcification rates of *M. capitata* appeared lower by 15%–20%. These and many other experiments helped raise the red flag of OA's damaging effects to coral reefs and their calcifying organisms. Similar field experiments addressed warming sea surface temperatures, pollution, reproduction, etc., at marine laboratories around the world (Valiela and Valiela, 1995; Fennel and Neumann, 2014).

Notable in the above examples are that coral reefs occur in relatively shallow waters. In situ experiments in deeper waters are more difficult. The effects of human perturbation of the benthic habitats have been considered, as attention turns toward the possible mining of deep resources. Few propose to mine or perturb coral reefs deliberately, but habitats at abyssal depths are typically far from human settlements and thus, "out of sight, out of mind." We will revisit mining as a specific threat to the benthos in Chapter 4. For now, I will review a few experiments to understand the potential ramifications of deep-sea exploration on the biology, ecology, and conservation of the habitats. For example, studies beginning in the 1980s were initiated as nodule mining simulations. This was immediately followed by assessment of the resident communities from the impacted site and adjacent undisturbed sites. The following studies represent some well-known large-scale deep benthic experiments: The Inter Ocean Metal Benthic Impact Experiment (IOM BIE) in the eastern Clarion-Clipperton Zone (CCZ) comprises one study initiated in 1995 (Brockett and Richards, 1994; Radziejewska et al., 2001). The Indian Deep-sea Environment Experiment (INDEX) in the Central Indian Basin (Indian Ocean) investigated total meiofaunal density 4 years after the disturbance was carried out in 1997 (Sharma, 2001; Ingole et al., 1999, 2005). Harpacticoids species composition within and outside the track 26 years after the experimental nodule dredging by the American consortium OMCO (Ocean Minerals Company) (conducted in 1978 in the CCFZ) was assessed by Mahatma (2009). Recovery of some taxa has not been observed (Miljutin et al., 2011).

On a more positive note, one successful attempt to better understand the impacts of deep-sea mining for polymetallic nodules was the DISturbance and reCOLonization (DISCOL) experiment, which was conducted in 1989 in the Peru Basin (Shirayama, 1999; Shirayama et al., 2001; Simon-Lledó et al., 2019; Vonnahme et al., 2020). The DISCOL project represented a massive undertaking, spanning at least 30 years (Thiel and Schriever, 1989; Thiel, 1991). DISCOL incorporated the most advanced technologies at the time: photo/video image systems are useful in detecting changes of epibenthic communities resulting from physical disturbance impacts. The fauna in the Peru Basin had been catalogued with sufficient detail to measure changes after the treatments (http://ccfzatlas.com; https://www.discol. de/megafauna). For example, Bluhm (2001) showed that the Peru Basin appeared dominated by the following taxa with highest relative abundances, arranged by tiers in descending order: Group 1: Crustacea, Porifera, Ophiuroidea, and Holothuroidea; Group 2: Actiniaria, Asteroidea, Osteichthyes; Group 3: Cnidaria (other than Actiniaria, Pennatularia, Gorgonaria, Ceriantharia). DISCOL also incorporated undisturbed "control" areas, as any robust experimental design would include to have reference sites for comparison. Problems arose with the scale of experiment—temporal variation, variable methodologies, and errors (Bluhm, 2001). Trawling an 8 m-wide plow-harrow towed 78 times for polymetallic nodules at a water depth of 3800–4300 m created a very large, disturbed area (~1100 ha). Disturbance to the natural habitat included direct plowing, as well as plume suspension and re-blanketing of the bottom by sedimentation. Dominance measurements appeared mostly driven by the relative abundance of Ophiuroidea. Unfortunately, the results indicated that the original biodiversity of this abyssal area never fully recovered, even after more than 25 years without disturbance (Vonnahme et al., 2020).

80 Chapter 2 Multiple approaches to understanding the benthos

Succumbing to the experimental procedures can be interpreted as fragility and low resilience in certain deep ocean, benthic habitats. These should provide alarming results and a harbinger for what could happen with unregulated mining of benthic regions. This also underscores the stability and incremental pace of change at many abyssal habitats. From these experiments, scarring of the seafloor appears too easy to induce and takes a long time to repair.

Although DISCOL provided one successful example of hypothesis testing in the deep sea, projects on this scale still remain exceptional. This will likely change as people extend a type of manifest destiny into the oceans or perceive a gold rush to take opportunistic spoils before it is too late. Only about 1% of the open ocean or high seas, which extend past the Exclusive Economic zones of countries (200 nautical miles (370 km) from any shore), is regulated. This is why passing a High Seas Treaty sponsored by the UN as soon as possible may be as crucial as ever (Helm et al., 2021), before further benthic biodiversity is lost, and we can fully understand its value. Efforts stalled to pass the text of Treaty in late 2022 for a fifth time due to intensive lobbying from extractive industries. However, through persistence, the treaty text finally was passed in March 2023, since the problem of climate change and mining affecting benthic biodiversity may increase with more commercial activity. The treaty is not finalized and has more steps to go before full adoption.

Closer to shore but still in deep waters are offshore oil rigs. Perhaps one of the most ill-fated large-scale "experiments" on the seafloor turned out to be an unplanned and uncontrolled trial, not formally carried out by a scientific entity. On April 20th of 2010, the BP-operated Macondo Prospect extracting station, Deepwater Horizon, located in the Gulf of Mexico (28°44′17.30″N, 88°21′57.40″W), exploded and discharged a total of 780,000 m^3 (210 million US gallons) of crude oil into the surrounding seawater at about 5000 ft (1500 m) until September 19, 2010 (87 days). The Deepwater Horizon oil spill (DWHOS) well rupture and platform explosion, in the Gulf of Mexico (GoM), may represent one of the largest, human-mediated accidents in US history. The Deepwater Horizon Oil Spill (DWHOS) in essence represented an unintentional experiment for how well the drilling and oil industry could respond to a deep well blowout, including a test of their equipment, which initially failed (Achenbach, 2012). Until the well could be capped, some of the remedies included traditional skimmers and booms, sometimes even composed of makeshift materials, e.g., donated shorn hair from salons. Hundreds of volunteer boats and thousands of Gulf residents mobilized to protect their shorelines from encroaching oil. As we now know the results of this new "large-scale" experiment were overall not always satisfactory to society, including many citizens and businesses living on the Gulf Coast (Nyankson et al., 2016).

Ad hoc experiments were added to the primary DWHOS accident during the 87-day well rupture itself. This is atypical but not unheard of, especially when new variables are recognized. The additional variable included testing the effects of the dispersants, such as Corexit 9500. This was warranted because the situation was dire. Known surface currents (e.g., the GoM Loop Current) were believed likely to transport oil and toxic pollutants to several parts of the Gulf or even up the Florida Straits. Transported oil could damage major marine ecosystems along the way such as beaches, mangroves, or reefs, so the dispersant was mostly applied at depth. However, many can argue that this procedure was mostly for expediency, as in preventing oil from reaching the surface where it could be measured and used for calculating fines against British Petroleum. Luckily, the oil was not taken up by the Loop Current. Nonetheless, the DWHOS had detrimental effects on multiple Gulf of Mexico species (Lewis et al., 2020).

Before the DWHOS, biomass at the bottom was analyzed at seven sites across a 3-km depth gradient of the continental slope of GoM by Rowe et al. (2008), through sediment community oxygen consumption (SCOC) rates. This provided a detailed analysis of carbon budgets and results that total benthic biota (mostly bacteria, Foraminifera, metazoan meiofauna, macrofauna, invertebrate megafauna) turns over on timescales of months on the upper continental slope, which becomes extended to years on the abyssal plain at 3.6 km depth. In general, the macrofauna biomass was greater than that of the meiofauna at depths less than about 2 km, whereas the metazoan meiofauna had higher biomass below 2 km.

Informative maps show a way

If the big experiments to better understand benthic biodiversity and its functions have many obstacles for practical feasibility and deployment, more attention could turn to the two other approaches—informative maps and leading, cutting-edge technological wedges (Eddy, 2013). Researchers (like our sea-faring ancestors) will always need maps for proper navigation. The most obvious maps display geographic landmarks for a traveler. We wish to know where the reefs, seamounts, or spreading ridges lie on the seafloor. Their extent, boundaries, and relation to each other are also important. A map allows the reader to know where they are and point to where they may want to go. It is typically two-dimensional.

The map is often a symbolic representation of a place, although modern technologies allow the user to interact more with digital maps in real time. Although map drawings can be found in prehistoric caves, accurate and informative maps can be viewed as relatively recent. The Mercator Projection was published in 1569. Notwithstanding the many maps used by military campaigns, the modern geographic information system (GIS) has its origins from the 1800s (Burrough et al., 2015). The map data are bound by the original scales of their creation. Fixed maps cannot be adjusted, while modern digital maps can adjust to almost any scale, which becomes relative due to the advances in technology. We can continue to map our planet and oceans at finer scales and also map outwardly beyond our solar system. For example, with the capability of space travel, the launching of geo-orbital satellites and recently the booting up of the new Webb telescope, human sight reaches into even more distant galaxies (Robertson, 2021). This glimpse also places our own solar system's coordinates into context, and in theory, we may be more accurately situated on the cosmic map.

Traditional physical maps of the ocean started as roadmaps for the sea, as people learned to navigate coastlines and then farther offshore. Seafloor mapping started with the development of bathymetric charts dating back to the 19th century BCE in ancient Egypt. In these early times, ocean depths were determined by lowering and measuring rope or cable from a ship. Contemporary methods have evolved to remote sensing from ships and drones using acoustics and laser systems at submeter resolutions. These high resolution seafloor depth measurements can be modeled into surfaces to identify hazards to shipping and topographic features such as reefs, sinkholes, seamounts, and sedimentary structures. The abyssal and deeper seabed is difficult and expensive to access, so the utility and value of maps as part of its characterization are obvious. As one example, a pioneering study made detailed bathymetric maps of the Portales Terrace of Southern Florida (Jordan et al., 1964). These maps were then refined through time with higher resolution bathymtry, geologic, and biological studies to gain a more complete understanding of the seafloor and associated communities (Walker et al., 2021). Lithoherms are structures rising from the seafloor, likely constructed by the subsea lithification of successive layers of trapped sediment and deposited skeletal debris (Neumann et al., 1977).

82   **Chapter 2** Multiple approaches to understanding the benthos

Another recent mapping example applied new technologies such as CTDs, multibeam echosounders (MBES), and computationally based predictive mapping of Whittard Canyon, North-East Atlantic at a maximum depth of 3400 m (Pearman et al., 2020) (Fig. 2.1). Overall, though the technology has improved, maps still often lack sufficient resolution because of our inherent limitations. Google Earth has provided multiple map layers including benthic topography to the lay person, and this is a start, but they are not intended for routine daily use. The resolution is not high enough. Most map users are not actually in the water but on its surface, where some boaters, fishers, or mappers also use sonar sounders to give a sense of the bottom as if they were reading braille. We are air breathers.

The difficulty of making and having underwater maps is illustrated by their relative scarcity. Routine benthic map use will happen only after we dramatically improve capabilities to submerge and spend more time in the oceans.

Currently, less than 20% of the seafloor has been mapped. However, seafloor mapping will take great leaps in the current decade as results from the ongoing Seabed 2030 accumulate (https://seabed2030.org/faq) (Mayer et al., 2018). This project was established to map the entire ocean floor by 2030 and is a joint venture of the General Bathymetric Chart of the Oceans (GEBCO) group and the Nippon Foundation. Seabed 2030 will use the latest technology, such as echo sounders, a mixture of single-beam, multibeam, and satellite altimetry-derived bathymetry to map 100% of the ocean at fairly high (depth dependent) resolution by 2030. A current estimate of US$3B is required to completely map the 93% of the world's ocean deeper than 200 m using existing sonar technology. Mayer et al. (2018) also interestingly point out that if better high-resolution bathymetry data had been available for the southeastern Indian Ocean in 2014 when Malaysia Airlines flight MH370 disappeared, search patterns and deployment for ships and AUVs could possibly have been more effective. However, even having a toporgraphic map would only represent the first step in the process of characteriing the seafloor. Geologic samples and biological data would also be required to create comprehensive habitat maps (B.K. Walker, personal communication).

Because this topic goes beyond my expertise, however, I will not dwell long on actual physical mapping in this volume. Obtaining topographic maps is only the first step in the seafloor mapping process. I point the readers to the dense literature and many reviews that cover decades-long progress on understanding benthic habitats (Dekker et al., 2011; Bayley and Mogg, 2019). Rather, the main idea of this section was to convince that maps come in a variety of sizes and scale. Maps are utilities with the commonality that they guide users along whichever path they choose to travel (https://tinyurl.com/y47gvzo5). With regard to studying benthic biodiversity, the traveler can use all of the help she/he can get. For example, maps of seagrasses vary according to locale and country, have many gaps, and thus most estimates of seagrass meadow coverage are not made with high confidence (McKenzie et al., 2020). We remain far off from fine-scale mapping of most benthic habitats.

Genetic and genomic maps

We could also envision different types of maps that go beyond the geographical surface and spatial. Thus, mapping more inwardly, we also have the capacity to work and produce at the nanometer scale (one billionth of a meter or a millionth of a millimeter). For example, since the invention of the microscope by Antonie van Leeuwenhoek in the 1670s, we view the myriad of activities that occur deeper

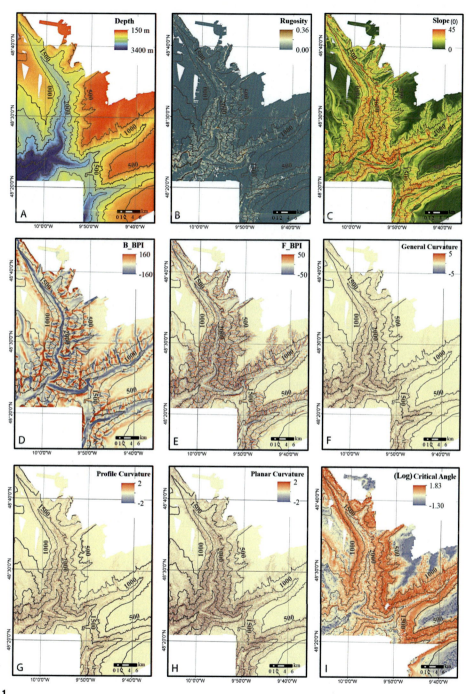

FIG. 2.1

Maps (50 m pixel resolution) of the bathymetric derivatives used as environmental variable proxies in the predictive models: (A) Depth (m), (B) Rugosity, (C) Slope (), (D) Broad bathymetric positioning index, (E) Fine bathymetric positioning index, (F) Curvature, (G) Profile curvature, (H) Planar curvature, (I) Log of bathymetric slope criticality to the dominant semi-diurnal internal tide.

Reproduced from Pearman et al. (2020).

84 **Chapter 2** Multiple approaches to understanding the benthos

within our own bodies or immediate environmental surroundings and communities (see the One Health Initiative in Chapter 5). This includes examining our hereditary material in more depth.

As the costs of high-throughput (HT) DNA sequencing of whole genomes markedly decrease (Goodwin et al., 2016), researchers in diverse biological fields will have better opportunities to sequence more complete genomes from non-model organisms including endangered marine and coral reef species. To date, very few genomes of the ocean's massive benthic biodiversity have been sequenced and studied. The large volumes of their data can be seen on the horizon and will depend on the computer sciences to store and analyze them. Along the same scales now being discussed, the silicon microchips that power and store information on today's computers represent some of the smallest structures ever produced, currently measuring 10 nm. These (logic) chips either process information to complete a task or act to store data as memory chips: DRAM (Dynamic Random Access Memory). Integrated circuits of transistors have to be mapped onto these chips, but I will not go into detail here. By 2025, the world will be producing 175 zettabytes (ZB) of data per year or roughly the equivalent of 1 billion terabytes (TB).

Thus, another excellent map metaphor for biology is the DNA sequence and protein molecules themselves. At the molecular scale of nanometers, the basic molecular structure of DNA is a linear double helix; it is versatile, flexible, and has the ability to fold, condense, coil, and follow all sorts of contortions (secondary structures). These three-dimensional shapes affect their function or may just allow the molecules to fit within the confines of a cell (usually only 50–100 µm in diameter in eukaryotes). In essence, the DNA guides the cells on what to metabolize, so they provide more of a protocol, which is a type of map. Cells, DNA, and several other biological molecules such as proteins have a landscape and orientation, and this can warrant the need for a map. Part of this need can be explained by Lewin et al. (2009). The benefits of determining whole genome sequences were described as such:

(1) "to discover evolutionarily conserved sequence motifs, particularly outside of protein-coding genes, which are responsible for regulatory and other critical genomic functions;
(2) to provide a framework for reconstruction of genome organization, content, and dynamics that have occurred during the mammalian radiations;
(3) to empower new models of human disease and heritable phenotypes;
(4) to provide a starting point for assessment of the expansion, contraction, and adaptation of gene families in different evolutionary lineages."

DNA and the genes they compose encode organismal "phenotypes," which generally have large impacts in the world and environment we can see. In addition, collectively the genomes (or complete hereditary material) of many organisms contain millions to billions of bits of information in the form of single nucleotides (Brown, 2018). Therefore, we need maps or any type of landmark to visualize and tell us where we are on long pieces of DNA or its chromosome. This is because DNA is essentially linear, so it is possible to move along the strands (which proteins and ribosomes do routinely). For decades, geneticists have been mapping our hereditary material, chromosomes in more and more detailed genetic maps (Rhie et al., 2021). This began with the first genetic and chromosome mapping in *Drosophila* fruit flies by Sturtevant (1913).

For the past decade, I have been fortunate to have become immersed in various consortia and communities of scientists dedicated to sequencing whole or large swaths of genome sequences of macro- and microorganisms. This began with the Genome 10,000 (10K) Project, which realized that rapidly advancing DNA sequencing methods should be applied to reading genomes beyond standard laboratory models (yeast, nematodes, fruit flies, mice, etc.). They set goals for sequencing a select group of species, while targeting where the most easily accessible species, and wrote a white paper for guidance (G10KCOS, 2009). Whole genome sequencing of non-model organisms caught fire, and the scope to sequence even more diversity and species across the phylogenetic spectrum of life grew even more. This resulted in plans to sequence the insects, by far the most speciose animal group, marine invertebrates, and also microbes (Robinson et al., 2011; Human Microbiome Project Consortium (HMP)-248 Collaborators, 2012; GIGA Community of Scientists, 2014; Koepfli et al., 2015). More recently, these efforts have culminated and converged onto the other coordinated genome consortia—e.g., the Earth Biogenome Project (EBP) and the European Reference Genome Atlas (ERGA) (Lewin et al., 2018, 2022; Formenti et al., 2022; Blaxter et al., 2022).

The EBP plan began around 2015 and has gained momentum in the ultimate goal to obtain quality whole genome sequences from the majority of taxonomically classified eukaryotic species, which number about 1.8 million (Lewin et al., 2022). To carry out this ambitious project, US National Academy member Harris Lewin along with John Kress at the Smithsonian Institution and Gene Robinson at the University of Illinois developed the idea of a collaborative "network of networks," a consortium of diverse scientific communities to accomplish the ambitious goal. The focus is both geographically and taxonomically based, and indeed, the EBP website now encompasses more than 50 distinct genome projects, from agricultural pests to fungi to ungulates to invertebrates. The overall plan is to fully read and eventually make all of the genetic and genome information highly accessible and useful to non-genomicist scientists and the wider lay community alike. For example, the spiny sea urchin (*Strongylocentrotus purpuratus*) represented one of the earliest whole marine invertebrate genomes (Sodergren et al., 2006). Not only are echinoderms a very close clade to chordates, but the sea urchin sequences revealed several surprises at the time. For example, some genes thought to be relegated to vertebrates were found in the sea urchin (deuterostome-specific). Other gene sequences were unique to sea urchin, suggesting a loss in the vertebrates. Expansion of some gene families occurred apparently independently in the sea urchin and vertebrates. Moreover as mentioned in Chapter 1, the *S. purpuratus* had a diverse and sophisticated immune system mediated by a large repertoire of innate pathogen recognition proteins (complement system). This led to the identification of one of the first extensive "defensomes." The sea urchin also had orthologs of genes associated with vision, hearing, balance, and chemosensation in vertebrates. This latter finding was consistent with then unknown sensory capabilities—photoreceptor genes expressed on tube feet! In addition, distinct genes for biomineralization exist in the sea urchin.

Fast forward 20 years and octopus genomes now reveal the basis of brain development supporting advanced behavioral features paralleling vertebrates. The octopus also has a well-developed eye and behavior that reflects high intelligence, emotion, and memory capacity. There are now at least three completed octopus genomes (*Octopus bimaculoides*, *Octopus vulgaris*, and *Octopus minor*), which show a large number (at least 50%) of repetitive elements (Albertin et al., 2015; Kim et al., 2018; Zarrella et al., 2019). Concordantly, multiple suites of protocadherin genes also appear in octopus genomes, which could permit the advanced neural development.

The EBP network also includes the Global Invertebrate Genomes Project (GIGA), which the author helped establish in 2013 on the heels of Genome 10K project (GIGA Community of Scientists, 2014, GIGA Community of Scientists (Voolstra, C as lead author), 2017). GIGA's mission is consistent with the mainline of whole genome sequencing project motivations: (i) to organize a community of invertebrate scientists dedicated to advancing whole genome sequencing in this taxonomic group; (ii) to develop a set of high-quality standards for whole genome sequencing; and (iii) to assist in the training of the next generation of genomicists and bioinformaticists who will be needed to analyze,

86 **Chapter 2** Multiple approaches to understanding the benthos

interpret, and "map" out the imminent large volumes of genome sequence data. More than likely, the EBP will eventually involve many shallow and deep benthic species, though these are likely well less well known or catalogued compared with their shallower relatives. I will delve a bit more into the molecular aspects of genome linkages to biodiversity later.

Leading wedge technologies for benthic assessments

What sounds most promising within Eddy's threefold paradigm are the new technologies that can serve as cutting-edge wedges. They have the potential to pioneer and attempt to illuminate forbidden ocean habitats. To start from the broadest bases of imagination, we could begin with the earliest forays to the ocean, perhaps even before the world was realized to be spherical (when few marine maps existed). The earliest floating devices, boats, and hollowed-out tree canoes may not sound like a technological wedge from today's perspective, but it is again relative. (The wheel was also once an innovation). For those first brave marine travelers who reached their furthest horizon and beyond, their boat more or less kept them dry and could go from point A to B. The areas of Oceania and Micronesia comprise about a quarter of the earth's surface yet were settled and populated by Asian origins with mostly dugout canoes and no instruments (Hutchins, 1983; Kirch, 2017). In the Atlantic, European expeditions were first exploratory (Vasco De Gama, Magellan, etc.) then commercial. Besides a better working knowledge of navigation, cartography, and introduction to new cultures, the journeys also likely spawned the fantastic tales of sailors and their sea monsters and sirens supposedly encountered during the journey. From those illustrious beginnings, our imagination could go to maritime stories rooted in science fiction. Jules Verne's *Twenty Thousand Leagues Under the Seas: A World Tour Underwater* (Verne, 1871) introduced the submarine plowing the oceans. This sparked imagination in the readers, possibly pushing some to actively pursue more underwater adventures and studies. And yet besides images, science fiction can also produce wish lists and predictions that benefit wide society (Beukes et al., 2017). As with any fiction, some ideas are grounded on contemporary strands of facts available to the public, while others are pure whimsy. Nonetheless, thoughts are free, as the saying goes.

Back to facts, the *HMS Challenger* expedition dutifully voyaged (1872–1876) with six scientists for the expressed purposes of characterizing the deep-ocean physical parameters as well as the biological inhabitants of these depths around the world across 70,000 nautical miles (130,000 km; 81,000 mi) surveying and exploring (Eiseley, 2011). The explorers discovered 4000 previously unknown species. The activities have been documented in the *Report of The Scientific Results of the Exploring Voyage of H.M.S. Challenger* during the years 1873–76. Their equipment may have been simple, but trawls and dredges are sometimes used today and are still effective. Some technologies for exploring the benthos have changed slowly, while others have had greater saltatory leaps (Thomson et al., 1889).

To reiterate this chapter's theme, a leading wedge according to Eddy is "a massed technology development effort, in an area where we need radically better methods…" We can all probably think of several examples that have served as leading wedges across science and technology—the pH meter, computer miniaturization leading to the personal computer, the polymerase chain reaction (PCR), solar panels, DNA sequencing, and SCUBA. With the urgency of marine biodiversity loss in the Anthropocene, society quickly needs cutting-edge technologies to get to the bottom and facilitate benthic assessments.

Submersibles and remotely operated vehicles (ROVs) are leading wedges that visit the ocean floor

Sometimes, leading wedges can be intertwined with the other two big science approaches—maps and experiments. The *Johnson Sea-Link* (JSL) submersible was first built in 1971 by Edwin A. Link, a pioneer in the field of underwater diving and underwater technology. The project began in the early 1960s, with the goal of designing a submersible capable of reaching depths of up to 2000 m. Link designed and operated the Deep Diver submersible, immediate forerunner of the JSL, and which could descend to 400 m (Link, 1973). Link and his team developed the first *Sea-Link* submersible, which was named after him. The submersible was designed to be a manned vessel and was equipped with a variety of scientific instruments and cameras to allow for the study of the deep sea. The first successful dive of the *Johnson Sea-Link* submersible took place in 1969, and it quickly became a valuable tool for oceanographic research and exploration.

The Woods Hole Oceanographic Institution's (WHOI) maintains the scientific deep submersible vehicle, DSV *Alvin*. This undersea vehicle was commissioned in 1964 and can carry two scientists and one pilot to dive for up to 9 h at 4500 m. To date, the DSV *Alvin* has carried out more 4500 scientific dives. The dives have produced more than 2000 papers. More details of the vessel can be found at this website: https://www.nationalgeographic.com/news-features/evolution-of-alvin/

When the first deep-sea hydrothermal vents were discovered and first observed in 1977, sophisticated submersibles such as Woods Hole Oceanographic Institution's (WHOI) DSV *Alvin* submersible were required. This vehicle's versatility went far beyond the technologies of the first bathyscaphs, such as the *Trieste* sphere, which visited the Marianas Trench and the Challenger Deep in 1960. Although relatively simple by today's standards, the *Trieste* safely carried (and returend) US Navy pilot Don Walsh and French copilot Jacques Piccard to a depth of 10,913 m in the western pocket of the Marianas trench, considered the deepest part of the oceans (Nakanishi and Hashimoto, 2011).

DSV *Alvin*'s 1977 encounter with teeming life and diversity at the vents (aka "black smokers") surprised the explorers so much that they were not prepared to sample (Ballard, 2000). Therefore, of course, they would return multiple times, and the habitat will be described later. Vents support profuse life (Thaler and Amon, 2019). We can imagine the thrill and awe upon seeing a thriving, new underwater ecosystem for the first time. I can relate to this feeling, although I still consider myself a relatively novice, inexperienced, deep diver, and even an irregular diver on shallow reefs.

Nonetheless, because most of the seabed remains an alien locale to many humans, each diver has the expectation that almost every dive and brief visit to the seafloor can reveal something new. This could apply to individuals, or if lucky, to wider society. The author had some of these feelings whenever I was fortunate to board and descend to depth in the *Johnson Sea-Link* (JSL) submersible, from Harbor Branch Oceanographic Institute (HBOI) in the early 2000s. This unique deep-sea faring vessel was designed and built by Edwin Link, who had risen to fame by inventing the flight simulator for fighter pilots. The JSL could probe oceanic depths to 3000 ft. (910 m) and hold three passengers and a pilot. The submersible distinguished itself by the 5-in. thick Plexiglas bubble, which allowed the pilot and front passenger nearly 330 degree clear vision of their surroundings. In contrast to WHOI's DSV *Alvin* described above, the JSL was also equipped with robotic arms and claws for collecting and handling delicate specimens and also a state-of-the-art video recording system. I foreshadow the video libraries catalogued for HBOI and which will be described in Chapter 5. Almost every view on the JSL from the bubble to the JSL's video cameras (or almost any other deep-sea camera device whether manned or

88 **Chapter 2** Multiple approaches to understanding the benthos

unmanned) would show a place on earth that probably had never been seen by humans before. JSL pilot Phil Santos has been on hundreds of dives and can testify to many of these encounters. There was great excitement when the first astronauts on Apollo 11 set foot and gazed at the lunar surface, which is even more barren. Yet the first views of the seafloor can be viewed quite routinely by our aquanauts with relatively little fanfare in spite of sudden discoveries. Other submersibles followed the JSLs, DSV *Alvin,* and include China's *Jialong*, which can dive to 3500m, and James Cameron's private submersible of the *Deepsea Challenger*, which dove 10,908 m at the Challenger Deep.

For the novices, and perhaps even some of the regular submersible divers, reaching and diving the seafloor continues to sustain one of life's great metaphors or allegory: we live on an expanse of monotonous topography with apparent barrenness, which only excites when punctuated by an anomaly. This may be represented by an outcropping, sea mount, underwater reef, or that elusive discovery. Overall, the metaphor can be extended to our daily lives—that we often have to wade through periods of silence, or noise and distraction before we get to the really valuable nuggets. On *Okeanos Explorer* cruises, there are hours of drifting over bland benthic areas. The "eureka!" moment in science that makes it onto highlight reels is generally a rare event for most researchers, who usually spend most of their time in focused concentration, preparations, and analyses; therefore, the parallel with benthic studies. Deep dives on the JSL lasted a minimum of 3–4h, depending on final depth of course. For a typical dive below 200m and the photic one, which took a 30min descent to the bottom, one expects to be occupied for a whole morning or afternoon. On a cursory view out the port hole of the deepwater submersible, the visiting researcher sees very little to write home about for the majority of the dive. Much of the seafloor appears as monotonic mud and sediments for many kilometers, with marine snow deposits accumulated over hundreds of years with little disturbance. However, to engage the lay audience, we choose the most exciting highlights to show in our B-roll footage. Fortunately, education documentaries on public television have become adept at portraying nature, since expected ennui does not provide incentives for tuning in. Therefore, when both expert and the lay public visit dedicated exploration project websites such as those by *Okeanos,* we are struck by the bizarre and dramatic: https://www.youtube.com/@oceanexplorergov (more in Chapter 5).

And then, the scientists from each specialty understand that there is much more complexity at sites that does not meet the eye. It will only be after days and weeks at the lab bench, microscope, or computer screen that the seemingly infinite diversity of benthic life exposes its awesome breadth and girth.

Because of the unexpected discoveries, early deep-sea explorers had to preserve specimens in stocks of vodka and other spirits, which although innovative and essential could not qualify it as a leading wedge per se. On the other hand, the DSV *Alvin* and the WHOI's program can be classified as one of the first manned submersibles to explore the mid-Atlantic ridge and as part of French-American Mid-Ocean Undersea study (FAMOUS) in 1974 (Ballard, 2000).

Also, in those early years, daring expeditions to study the benthos did successfully launch. Going back to a more realistic biological scale, simple sampling of the benthos remains a nontrivial effort. It is possible that society collectively spends more funds and energy trying to sample the Martian landscape for life than the ocean floor. For example, previous estimates for the cost of the Mars One expedition to send four people to the planet were too low at least $6 billion, while the Artemis program for sending astronauts back to the moon may cost >$90 billion in total (https://www.space.com/artemis-program.html). These expenditures exceed the total 2019 budget for NOAA, which had an annual Operations, Research and Facilities (ORF) budget of about $5.5 billion US dollars.

What if the same or even only 50% of the investments into technology spent on the US NASA moonshot also went into understanding our deepest oceans and unseen benthic habitats? We would likely be in a different place and know much more about our own backyard, or more precisely our backyard pools, including brine pools found on the ocean floor. We will revisit the latter further down.

As dive bottom time is limited by human physiology, typical SCUBA parameters, and expenses, automated exploration becomes an important alternative approach. This is common sense. We acknowledge that deep-sea habitats have high pressures, making most inhospitable and dangerous environments. One way to increase precious "bottom time" is to actually live in the habitat. Surprisingly in this age of planning new lunar or Martian voyage ambitions, including colonization, few bona fide underwater habitats have been manufactured or used. Perhaps there have not been sufficient incentives to coax people to try and live in an artificial habitat within dark oceans, but at least this would be closer than the moon.

One such example of a continuing attempt at underwater living is the Aquarius Reef Base underwater habitat set up in Key Largo. Inhabited by marine scientists, the base allowed these "aquanauts" to spend extended periods of time, from days to weeks, at depth. The concept of living underwater was envisioned by Dr. George F Bond, which was also the original name for the *Aquarius* Sea Base originally designed by Perry Submarine Builders of Florida and constructed by Victoria Machine Works, in Texas. Bond had originally initiated and commandeered the US Navy's Sealab underwater vessels, which directly preceded the more successful Aquarius habitat. The pioneering ways of this physician could actually fit into a fictional James Bond story, as the doctor experienced life and death situations such as emergency ascent dives from over 100 m (Hellwarth, 2012). Therefore, we know that exploring the depths requires some fortitude and bravery, and the tests can be demanding. The absence of light and warmth is not a welcoming feature to most habitual land dwellers.

These examples underscore the difficulties of studying the benthic habitats in general and deep sea in particular. In spite of the unexpected abundance of life and diversity, the habitats remain unwelcoming, cold, and foreboding. Few seek to find comfort and shelter in the deep oceans, even if one were to consider them as the cradle for life and for some human burials, a final resting place. And yet, some of us are driven to finding better ways to explore it.

From the preceding discussion, we know that sending people down to the seafloor can be labor-intensive, expensive, difficult, dangerous, and not always desirable for long periods, so utilizing proxies becomes an option. These include unmanned, robotic remotely operated vehicles (ROV) and autonomous underwater vehicles (AUVs) equipped with recording cameras. Marine ROVs revolutionized the field of ocean exploration since their invention in the 1950s. The first ROVs were primarily used for military and industrial purposes, such as inspecting pipelines and performing deep-sea salvage operations. In the 1960s and 1970s, ROVs began to be utilized for scientific research, allowing researchers to explore the depths of the ocean and study previously inaccessible ecosystems (Ramirez-Llodra et al., 2011). In the 1980s and 1990s, ROV technology continued to improve with the development of tether management systems and more advanced imaging and sampling capabilities. The advances included the *Argus/Jason* ROV and robotic documenting system, which were originally designed for deep-sea scientific research, but also played a part in the exploration of the *Titanic* wreck. More recently, the *Global Explorer* can reach 2700 m. ROVs will continue deep-sea exploration sometimes for oil and gas exploration, but also for ocean conservation and monitoring. Mallet and

90 **Chapter 2** Multiple approaches to understanding the benthos

Pelletier (2014) provided a summary of biodiversity video monitoring from 1952 to 2012, indicating that during that time, biodiversity fell into either extractive (e.g., fishing, dredging), acoustics, or Underwater Visual Censuses (UVC).

Thus, with our increasing surveillance tools on land, can we provide better visualizations of life under the water? Better video would benefit many studies and help address hypotheses and questions. I have already alluded to the success of many public educational videos and will continue to emphasize this in later chapters. One example of a dive to near hadal depths (5862 m) in the Phoenix Islands Protected Area (PIPA) by the *Okeanos* Explorer ROV is documented by Kennedy, Brian R.C.; Pawlenko, Nick, in March 2017 (EX1703, Dive 16, https://repository.library.noaa.gov/view/noaa/ 15844). Entries of dive experiences such as the following underscore the excitement and novelty that can be easily encountered with the new technologies and the deliberate focus on heretofore newly explored sits:

> This dive should be nicknamed "white house, black market". Throughout the dive, both on the steep slope and sedimented ledge at the end of the dive, we encountered taxa that lacked pigmentation, which contrasted with the dark color of the seafloor. We observed white anemones, holothurians, brisingids (Freyastera—some with swollen arms filled with gonads), cladorhizid sponges (2 species)...

Terrestrial studies and hypotheses can be based upon direct, long-term observations, which is not a luxury afforded to most marine biologists. One of the most well-known terrestrial studies and proclamations was conducted by Terry Erwin in the 1970s and 1980s (Erwin and Scott, 1980). Working in the tropical rainforests of Panama, Irwin and colleagues counted beetles and other insects, sometimes by smoking them out of the treetops. In the end, they extrapolated that over 30 million organisms could exist globally when they considered rainforest cover at that time. This estimate was well over the current 1.5 million total species (Erwin, 1982, 1997). Of course, insects compose the bulk of the species total, Erwin on beetles and E. O. Wilson's focus was on ants. However, as Wilson wrote, [insects] are "the little things that run the world" (Wilson, 1987). The implications to their pivotal roles in ecosystem functions can be compared with those of microbes and nematodes and other benthic invertebrates to be discussed later. The other major point of this terrestrial example is that the same concepts have since been applied to estimating the cryptic biodiversity of marine ecosystems. The experimental approach will be inverted for the seafloor, since much of the biomass is on the bottom and not dropping like tree leaves or the surface (whale carcasses do drop, but sporadically). However, we can now discuss modern methods that have proved effective in discovering new species or erecting stable community profiles across specific biomes (reefs, deep sea sediments, etc.) from different geographic locales (Leray et al., 2019; Lejzerowicz et al., 2021).

Visual technologies in the deep sea have made significant leaps in the last few decades. Automated underwater vehicles (AUV) and remotely operated vehicles (ROV) have helped facilitate, but the still need to become more widespread, reliable, and less expensive. The expenses do accrue based on the size of these instruments and the amount of trained personnel required to operate.

Occurring in the twilight zone of the coastal oceans, mesophotic coral ecosystems (MCEs) represent excellent subjects for ROV, AUV, and other underwater visualization applications (Baker et al., 2016). Armstrong et al. (2019) describe that these methods actually have many different flavors. For example, methods such as acoustic mapping with side-scan or multibeam sonars, assist tethered vehicles (towed sleds and drop cameras). Untethered vehicles employ manned submersibles and

Lagrangian buoys. Hybrid remotely operated vehicles (HROVs) combine both AUVs and ROVs into one vehicle. Focus on specific mesophotic reefs will be discussed in Chapter 3.

The US National Oceanic and Atmospheric Administration (NOAA) has been at the forefront of US undersea research, outfitting vessels such as *Nancy Foster*, *Ronald Brown*, and *Okeanos Explorer* for benthic-focused expeditions. Because of such advances, many benthic fauna have been observed and monitored. Deep-sea mollusks, monoplacophorans, which are considered living fossils, were recently visualized by high-definition (HD) video (Sigwart et al., 2019). The deep-sea monitoring activities have been updated and exemplified by NOAA research vessel *Okeanos Explorer* (R 337), converted from a United States Navy ship. NOAA has used this approach since the early 2000s (McNutt et al., 2000). On every research cruise, the public is invited to engage with the scientists via the internet, which is dubbed as "telepresence" (Martinez, 2009) (https://oceanexplorer.noaa.gov/technology/telepresence/telepresence.html. *Okeanos Explorer* launched the ROV Deep Discoverer and Seirios. This used five HD and five standard-definition video cameras, by 24 LED lights (144,000 lm) and paired lasers (10 cm apart) mounted on the main high-definition video camera for scale. Relying on "modern computer networks and high-bandwidth satellite connections to enable remote users to participate virtually in ocean research and exploration expeditions," this Ocean Exploration and Research (OER) program has amplified our reach into the benthos. The NOAA research vessel *Okeanos Explorer* operates in situ with remotely operated vehicle (ROV) dives and sends back images to experts at the expedition headquarters on land. This hybrid expedition in some ways mimics a longer range submersible, less dependent on the mothership. The *Okeanos Explorer* may be more independent, but the shore team can conduct highly complex data processing and telepresence mapping and relay it back to the ship. These studies show how video can be a sufficient substitute to in-person visits to the benthic study sites. Mapping from afar can also be performed through the use of "telepresence" methods. More of this will be expanded in Chapter 5.

For example, NOAA launched a 3-year exploration campaign in 2015 entitled CAPSTONE: Campaign to Address Pacific monument Science, Technology, and Ocean Needs. Led by NOAA's Office of Ocean Exploration and Research (OER) to map and characterize Exclusive Economic Zones (EEZ) in the Pacific (Kennedy et al., 2019). Of the $363,526 \, km^2$ mapped during CAPSTONE, 60.86% was within the US EEZ, which was 2.99% of US EEZ total. Overall, this expedition on the NOAA Ship *Okeanos Explorer* conducted 187 dives with the dedicated two-body ROV system, *Deep Discoverer* (D2) and camera platform *Seirios* (6000 m depth rating). Acoustic mapping during CAPSTONE was performed with EM302, EK60, and Knudsen sub-bottom profilers (SBP). The CAPSTONE effort totaled 891.5 h of ROV benthic imaging time, making 89,398 annotations of biota over an estimated $29.3 \, km^2$ of seafloor imaged and >347,000 individual documented organisms. Identifications were performed visually by expert taxonomists, and raw image data can viewed at the OER Digital Atlas (See Table 5.4 and https://www.ncddc.noaa.gov/website/google_maps/OE/mapsOE.htm). This project shows the potential for generating large amounts of coverage and data, and excellent applications of taxpayer funds, but should receive more exposure.

Another solution is if we could leave observational cameras and equipment on the seafloor in our place. These would watch, record, and monitor their surroundings, continuously, through long periods, of inactivity until the time of visitation by curious benthic residents. Again, people are getting better at voyeuristic occupations on land (surveillance cameras on the streets and airports, due to domestic, national security concerns, social media, etc.), so it should not be so far-fetched that long-term visualization and monitoring be extended to the oceans for the purposes of scientific enterprises or plain curiosity. Some animals get easily spooked by humans and other large animals, and so we would all benefit from a "candid critter camera" of peaceful unobtrusiveness.

92   **Chapter 2** Multiple approaches to understanding the benthos

One popular example of this was in the 2017 film "Chasing Coral." Due to the serendipitous convergence of curiosity and unexpected major bleaching event, a unique opportunity for benthic monitoring was innovated and captured in real time. Sadly, the documentary focused on the partial ecosystem collapse of the Great Barrier Reef in 2016, when almost a third of the habitat bleached and eventually died (Hughes et al., 2018). This was not foreseen even by experts in our lifetime. It would be much more pleasing to have footage of reefs growing by accretion (which is too slow of a process to witness). As noted by many, some people may rarely take notice of the regular and mundane, but many cannot turn their gaze away from the signs of a horrific accident. Fortunately, there is now some evidence that the northern third of the GBR, which bleached and crashed in 2016–2018, has shown some signs of recovery (Australian Institute of Marine Sciences (AIMS), 2022).

The author recalls this point of proxy raised several times by pioneering marine biologist and deep diver, Edie Widder. Dr. Widder is a renowned marine scientist, previous recipient of a MacArthur genius grant. She has recently published her memoirs about studying bioluminescence in the deep oceans (Widder, 2021). With her forte in bioluminescence, I recall her discussing how she mounted underwater "splat" screens (equipped with a dichroic beam splitter, strobed IR illumination, and a video mixer, mounted on submersibles), which could capture the shapes of bioluminescent organisms as they hit the screen similar to insects on the windshield of a car driving down a highway meeting their maker (Widder, 1992). For years, the Widder laboratory has created innovative marine technology to monitor underwater systems and organisms. For example, Widder has developed various sensors and cameras to provide more realistic glimpses. The "Eye in the Sea" was a system deployed at Gouldings Cay in the Bahamas at 488-m depth. Equipped with two LED illuminators, the system considered far-red illumination, which would allow animals to approach the camera without fear (Widder, 2007). In another example, Widder founded the Ocean Research and Conservation Association's (ORCA) to create new innovations in ocean engineering. One of their efforts produced the water quality Kilroy instrumentation, which could be mounted on solid benthic substrates such as pilings to measure various parameters such as Chl*a* and Blue Green Algae phycoerythrin at 30-min intervals (Thosteson et al., 2009). Much of her work qualifies as technological wedges.

Remote Sensing, Lidar (Light detection and ranging), and other technologies (Bayley and Mogg, 2019) represent sophisticated types of visualization in the sea and are applied for measuring distances and high- resolution ocean floor mapping. Within the author's institution, Brian Walker and colleagues have put together fine-scale resolutions of the uppermost (Martin/Broward/Miami-Dade counties) Florida Reef tract benthic topography (Walker et al., 2008; Walker, 2012; see also Fig. 3.1). Using a combination of technologies, such as LIDAR, GIS, this characterization revealed physical evidence of benthic gradient characteristics that reflect the transition from tropical to subtropical biogeographic partitioning.

The technologies are broader in scale than we have portrayed in this volume, but they are mentioned because of context. Just as topographies on land affect biological patterns, the same mesoscale structures of sea mounts, ridges, and slopes affect deep ocean currents, thermal gradients, upwelling, local eddies, etc. The relative size of benthic structures will probably reflect these effects proportionally.

Benthic monitoring—All eyes on the sea

For effective assessments, we first have to know what species or taxa are present at any specific site. This begins with effective monitoring. Monitoring of marine habitats has been going on for quite a while. Of course, by necessity it started in the most shallow regions. Various and multiple monitoring programs around the world aim to assess biodiversity of benthic habitats, especially of threatened coral reefs. In addition, Thrush and Lohrer (2012) discuss the various benefits and requirements for establishing benthic and oceanic monitoring programs with a biodiversity-ecosystems functions (BEF) approach. They emphasize careful consideration of temporal and spatial variability, which can affect overall assessments. The episodic but irregular El Niño-Southern Oscillation (ENSO) is a good example of the former.

For an initial example, the Atlantic and Gulf Rapid Reef Assessment Program (AGRRA) (https://www.agrra.org/resources/) was formed in 1997 under the guidance of Robert N. Ginsberg from the University of Miami. With an international consortium of reef scientists, managers, and supporters, and eventually the formation of the non-profit Ocean Research and Education Foundation (ORE), AGRRA has provided important baselines from 3000 site surveys in 29 countries or territories throughout the Caribbean describing reef "structural and functional indicators that could be applied to reveal spatial and temporal patterns of regional reef condition." The data, currently in a public data explorer (see Table 5.3 in Chapter 5), include coral and benthic cover at specific locales for regional-scale interpretations. The AGRRA was meant to project future shifts in finfish resources and reef health. For example, applying AGRRA data sets collected between 2011 and 2015 at 328 fore reef sites across the Caribbean, Lester et al. (2020) found reasons for optimism from 10 key ecosystem metrics: vertical relief, calcareous macroalgal cover, coral recruit density, coral calcification rate, mean coral species richness, herbivorous fish biomass, potential fishery value, density of large fish, parrotfish, bioerosion rate, and mean fish species richness. The investigators also noted a poor correlation of coral cover with numerous non-coral ecosystem metrics, including herbivorous fish biomass, density of large fishes, fishery value, and fish species richness, suggesting that various ecological states are possible in the absence of high coral cover. AGRRA is now run by Judith Lang.

Other monitoring systems have been organized such as the Global Ocean Observing System (GOOS—https://www.ncdc.noaa.gov/gosic/global-ocean-observing-system-goos). The GOOS system was established in 1991 by the Member States of the Intergovernmental Oceanographic Commission (IOC) of the United Nations Educational, Scientific, and Cultural Organization (UNESCO), with the World Meteorological Organization (WMO) to coordinate, standardize, and facilitate global ocean observing efforts (Tanhua et al., 2019). The effort expanded even further with the Framework of Ocean Observing (FOO), which had an even more technological approach for the data and its readiness for stake holders. Another program for the deep sea has also been set up (Levin et al., 2019), which has tried to recognize the most "essential ocean variables" (EOVs). For coral reefs and other shallow benthic habitats, the Marine Biodiversity Observation Network (MBON) of the Group on Earth Observations Biodiversity Observation Network (GEO BON) aims to integrate efforts and ensure that Essential Ocean Variables (EOVs) through GOOS and Essential Biodiversity Variables (EBVs) from the Group on Earth Observations Biodiversity Observation Network (GEO BON) are complementary, consistent and can apply the common set of agreed upon scientific measurements (Muller-Karger et al., 2018; Obura et al., 2019).

NOAA established the Coral Reef Conservation Program (CRCP) in 2000, as a partnership between NOAA Line Offices that focus specifically on coral reef issues. As part of a broader monitoring effort, CRCP implemented the National Coral Reef Monitoring Program (NCRMP), which is "a strategic framework for conducting sustained observations of biological, climatic, and socioeconomic indicators in U.S. states and territories. NCRMP stems from the NOAA Coral Reef Conservation Program (CRCP) established as an integrated and focused monitoring effort for the US. The resulting

Chapter 2 Multiple approaches to understanding the benthos

effort provide a robust picture of the condition of U.S. coral reef ecosystems and the communities connected to them." Since its inception, the program has conducted monitoring activities to provide long-term data, status, and trends of coral reef ecosystems across seven US jurisdictions, including Puerto Rico, the US Virgin Islands, Florida, Hawaii, American Samoa, Guam, and the Commonwealth of the Northern Mariana Islands. NCRMP makes its data available through reports and CoRIS (the Coral Reef Information System), the official CRCP information portal is CoRIS (https://www.coris.noaa.gov/welcome.html), where data are collected and stored and publicly available for use in research and management activities (https://www.coris.noaa.gov/monitoring/data_summary_report_2018/).

Over the years, the NCRMP has expanded its activities and collaborations to improve the quality and relevance of the collected data. For example, in 2013, the program partnered with the US Coral Reef Task Force to develop a national framework for coral reef conservation, which led to the inclusion of new indicators for monitoring coral reef resilience and adaptation. Additionally, the NCRMP has collaborated with other monitoring programs, such as the Pacific Islands Ocean Observing System and the Atlantic Oceanographic and Meteorological Laboratory, to incorporate oceanographic data and improve the understanding of the environmental drivers of coral reef dynamics.

The NCRMP implements standardized protocols to collect data on coral demographics, coral reef structure, benthic cover, fish populations, and oceanographic parameters. Part of these activities include a nondestructive, visual assessment of fish and benthic organisms using line point-intercept transects (LPI), coral size structure and demographic surveys, and underwater visual censuses to determine fish biomass and abundance. During the assessments, measuring tapes are deployed gently along the bottom to define the assessment area. Divers measure coral colonies with tapes or PVC sticks and take pictures of the bottom and surrounding area. Organisms are counted and measured. Sea water is collected at a subset of sites for ocean acidification analyses. New survey methods are continually improved. Through concerted efforts, the NCRMP has become a critical resource for coral reef management and conservation in the United States.

In this context, the Nova Southeastern University Coral Reef Restoration, Assessment & Monitoring (CRRAM) laboratory run by David Gilliam has been conducting long-term monitoring of the South Florida reef track (FRT) for almost two decades (https://tinyurl.com/2s47kukh). As part of the Southeast Florida Coral Reef Evaluation and Monitoring Project (SECREMP) as an expansion of the FWC managed Coral Reef Evaluation and Monitoring Project (CREMP) in the Florida Keys.

CRRAM runs standard transects and counts specific indicator species abundance. One drawback of this approach is that only a handful of taxa are analyzed as broad indicators for the habitat, and these may not even include keystone species. This includes identification to species (barrel sponge *Xestospongia muta*, (e.g., stony corals, *Gorgonia ventalina*), genus (e.g., *Dictyota* spp., *Halimeda* spp., and *Lobophora* spp.), or higher taxonomic levels (e.g., encrusting or branching octocoral, crustose coralline algae, zoanthid, sponge, and macroalgae). Uncolonized substrates are also marked as sand or substrate (consolidated pavement or rubble). By definition, the monitoring efforts are labor-intensive. They require divers on the scene that may or may not be able to identify species. Many of the videos and photos should be ground truthed to verify taxonomy and identification.

For this reason, a helpful software program called coral point count or CPCe was created to help scientists analyze video and images back in the laboratory (Kohler and Gill, 2006). Needless to say, this could be tedious work, but it has driven several master's projects and publications. When one square meter of space on a thriving call reef can potentially hold hundreds of species, trying to assess biodiversity across tens, or even thousands of square kilometers will require great efforts.

Reef fishes are also among the taxa counted by the CRCP, in conjunction with local Florida agencies such as Florida Department of Environmental Protection-Coral reef Conservation Program (FDEP-CRCP) and other multiple Southeast Florida Coral Reef Initiative (SEFCRI) partner agencies. To portray the scope of the monitoring, in the 2018 report, >1.2 million individual fish representing 305 species and 70 families were recorded along the Florida Reef tract from Martin County to Government Cut in Miami-Dade County (Kilfoyle et al., 2018). The total mean density for all sites and strata combined for all 5 years was 176 fishes/SSU ($\pm$4.6 SEM) (Second-Stage Sample Unit = SSU or site, 177 m^2). Multivariate analyses showed patterns in the reef fish communities associated with benthic habitats. Topographic complexity has been shown to affect local fish distributions (Walker et al., 2009). One example for the utility of this monitoring can focus on benthic-dependent fish species, such as gobies and blennies, as their presence and relative abundances can reflect the status of the habitat. This 2018 CRCP report indicated an order of abundances of the Seaweed Blenny (*Parablennius marmoreus*), Rosy Blenny (*Malacoctenus macropus*), and Saddled Blenny (*Malacoctenus triangulatus*) species, with all increasing from the Florida Keys > SEFCRI > Dry Tortugas. All goby species such as the Bridled Goby (*Coryphopterus glaucofraenum*) and Goldspotted Goby (*Gnatholepis thompsoni*) had a slightly different order increasing steadily in abundance from the FRT to the Keys, then Dry Tortugas: The results likely reflect the quality of substrate, which these low dispersal fish species depend upon.

Underwater soundscapes, landscapes, and unexpected sources of innovation

With the deep sea and benthos lying mostly in darkness, sound becomes another alternative for probing the landscape. Actually, acoustic listening of marine animals has become specialized in only a few lineages, and we may ask "why not more?" For many years, divers could "hear" they were near viable coral reefs because of the sound of snapping shrimp. For our purposes of understanding the benthos, sounding boards have long been in use. The *Challenger* expedition carried several with them. Acoustic methods have long been used for detecting pelagic organisms and thus enhanced fishing catches. Acoustic mapping provides broader spatial configuration of the seafloor biophysical characteristics (Brown et al., 2011). In this context, a recent application of Remote Sensing Object-Based Image Analysis (RSOBIA) of object-based segmentation of the images (backscatter data, bathymetry, and its derivatives) was performed in the southern Adriatic Sea (Mediterranean Sea). This area is juncture of Africa and Europe tectonic plates. The study successfully created a benthic habitat map based on 20 acoustic data sets collected within 12 cruises and 255 bottom samples collected during 19 cruises (Prampolini et al., 2021).

With regard to mapping the seafloor, innovative soundscapes have recently been implemented, and some could be considered technological wedges. Acoustic layers and terrain attributes have proven effective in isolating and characterizing specific habitats or features, for example, coral mounds (Savini et al., 2014). Mooney et al. (2020) have recently reviewed the novel methods in characterizing biodiversity with sound or bioacoustics. They distinguish biological sounds (biophony), geophysical sounds (geophony; natural abiotic sources such as wind, waves, rain, water flow), and anthropogenic sounds. Moreover, the ocean is elevated as a prime medium for soundscaping marine biodiversity, as sounds travel easier through water than light, including times of low visibility. Long-term, broadband recordings may also be performed with relatively low cost, advanced passive acoustic recorders to monitor biodiversity over time, after disturbances or in marine-protected areas (Mooney et al.,

2020). This Soundscape Ecology has also been termed *Eco-acoustics* and profiles environmental characteristics and variation within them (Sueur and Farina, 2015). Lin et al. (2019) posit that "deep-sea larvae may use habitat-specific soundscapes to detect their habitats." In these examples, we can see how these new approaches combine both mapping and leading wedge aspects.

Probing the depths introduces us to new forms, shapes, and phenotypic capabilities, since the organisms at depth have adapted to their unique environment. With regard to the deep sea, this means higher pressure, zero surface light replaced by bioluminescence and colder temperatures (Yancey, 2020). These conditions could yield ideas and resources to apply to humanity's extraterrestrial ambitions to explore other realms that may exist beyond earth (Lopez et al., 2019).

Sometimes, we cannot improve on the innovations of nature, which has experienced millions of years of natural selection, resulting in some unique adaptations. Yet during times of stress and strain, organisms can find ways to adapt. For example, humanity does not shy away from its own type of biomimicry to borrow from nature to meet its own needs. The borrowing resulted in the commercial inventions of Velcro (based on some plant seeds' ability to apply a hook and loop attachment mechanism for greater dispersal after attaching to a mobile animal), synthetic or derived antibiotics such as penicillin based on biochemicals encoded and metabolized for purposes unknown, but possibly for protection from competitors. Then there are the modern Japanese bullet trains, which were modeled after Kingfisher bird beaks to become faster (170 mph) and quieter over the older models.

Exploring and surviving depths such as the 10,900 m Challenger Deep of the Marianas Trench has now borrowed from deep-sea inhabitants themselves (Li et al., 2021). In this example, nature has inspired the creation of a soft-bodied robot to probe the depths. Most of the time, strong rigid vehicles, such as the JSL or DSV *Alvin*, are constructed to withstand the high pressures. The engineering solution was necessarily to resist and confront more than 1082 atm pressure at trenches. However, the opposite strategy would involve the construction of a vehicle, which could flex and give to the increasing pressure.

Enter the deep-sea snail fishes. These liparid snail fishes display an astounding versatility by range wider than any marine fish family from intertidal to depths exceeding 8100 m (Wang et al., 2019). Snail fish adaptations to hadal living (from 6000 to 11,000 m) include an incompletely closed skull, thin, lower bone hardness due to incomplete ossification, unpigmented skin and scales, inflated stomachs, and higher weight percentage of unsaturated fatty acids allowing greater fluidity in its cell membranes. The genome of the species *Pseudoliparis swirei* has been recently sequenced, and the sequences confirm these adaptations while phenotypically the fish thrives at depth. For example, *fatty acid metabolism* gene families exhibited an expansion of gene number in *Pseudoliparis swirei*, which would facilitate more unsaturated fatty acids that assist membrane fluidity, while the piezolyte-generating enzyme flavin monooxygenase 3a (*fmo3a*) gene exhibited positive selection. Many of these are common adaptations that occur across multiple organisms from microbes to metazoans (Allen et al., 1999; Winnikoff et al., 2021; Li et al., 2022; Parsekar and Jobby, 2022). The novel artificial soft-bodied robot produced by Li et al. has flapping arms and has now been successfully actuated and tested at the Mariana Trench down to a depth of 10,900 m and to swim freely in the South China Sea at a depth of 3224 m (Li et al., 2020).

In a provocative review, Hand and German (2018) proclaim the advantages of downsizing oceanographic enterprises, "small is better," and also describe the use of "NASA large flagships to the smallest CubeSats which have pushed in situ instrumentation to be lighter, smaller and more capable." By these recommendations, this viewpoint argues against the big science paradigm introduced earlier and for a nimbler one instead. Several such instruments include mass and Raman spectrometers and 'lab-on-a-chip' systems." As mentioned earlier, having eyes on the ground, video recordings in situ and in real time provide important advantages and compelling observations for new hypotheses and testing.

In another example, California Institute of Technology's Jet Propulsion Laboratory has a group specifically focused on "Ocean Worlds." Led by Kevin Hand, this lab aims to "pursue experiments and numerical modelling that simulate the surface and subsurface physical and chemical conditions on ocean worlds" (https://oceanworldslab.jpl.nasa.gov/). The laboratory's focus also acknowledges that chemosynthesis is the driver of deep-sea biology and interactions and therefore is a key to the success of their program. This also harkens back to our extensive discourse of symbiosis as a key to biological survival and innovation.

Autochemosynthetic pathways underpin many physiological innovations of deep-sea creatures that live without the sun and allow them to thrive. These metabolic processes, which do not depend on the direct input from sunlight, appear to have most of the essentials—e.g., energy sources, electron donors and receptors, and organic materials. Carbon may be limiting, but constantly sinking marine snow provides an important input.

One area where these biological principles and capacities could be applied is in the field of energy storage for "batteries." Batteries are basically any physical or chemical source for potential electrical power. Yet most biological systems carry out reactions that can also harness or release energy in the form and transfer of electrons. Lee et al. (2018) provide an interesting review of the possibilities, driven by the inherent reversibility of biological reactions, such as reduction-oxidation (redox) reactions. For example, "In particular, biomolecules are usually composed of light and cost-effective elements such as carbon, hydrogen, oxygen, and nitrogen, implying the possibility of achieving a high gravimetric energy density for biomolecule-based electrode materials" (Lee et al., 2018).

Based on their high redox potential and activity, various organic materials could serve as electrodes. For example, quinone and its derivatives have structures that hold electrons easily. These molecules are part of the electron transport chain of mitochondrial, the eukaryotic organelle known as the energy source due to its production of ATP. A drawback of quinones is their relatively low life span to just 50 cycles, due to dissolution of the active materials. Other compounds that can store energy are the riboflavin (also known as vitamin B2) derivatives. With a chemical structure based on flavin adenine dinucleotide (FADH2) also built to simultaneously release protons and electrons, they could provide energy storage functions.

Benthic habitats such as sediments on shallow escarpments and even the deep sea likely have sufficient organic materials to provide resources for these processes (e.g., hydrothermal vents), and why their communities thrive in what appears to be desolate or remote locations (Kennicutt et al., 1985). More importantly, certain keystone species of these surprising deep-sea communities have evolved traits to capture, store, and use energy. These adaptations can be instructive to the more shallow residents dependent on sunlight.

The rise of artificial intelligence (AI)

When it is not possible to directly see and observe our intended subjects of study, then other tools may be developed to fill in the gaps through inference or other means. This is where AI, deep and machine learning approaches can be very helpful (Salvaris et al., 2018). Where could intelligence be more

Chapter 2 Multiple approaches to understanding the benthos

welcome than for shining as much light onto the mysterious deeps? Moreover, at the time of writing, AI in the form of online software such as ChatGPT is being applied to write basic code and touted to carry out mundane tasks for the public. AI has been projected to have as big an impact on society as when the internet was introduced (Ghaffary, 2023), while others raise ominous warning signs of AI's potential power and influence (Nath and Manna, 2023; Roose, 2023).

Linked to our various references to the much heralded "information age" and our reliance on computation to process large data sets, AI could be seen as a tool and logical key to help understand the vast and hidden oceanic benthos. AI can be defined as the capacity of non-human entities (e.g., a computer) to simulate process distinctive of human cognition, such as "learning" and "decision-making," to autonomously accomplish a specific task (Russell and Norvig, 2009). The need for AI arose from the information revolution and impossibility of humans to analyze the large caches of big data some fields were generating discussed earlier.

For example, with better mapping tools, geographic information system (GIS) mapping of the seafloor is possible, and thus processing, analyses, and visualization of the data invoke some level of high-performance computing (HPC). AI has now been applied to assist in various large-scale tasks such as digital habitat monitoring, remote sensing (Malde et al., 2020), data processing for marine monitoring, including reefs (Beijbom et al., 2015; Gonzalez-Rivero et al., 2020), and integrating different types of ecological information (Aguzzi et al., 2020). A cautionary note here is that "ground truthing" AI additions should be conducted.

A coral reef study by Gonzalez-Rivero et al. (2020) used Convolutional Neural Networks (CNN), a class of deep learning networks commonly applied to analyze visual imagery, to study coral reef habitats and compare results with manual expert observations. This study took observation data from various oceanic regions (Atlantic, Australia, Indian Ocean, Pacific, and Southeast Asia) and taxonomic groups (e.g., algae, hard corals, soft corals, other invertebrates) to compare the findings between experts or deep learning algorithms. An interesting finding of Gonzalez-Rivero et al. (2020) was that there were no statistical differences in the amount of error and repeatability of benthic abundance estimations when comparing between manual and deep learning methods. This indicates the efficacy of relying on in silico methods to plumb the depths of data, so to speak.

Starting to gain momentum as a field only around 2012, machine learning is a constantly evolving field. Saving in annotation costs is one reason for applying AI in these tasks. For example, assessing photos of coral reef cover for benthic organismal composition or taxonomy requires real expertise, is still laborious, time-consuming, and thus expensive. According to Gonzalez-Rivero et al. (2020), annotating a single image by an expert is certainly feasible and estimated to cost US$ 5.41, while using machine learning runs comparatively less at only US$ 0.07.

Machine learning has been applied to improve statistical methods such as the Random Forest methods (Hastie et al., 2008). Recent monitoring work of the Southeast Florida Reef Tract (SEFRT), a high-latitude reef system offshore of a highly urbanized coastline of Broward County, has identified some key factors affecting biodiversity (Jones et al., 2021) and indicates that temperature has a large effect on benthic species composition. This does not fully discount the effects of other factors such as sedimentation and nutrient loading. However, by applying Random Forests methods on a long-term data set and well-studied transects and study sites, temperature appears to have even greater effects than nutrient or sedimentation on the abundance and resilience of hard coral density in this important reef habitat.

Biotechnologies applied to the seafloor

We began this chapter discussing a three-pronged framework for addressing the scientific problems and especially large-scale issues, such as combing the oceanic depths. That framework depends heavily on technology, broadly defined as the application of science toward specific industrial or commercial objectives. "Biotechnology" is the application of scientific methods to generate products from living organisms. I first came across the term as an undergraduate at the Georgia Institute of Technology in the early 1980s, when I saw it as a course offering listed as "Genetic Engineering." I was still an undecided engineering major at the time, but as my interest in introductory architecture and chemical engineering courses waned, the new discipline intrigued me even more. The course covered the relatively nascent discipline of molecular biology. This has a whole fascinating history to itself, so I refer readers and non-molecular biologists to well-documented stories such as Judson's *The Eighth Day of Creation* (1979) or more recently *The Gene* (Mukherjee, 2017). Eventually I would switch majors from generic to genetic engineering, in which case meant studying to be a biologist.

Can biotechnological leading wedges be applied in a constructive way to conserve, protect, and characterize benthic habitats? Does biotechnology, molecular biology, and current genomics methods assist in the effort toward characterizing benthic ecosystems so far discussed? The answer is affirmative, and there are many examples. I cannot list all of the noteworthy advances here, but I can recapitulate some of the most relevant and interesting examples.

Firstly, for background, the universal genetic material of deoxyribonucleic acid (DNA) and its ability to code for RNA and eventually proteins undergird much of modern biotechnology. DNA also is the fundamental basis of all biodiversity, and we will later see the convenience of having a few molecules illuminate complex phenomena. These include species boundaries, population subdivision, and organismal development of traits such as brain coral patterns or interlocking spicules of rock-like "lithistid" sponges found in deeper waters. The field of molecular genetics can be traced back before 1970s (Judson, 1979), and in a nutshell covers the cloning of recombinant DNA, gene amplification via the polymerase chain reaction, high-throughput sequencing (HTS) of DNA and RNA, the application of computational bioinformatics to store and analyze all of the sequence data, DNA and RNA sequences, and then all of the many experimental and computational innovations in between these milestones (Pevsner, 2015; Brown, 2018). The pace has not slowed, as inexorable improvements in long-range DNA sequencing methods will continually be explored (Shih et al., 2023). Moroever, one of the most significant innovations in biology over the last few years has been the application of bacterial CRISPR (clustered regularly interspersed palindromic repeats)-CAS methods, which allow researchers to edit DNA sequences fully and specifically by design in almost any genome (Doudna and Charpentier, 2014; Shivram et al., 2021). These repeats were first discovered in bacteria, which use them as regular anti-viral (phage) defenses. Novel adjustments such as adding anti-CRISPR components also found in bacteria will enhance the system even further (Nakamura et al., 2019). Application of CRISPR has already commenced on some reef organisms to investigate the molecular basis of heat tolerance and responses to stress and other aspects of their biology (Levin et al., 2017; Cleves et al., 2020; van Oppen and Oakeshott, 2020). Perhaps still in the realm of science fiction, but CRISPR-CAS methods could possibly be used to introduce genetic variation in a species, which have high levels of genetic homozygosity that can lower organismal fitness. CRISPR-CAS could one day solve or reverse the problems associated with low genetic variation and inbreeding depression effects, often found in isolated or decimated wild population numbers. CRISPR would be done in a laboratory setting first for proof of principle.

100   **Chapter 2** Multiple approaches to understanding the benthos

The ability to read and even manipulate DNA sequences, the basic hereditary material of the planet, should never be regarded as nontrivial. Harkening to the start of this chapter, our advances to sequence DNA, RNA, and proteins belong in the pantheon of technological breakthroughs. DNA sequences have also allowed researchers to use the code to determine taxonomic identities, organize, classify, and subsequently construct genealogies or "phylogenetic trees" to portray deeper relationships. Many know that Charles Darwin drew the first phylogenetic tree, which appeared in his Origin of Species, but it was not very sophisticated nor applied any large data sets. The first use of molecular data for tree building was ostensibly produced by George Nuttall in 1904, who was studying immunology and animal blood proteins as one method for inferring genetic distances (Nuttall, 1904). Another major landmark came about with Linus Pauling and Zuckerkandl's 1965 description of a molecular *clock* correlated to evolutionary distances, stating that most sequences appeared to accrue countable mutations in a random and regular manner (Zuckerkandl and Pauling, 1965). From these modest beginnings, we can then explore how molecular systematics, which spans diverse genes and gene families with varying utilities, can answer evolutionary questions. For example, animal mitochondrial DNA sequences tend to mutate quickly in the oxygen-rich environment of the organellar lumen, and thus most its sequences best track populations or recently diverged species. Genes that are required to code for basic cellular functions such as protein synthesis in ribosomal organelles or the cytoskeleton are more conserved, and thus, the sequences evolve slowly even across long evolutionary time periods. An example of a conserved gene with high utility to resolve relationships is the small subunit ribosomal RNA, such as 16S rRNA. Again, these are universal genes, found in organisms as different as yeast and killer whales. This ubiquity across diverse taxa makes the genes very useful, as they can be identified and extracted, with another revolutionary biotechnology tool, such as the polymerase chain reaction (PCR) (Brown, 2018).

Phylogenetic trees can be viewed as one way to map out temporal pathways of species and their lineages. An accurate phylogenetic tree of biological species provides a way to display biological diversity as the members radiate across both time and space. When phylogenies are accurately rooted with fossil or other dating evidence, the bifurcations can provide a way to map speciation over time. This approach aims to eventually point to significant events (sea level rise leading to isolation, earthquakes) that go beyond a single individual organism.

The novel large sequence data sets have been educational, as they have allowed a reassessment of the most relevant evolutionary models and algorithms. For example, this animal root debate goes back over 500 million years into the deep past. This time span is long enough for multiple mutations to accumulate over each, which essentially erases most phylogenetic signal found in any sequences, no matter how ancient or primordial the organismal lineage. The recent debates around sponge or ctenophore basal hypotheses are based on reevaluating these models and identifying artifacts that could erroneously affect the placement of taxa on phylogenetic trees (Kapli and Telford, 2020). This paper also argues that more gene sequence data, which are getting easier to generate and precipitated the initial reassessments, may not be necessary to clarify the basal taxon placement questions. As more data and better models accrue, a metazoan "tree of life" will continue to be modified, pruned, and hopefully reflect historical radiations. For example, the phylogeny in Fig. 2.2 shows yet another possible revision of placements, this time involving the organization of Lophotrochozoa and whether chaetognaths are a sister taxon to Spiralia phyla or protostomes (Marlétaz et al., 2019).

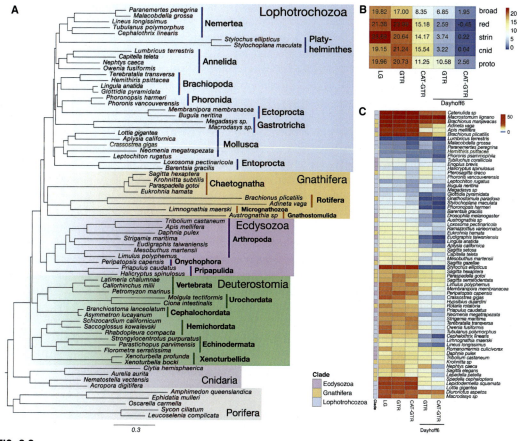

FIG. 2.2

(A) Bilaterian phylogeny reconstruction with Phylobayes using the CAT+GTR model using only taxa with slowest evolutionary rates and steady deviating amino acid composition. All posterior probabilities are maximal. (B and C) Z score statistics of posterior predictive analyses (PPAs) to assess compositional heterogeneity. Global scores for different datasets and models (B) and the detail of the Z score for each species in the "proto" datasets (see https://doi.org/10.5281/zenodo.1403005). Species order was derived by k-means clustering based on PPA Z score. The scale bar indicates the inferred substitution per site.

This diagram to reflect the impact of compositional heterogeneity and reproduced with permission from Marlétaz, F., Peijnenburg, K.T., Goto, T., Satoh, N., Rokhsar, D.S. 2019. A new spiralian phylogeny places the enigmatic arrow worms among gnathiferans. Curr. Biol. 29 (2), 312–318.

Bioprospecting for new natural products and ideas

It may seem ironic that we could deem an animal as simple as a sponge a possible technological wedge. However, the durability of this group over millions of years cannot be denied and can be utilized for a better understanding of the benthos. At this juncture, the interesting features of how sponges play a role in another biotechnology now come into focus. Sponges, other sessile invertebrates, and marine microbes exemplify how biodiversity can be the potential source of new medicines, foods, and technological innovations. This includes the search for new therapeutic compounds also known as natural products or "secondary metabolites" (SMs) (Newman and Hill, 2006; Leal et al., 2020), because they constitute ancillary products to an organism's survival, compared with a DNA polymerase or having a functional Kreb's (tricarboxylic acid) cycle. Nonetheless, SMs represent nature's tinkering with exquisite biochemistry, and organisms that encode for SM capacity also could have an adaptive advantage to other species that lack them. Thus, SM utility for the source organisms would probably encompass enzymes that could biosynthesize novel compounds involved in chemical signaling, virulence factors or toxins, defense again predation, or microbial infections. There are more than 200,000 described species of marine invertebrates and algae, and it is likely that there is an equal number still could be discovered or catalogued. Estimates of the number of marine microbial strains range into the millions, considering that any individual organism may host hundreds or more of its own unique symbiotic microflora. Although SMs can be produced by any organism, a large biosynthetic potential occurs in sessile invertebrates such as sponges and bryozoans. Over the past several decades, investigations by natural products chemists have revealed that sponges and other benthic organisms produce a significant source of biochemical diversity with over 2500 compounds having been described in the 1990s alone (see for example, Faulkner, 2001, Bewley and Faulkner, 1998; Smith et al., 2018). Phylum Bryozoa also produces a large number of potential candidates (Garson, 1994; Figuerola and Avila, 2019). For many years, John Falkner updated lists of marine natural products (Faulkner, 2001), and this has continued to the present (Carroll et al., 2023).

Bioprospecting within this context is defined as "the discovery from natural sources of medicines or genes of use for human health, agriculture, industry or other applications (Kate and Laird, 2019)." Although synthetic chemistry and more recently synthetic biology have provided hundreds of chemical products with significant utility and value, we can still leave it to nature to hold many of the original patents. Because the discovery of new natural products or SMs also spurs new multidisciplinary collaborations, analyses into their mode of action and metabolic functions, and potential new (extractive and sustainable) technologies and policies, I classify the field as a leading wedge. It can provide methods and products that can be used widely.

Natural products biosynthesized by sessile and benthic marine macro- and microorganisms represent an excellent source of potentially new therapeutic drug candidates, and these are plentiful in benthic habitats. Some of the novel biochemical structures of SMs originate from microbes living in sediments (soils on land). It has been estimated that for the period 1983–1994, 78% of the antibacterial agents and 61% of the 31 anticancer agents approved for use were derived from terrestrial natural products (Cragg et al., 1997). A majority of these derived from plants. By comparison, marine environments still hold much to be explored, with only eight compounds derived from a marine source currently approved for clinical use (one is a drug used to treat herpes infections; the other is a drug used to treat leukemia and non-Hodgkin's lymphoma). Many of the invertebrates are sessile as adults and have evolved in habitats that are often more extreme and self-limiting compared with their terrestrial counterparts. This increased evolutionary pressure has resulted in organisms that produce chemically diverse compounds that have a wide variety of possible ecological roles. These include, but are not limited to toxins, which can reduce predation, larval settlement and overgrowth by neighboring organisms; compounds that reduce palatability or nutrient uptake in predators; and compounds that direct larval settlement and reproduction.

Many SM compounds have potent biological properties and multiple compounds are currently under clinical investigation for their use as anticancer and disease agents. Several examples of secondary metabolite innovation from benthic organisms have been well documented: the nucleosides were isolated in the 1950s from a Caribbean sponge, later used as scaffolds for commercial antiviral drugs [Ara-A (Acyclovir) or Ara-C], the analgesic ω-conotoxin analog ziconotide (Prialt) from a *Conus* mollusk used for chronic pain, ecteinascidin 743 (Yondelis) from a tunicate for cancer chemotherapy (sarcoma), and eribulin mesylate (Halaven), a derivative of halichondrin B from the cold-water sponge *Halichondria okadai* as an anticancer drug (Gerwick and Moore, 2012; Jaspars et al., 2016; Giordano et al., 2018). There are currently least eight anticancer drugs derived from marine sources, and at least 20 other marine compounds have advanced to phase I clinical or advanced pre-clinical investigations (Dyshlovoy and Honecker, 2019; Ghareeb et al., 2020): Pliditepsin (Aplidin), PM00104 (Zalypsis), Kahalalide F, Hemiasterlin (E7974), Spisulosine (ES-285), Pseudopterosin A, Marizomib (Salinosporamide A; NPI-0052), Tetrodotoxin (TTX), and Neovastat (AE-941) for cancer treatments. Conotoxin G (CGX-1160), IPL-576092, DMXBA (GTS-21), Bryostatin 1, and Plinabulin (NPI-2358) for the treatment of pain, anti-inflammatory, Alzheimer's, wound healing, and schizophrenia, respectively.

In spite of these examples, we are still beginning to scratch the surface in terms of identifying and controlling how these products are biochemically constructed or evolve. Even more biochemically interesting, is that most metabolically potent SMs have complex structures, spanning alkaloids, cyclic peptides, or polyketides, and so are actually not composed of protein. However, proteins in the form of enzymes are responsible for non-protein SM biosynthesis. These biochemical aspects of SMs are interesting because they are part of unique benthic phenotypes and are directly related to biodiversity (Hill, 2004).

However, even more metabolic diversity and unique products can be produced when in tandem with specific partners. If individual organismal natural histories can be imagined as capillary like single lane dirt roads and two-lane highways, then symbiotic and epibiotic partnerships of >1 species would be the interstate or mega highways. Thus, generation of some marine SMs involves metabolic intermediates from microbes.

The development of the potent anti-tumor compound, bryostatin, has a very interesting story, which includes the association between the specific bryozoan host, *Bugula neritina*, and its bacterial symbiont *Endobugula sertula* (Davidson et al., 2001). As a class of macrocyclic lactones (polyketide) with a unique polyacetate backbone, Bryostatin-1 showed strong activity against the murine P388 lymphocytic leukemia by activating and eventually depleting the regulatory protein kinase C (PKC) via autophosphorylation (Isakov et al., 1993). Subsequently, a γ-proteobacterium, Candidatus *Endobugula sertula*, was identified to code for polyketide biosynthetic clusters for bryostatin (Haygood and Davidson, 1997; Davidson et al., 2001; Lim-Fong et al., 2008). Because the *E. sertula* and *B. neritina* did not seem to co evolve, bryostatins did appear to provide *B. neritina* larvae some protection against predators, indicating one chemical ecology explanation for this unique natural product (Schmidt, 2008). An issue with many natural products is having a sustainable source of the target compound. Currently, the symbiotic "*Ca.* Endobugula sertula" bacterium cannot be cultured, even though it seems capable of independent growth, so only aquaculture of *B. neritina* and total chemical synthesis can yield a sufficient supply of bryostatin (Miller et al., 2016). More recently, the first draft whole bryozoan genome sequence has been elucidated and could eventually be used to help meet supply issues through heterologous expression (Rayko et al., 2020).

104   **Chapter 2** Multiple approaches to understanding the benthos

As elaborated, we can see how the process of marine bioprospecting for any particular therapeutic often manifests as a long journey. I had the pleasure to work in one unique drug discovery program of the Harbor Branch Oceanographic Institution (HBOI) in Ft Pierce FL for 10 years. Harbor Branch Oceanographic Institute (HBOI) became part of Florida Atlantic University (FAU) in 2007. In the Institute's own words recently commemorating their 50th anniversary in 2022, HBOI's efforts "have produced a unique collection of specimens designated as the Marine Biotechnology Reference Collection (MBRC). The MBRC is a taxonomically and geographically diverse collection of more than 31,500 preserved benthic marine invertebrates and macroalgae (seaweeds). This collection is unique in that nearly 30% of the samples were selectively collected by the manned Johnson-Sea-Link submersibles at depths of more than 150 m, from locations that are otherwise difficult to access by trawling or dredging due to rugged topography or reef habitats. Collection by submersible means that these specimens are relatively intact compared to those traditionally collected with trawl or dredge. Collections that were made over a period of 30 years or more provide an opportunity to study changes over time. Research on the specimens, data, images, and videos has already contributed to a better understanding of evolutionary biology, ecology, characterization of marine habitats, and resource management, and has been vital to HBOI's drug discovery program. A goal of this project is to provide the scientific community access to these unique samples and accompanying data. The excellent condition of the specimens makes them useful for molecular analyses, and the videos provide important data to characterize poorly known marine habitats, some of which support commercially and recreationally important fish and benthic invertebrate species. (D Hanisak, S Pomponi, J Reed, A Wright, pers comm)" Many of the deep-sea samples have been maintained at −20°C storage at HBOI. The associated metadata of in situ photos and videos are available online, and the database will be discussed again in Chapter 5. Most of the invertebrate and microbiology collections are available upon request. The Division of Biomedical Marine Research (BMR) at HBOI has been led by sponge biologist Shirley Pomponi, who was also one my postdoctoral advisors. The author participated in several expeditions on the R/V *Edwin Link* to Caribbean locales such as the Bahamas, Turks and Caicos, Honduras, and San Salvador, to find potential new therapeutic drugs from the sea. Collections were made by both scuba and the JSL submersibles. Pomponi was able to make expert identifications of most of the sponge species collected, and she also has led a team to work on sponge cell cultures for a number of years (Pomponi, 1999). Recently, her team reported successfully establishing cell cultures of several sponge species, a long-term goal (Conkling et al., 2019; Hesp et al., 2023).

One of the most promising leads from this group in the 2000s was the polyketide compound discodermolide, found only in a single genus of lithistid sponges (Gunasekera et al., 1990; Longley et al., 1991). The discovery went through detailed biochemical structure analyses, and multiple bioactivity assays at HBOI, which indicated its strong binding to tubulin microtubules. This was a promising anticancer agent, since discodermolide effectively stopped cancer cell replication. Because of the compound's structure, BMR chemists also postulated that part of the biosynthesis involved a symbiotic microbe within the sponge *Discodermia dissoluta*, so the microbial fermentation team led by Peter McCarthy carried out hundreds of bacterial and fungal isolations. Several times discodermolide was detected in culture and larger fermentations, but the amounts waned after multiple passages and time. A microbial mode of production would have helped provide sufficient quantities of a compound, which can be a limiting factor when running numerous trial experiments (Cuevas and Francesch, 2009). Unfortunately, these types of result occur with many laboratory fermentations of natural microbial isolates, since at the time we did not understand the regulatory mechanisms for the discodermolide pathway. Discodermolide advanced to Phase I clinical trials in patients with solid tumors. Unfortunately, like many other potential NPs, the compound also caused associated pneumotoxicity (Mita et al., 2004; Liang et al., 2019).

Secondary metabolism in sessile benthic fauna likely arose out of necessity, possibly as a defensive solution and protection from predators. The life histories of several aquatic invertebrate species, such as corals, tunicates, and bivalves, involve a free swimming, planktonic larval phase that can last from days to months depending on the species. The logistical problem is not only finding a suitable place to settle but also how to survive before and after settling. Potent SMs may have provided some defense. Both pelagic and benthic habitats carry their own inherent jeopardies, but the choice was made evolutionarily. The chemical ecology of several marine SMs has been discussed recently (Paul, 2020).

Marine discovery and bioprospecting have provided newer infusions of thoughts (Sigwart et al., 2021). The suggestion here is based on widening the potential connections between public and private enterprises: "(1) investing in fundamental research, even when the links to industry are not immediately apparent; (2) cultivating equitable collaborations between academia and industry that share both risks and benefits for these foundational research stages; (3) providing new opportunities for early-career researchers and under-represented groups to engage in high-risk research without risking their careers; (4) sharing data with global networks; and (5) protecting genetic diversity at its source through strong conservation efforts."

Secondary metabolites from marine microbes

Microbes can carry out a great deal of unique metabolic production on their own. Because they have had more than a 2 billion years head start to perfect the pathways, as well as using that time to investigate and adapt to all of the possible habitats on the planet—air, land, sea, and even subsurface.

A large portion of the biochemical diversity found in nature discussed above can often and ultimately be traced to microbial organisms and communities, which are poised to provide an enormous boon to biomedicine (Bull et al., 2000; Jensen et al., 2005). It is well known that many secondary metabolites with bioactive effects have been derived from microorganisms. Thus, large-scale efforts involving more sophisticated biotechnologies have been initiated in recent years to characterize and exploit the rich biochemical diversity within many microbes (Handelsman et al., 1998; Engel et al., 2002; Piel et al., 2004). Though less characterized, these early observations and strategies were realized to be applicable to microbes living in marine habitats (Imamura et al., 1997; Faulkner, 2001; Engel et al., 2002; Costantini, 2020). The awareness of the biochemical cache within many microbes was realized relative early (Chadwick and Whelan, 2008), but full harnessing had to wait for genomic methods ("leading wedges") to blossom past theoretical models (Schorn et al., 2016). Part of the reason for this metabolic wealth is that microbial genomes have no nonsense DNA coding, with few wasted residues on introns, repetitive sequences, or intergenic spacers. That is, almost every part of bacterial genomic DNA has biological meaning, which produces a functional product (protein or RNA).

Specific and interesting examples of useful microbial secondary metabolites are the polyketides, created through successive condensations of simple carboxylic acids (Hopwood, 1997; Staunton and Weissman, 2001). Polyketide metabolites exhibit versatile antibiotic, antitumor, anti-inflammatory, and immunomodulatory activities. Sales of the more than 40 polyketide drugs currently on the market (including lovastatin, FK506, erythromycin, avermectin, and amphotericin B) exceed $15 billion a year.

106 **Chapter 2** Multiple approaches to understanding the benthos

In the environment, polyketides may be involved in organismal defenses, such as with brevetoxin, a potent red tide agent affecting human and marine ecosystem health (Baden et al., 2005; Chekan et al., 2020). Yet in spite of having one of the largest genetic databases available among SM biosynthetic loci for genomics and proteomics studies (Staunton and Weissman, 2001) and sharing ancestry with fatty acid synthases (FAS) (Hopwood, 1997), most manipulation of polyketide (and in general, other SM) biosynthesis via genomics and recombinant DNA has been difficult (Kornfuehrer and Eustáquio, 2019; Zhang et al., 2017). Fungi are eukaryotic but also mainly microscopic, yet they are phylogenetically close to the metazoan clade. Thus, their metabolic capacities have also not been overlooked by bio-prospectors of novel natural products. For example, fungal species have yielded some very important antibiotics and SMs such as penicillin, cephalosporin, ergotrate, and the statins.

The most significant challenge to the development of marine natural products is the need to provide adequate supplies of these chemically complex metabolites to allow experimentation, diagnostics, and clinical use. Possessing the coding sequences of biosynthetic genes and proteins does not guarantee or ease the production of the target metabolites. This need has led investigators to begin to study how and why sponges and other marine organisms can produce such a broad library of chemical diversity and what role if any their diverse microbial communities may play in production. Coaxing organisms to become manufacturing tools or factories for human needs is not an uncommon occurrence in history (Huo et al., 2019).

Newer studies have now shown the successful application of molecular biology and functional genomics (transcription profiling) methods to the discovery of SM genetic regulatory factors (Keller, 2019; Costantini, 2020). For example, in up to 50% of biosynthetic gene clusters (BGCs), specific transcription factors (TFs) such as C6-zinc cluster proteins, which regulate gene expression via the recognition of palindromic motifs in cluster gene promoters. Abbasi et al. (2020) have also recently begun to use recombinases RecE/RecT and homologuos recombination for the direct cloning of large DNA sequences up to 100 kb that encode complex SM biosynthetic pathways.

In terrestrial soil ecosystems, the microbial group *Streptomyces* has been a target of natural product searches for many years. Because the capabilities are built into their genomes, the same focus continues to hold true for marine *Streptomyces* (Dharmaraj, 2010; Yang et al., 2020). Unfortunately, even to the present day, inducement of BGC natural cultures of microbes in vitro remains difficult and beyond our knowledge in most organisms, as the BGCs are often transcriptionally silenced (El-Hawary et al., 2018). Part of the problem appears to be unknown quaternary protein-protein interactions that regulate expression. In many organisms, secondary metabolites-associated genes are expressed in response to biotic and abiotic environmental factors (Shwab et al., 2007).

One biotechnological solution to genetic regulation issues has been to use recombinant DNA and high-throughput sequencing technology, which can apply both cloning and sequencing of complete biosynthetic loci. Expression of the secondary metabolic pathways could then be performed in vitro in a foreign, heterologous microbial host. This strategy is termed "combinatorial biosynthesis" (Hutchison, 1998; Pfeifer and Khosla, 2001; Liang et al., 2019; Ruijne and Kuipers, 2021). Several problems with this approach remain, however, which ultimately stymie consistent production of the desired secondary metabolite compound:

"Firstly, combinatorial biosynthesis harnesses the promiscuous substrate specificity of biosynthetic enzymes, but many enzymes do not have sufficient substrate flexibility. Second, a general programming rule for recombining enzymes from different sources lacks the ability to ensure efficient substrate channeling along the assembled biosynthetic machinery. [Genetically] Engineered assembly lines (e.g., domains or modules) could result in the disruption of overall enzyme integrity and

> incompatible enzyme kinetics. Third, lower production of novel natural products in heterologous hosts is often encountered. Furthermore, there is no optimized universal heterologous host capable of functionally expressing the diverse metabolic pathways (such as polyketides (PK), nonribosomal peptide synthase (NRPS), terpenes etc.) derived from disparate sources. Finally, another problem arises from the lack of facile genetic tools available for the rapid assembly of large gene clusters and the engineering of the production host. Consequently, a new multidisciplinary synthetic biology approach aiming to design and construct new biological components (e.g., enzyme, genetic circuits, or cells) or to remodel the natural biological systems is needed to overcome these challenges."
>
> **(Ruijne and Kuipers, 2021).**

Modern genome science again fits the bill for one solution (van Oppen and Coleman, 2022), by providing methods to tailor gene sequences to fit changing environmental conditions. Previously mentioned CRISPR-cas gene editing systems fall into this bin and are borrowed from the bacterial quiver of defensive arrows (Doudna and Charpentier, 2014; Barrangou and Doudna, 2016). Restriction enzymes fall into this same category. Another alternative to cloned expression stems from computational approaches of bioinformatics and functional genomics research, which continues to expand the knowledge of metabolic regulatory networks (Medema and Fischbach, 2015; Khalifa et al., 2019; Costantini, 2020). For example, a new genetic unit under coordinated regulation (regulon) was discovered by Rigali et al. (2004), who used in silico computational approaches to find TFs to the new HutC regulon that comprises genes of the phosphotransferase system (PTS) from anonymous gene sequences. Newly generated cDNA and RNA sequences can also be queried against the growing databases of transcription and sigma factors (Park et al., 2017). A more extensive discussion of databases for biodiversity purposes follows in Chapter 5.

Before fully leaving the subject of bioprospecting, I can introduce one other aspect that these natural products and sciences (chemistry, physiological responses) can offer broader societies in terms of a leading wedge. Bioprospecting has interfaced with the humanities aspects in which the natural products are embedded. That is, the field can continue to take lessons from the social sciences and humanities with regard on how to negotiate royalty contracts, sustainably collect or extract samples, from sovereign nations, which hold the biodiversity sources (species or specimens) (Shiva, 2007). Sensitivity to benefit sharing between providers and end users was introduced with the passing Convention for Biological Diversity (CBD) in 1993, which then led to the later Nagoya Protocol. Both will be discussed in more detail in Chapter 5. One of the first successful implementations of CBD ideals occurred in Panama where International Cooperative Biodiversity Groups (ICBGs) were established (Rosenthal and Katz, 2003; Kursar et al., 2007). Besides the NP research, Panama benefited from the NP research through the training of students, retention of the human, technological, and institutional resources and skills necessary to independently conduct scientific work. This partnership also fully implemented capacity-building between Panama institutions and foreign or commercial entities. Success was measured by the number of publications, successful grant applications, career progressions, patent claims, financial outcomes, technology transfers, and sustainability (Leal et al., 2020).

Molecular ecology, conservation genomics, and genome sequencing

Bioprospecting along with other biotechnologies belongs to growing waves of innovation dubbed as "blue biotechnologies," which involve aquatic bioresources and contribute to a wider bioeconomy (Vieira et al., 2020). The application of biotechnologies including genomics and metagenomics

108 **Chapter 2** Multiple approaches to understanding the benthos

(parallel analysis of total community genomic material) has gained momentum in approaches to improve our understanding of many natural ecosystems (Hiraoka et al., 2016). In the last several years, these studies have been driven by the rapid advances of DNA sequencing technologies, which have paralleled great strides in computational applications (bioinformatics) required to reconstruct the systems biology of each ecosystem, including human and environmental microbiomes (Mardis, 2011; Human Microbiome Project Consortium (HMP)-248 Collaborators, 2012; Thompson et al., 2017). The rapid advances in sequencing and data analysis methods have enabled the design of more appropriate experiments and observational studies to characterize ecosystem dynamics and microbiome interactions (Knight et al., 2012; Hamdan et al., 2013; Cho et al., 2020).

In addition to metagenomic techniques, the second messenger nucleic acid of RNA provides snapshots of cellular function. RNA is a more fragile molecule than DNA and is expressed and transcribed only when required by cells (Brown, 2018). Moreover, since their protein coding exons is represented by only a small fraction (<10%) of most eukaryotic nuclear genomes, less sequencing coverage is required for functional profiling. In this context, meta-transcriptomic (cDNA, EST) or RNAseq sequencing approaches can provide accurate, instantaneous profiles of host and microbial communities (Riesgo et al., 2014; Keeling et al., 2014; Shakya et al., 2019; Treibergs and Giribet, 2020; Yoshioka et al., 2021). This approach can identify rare or uncultivable species, genes, and active phenotypes such as potential metabolic activities, gene expression, and the possible interplay between species in response to environmental drivers (Penn et al., 2014; Sunagawa et al., 2015). Microbial metabolic diversity has been acknowledged to greatly exceed the metabolic capabilities of eukaryotes (Stone and Williams, 1992; Cowan, 2000; Stincone and Brandelli, 2020; Paoli et al., 2022), and thus marine microbes probably contain a large untapped reservoir of genetic, biochemical, and metabolic capabilities. Biochemical diversity and the coculture experiments can also be viewed in a microbial ecology context, a relatively understudied area.

Genomic technologies and their applications are relatively recent, and many qualify as leading wedges useful for our understanding and conserving benthic habitats. The pace of progress in the molecular sciences in itself cuts through expectations, if not fully through to understanding. This section attempts to concisely synthesize some of the historical and recent efforts toward these goals. The endeavor attempts to balance both theoretical and applied sciences.

For example, measuring and conserving genetic variation such as the previously mentioned topic of connectivity are rooted in fundamental population genetics principles, while modern genomics sciences continue to steam ahead with a technological momentum driven by biomedical imperatives or external factors (industry, data gathering) that may have fewer attachments to pure academics or hypothesis testing. To connect these various disciplines, hybrid fields such as *molecular ecology* and *conservation genomics* have helped bridge the gaps. Allendorf et al. (2010) provided a comprehensive list of benefits that genomics will bring to these fields:

- Enormously increase the precision and accuracy of estimation of crucial parameters that require neutral loci (e.g., effective population size and migration rate).
- Reveal the molecular basis and genetic architecture of inbreeding depression.
- Genomic approaches will allow the identification of adaptive genetic variation related to key traits for the response to climate change, such as phenology or drought tolerance, so that management may focus on maintaining adaptive genetic potential.

Molecular ecology, conservation genomics, and genome sequencing

- Assess differential rates of introgression across different genomic regions following hybridization between native and introduced species.
- Assess the role of epigenetic processes in species hybridization.
- Attempt to find the connections between individual fitness and population growth rates.

These principles have been elaborated and expounded further in a recent textbook update, where multiple population genomics methods are described (Allendorf et al., 2022). Because population genomics varies widely between benthic species and their life histories, locale, and depth, I do not extensively summarize all of the possible scenarios in this volume. Moreover, the population data can be rare and deficient, revealing gaps in our basic knowledge. For example, in a meta-analysis, Taylor and Roterman (2017) find a relative paucity of invertebrate population studies for most deep-sea habitats—e.g., "only nine papers account for ~50% of the planet's surface (depths below 3,500 m)... and used only single locus, mitochondrial surveys." Microevolutionary factors of course are relevant to much of our discourse on speciation, connectivity, endemism and thus biodiversity conservation and assessments. Population genetics/genomics data and studies have been and will be exhibited when necessary and supportive.

DNA sequencing can trace its origins to Walter Gilbert's chemical sequencing, but this proved to be hazardous and slow. The major advancement was made by Nobel Prize winner Frederick Sanger in 1977, who invented the dideoxy chain termination sequencing technique (Sanger et al., 1977). These were considered first-generation methods. The "next-generation" methods began in the early 2000s. For example, DNA pyrosequencing technology was developed by 454 Life Sciences Inc. ("454 sequencing") and relied on an emulsion method for DNA amplification and an instrument for sequencing by synthesis using a pyrosequencing protocol optimized for solid support and picolitre-scale volumes (Ronaghi, 2001; Margulies et al., 2005). It was one of the first massively parallel sequencing systems that could read millions of nucleotide sequences in each run. This raw throughput significantly exceeded that of capillary electrophoresis and permitted the simultaneous generation of at least 20 Mb of sequence information per 4-h-long run, comprising reads of approximately 100 bp in length, at 99% or better accuracy.

Multiple studies have applied genomics methods to addressing the ecological components of biodiversity conservation issues, with some already mentioned earlier (Allendorf et al., 2022; Narum et al., 2013; Steiner et al., 2013). At this point, I will not go too deeply into molecular evolutionary principles except to emphasize the importance of characterizing levels of genetic variation in our discussion of the benthos.

One key to understanding genetic (now often referred to as "genomic") changes in populations below the species level utilizes DNA sequences that evolve fast enough for detection with current methods. Not all genetic loci behave similarly. Some vital sequences like those that are regulatory or affect protein synthesis cannot tolerate too many mutations, so they have slower nucleotide and amino acid substitution rates. For example, DNA codons that translate into cytoskeletal proteins such as actin or tubulin have barely changed over hundreds of millions of years. A pairwise comparative alignment of tubulin genes from *Homo sapiens* can be constructed with many species of unicelluar protists such as the common freshwater ciliate *Euplotes* using a standard BLAST tool (Altschul et al., 1997). The results are usually the same—82% of the roughly 435 amino acids match identically, in spite of more than 1 billion years of evolutionary separation. The high percentage matches across a

110 **Chapter 2** Multiple approaches to understanding the benthos

regularly sized protein precludes that the two sequences converged to the same result, but rather likely derived from a distant common parental ancestor. Similar comparisons with other long diverged taxa would produce the same result for this gene. These "slow-evolving" protein and gene sequences show no or few changes within a population and thus not be useful for population genomics studies. The sequences were conserved across hundreds of generations, multiple lineages, and evolutionary time, because their functions in the cell or physiology are vital and fixed. They outlast any individual and embody the idea "…the organism is a *survival machine* built by short-lived confederations of *long-lived* genes" (Dawkins, 1976). Any major changes in the sequence would not necessarily improve function and more likely reduce the fitness of the organism. Thus, certain sequences live on, while individual organisms pass away.

On the other hand, many sequences exist within genomes that have high propensities to accumulate mutations when compared between conspecific individuals. These sequences appear to accumulate more mutations because they typically do not code for vital proteins, or have no (known) cellular functions, occur in between coding sequences or exons, and thus do not appear under any major selective constraints. Because these sequences are thus newly arisen and relatively unique, they are used as genetic markers for *genotyping* populations. Non-coding sequences can compose 25%–40% of a total eukaryotic genome. Examples of rapidly changing sequences include introns (non-coding sequences between protein coding exons) and intergenic regions (sequences between essential coding genes). Other genetic loci with relaxed constraints are various classes of repetitive DNA and genes such as cytochrome B and NADH within mitochondrial DNA (Britten and Kohne, 1968; Lopez et al., 1997; Brown, 2018; Allendorf et al., 2022). For the latter, most animal mitochondrial genomes have been known to have a substitution rates as high as $10\times$ faster than nuclear genomes. This partly stems from the oxygen radical rich environment of the mitochondrial lumen. In addition, the third position of codons often follows neutral patterns of substitution, as multiple nucleotides can be substituted without changing the encoded amino acid (Graur and Li, 2000; Brown, 2018). These mutations are considered "neutral" or nearly neutral in the genome (not under selective pressures to be removed or increase) and posed a surprise to evolutionary biologists when first discovered with molecular biology techniques such as electrophoresis in the 1960s (Hubby and Lewontin, 1966; Lewontin and Hubby, 1966; Kimura, 1983). Explaining the maintenance of the unexpected amounts of neutral mutations and variation in most organisms at the time propelled *neutralist* vs *selectionist* debates (Ohta, 1992; Graur and Li, 2000). The contrasting views were eventually reconciled by the finding that both types of DNA sequence evolution occur and vary based on genomic location and context.

As mentioned above, another significant portion of eukaryotic nuclear genomes found to be hypervariable across many taxa are repetitive DNA or elements. Some repetitive DNA have cellular functions such as sequences at the termini (telomeres) and central regions (centromeres) of chromosomes, which regulate structural integrity and chromosome segregation into daughter cells, respectively. Other repeats can be classified into families based on their sequences or length. For example, SINES (short interspersed nuclear elements), LINES (long interspersed nuclear elements), and tandem repeats, which can be long or short, can compose up to 40% or more of many nuclear genomes. For example, the 707 Mb draft genome of the photosynthetic amoeba *Paulinella micropora* has 76% repetitive sequences (Lhee et al., 2021). Gymnosperm plants boast some of the largest eukaryotic genomes ranging from 17 to 35 Gb (the human genome is 3.3 Gb), but 70%–80% of the plant genomes comprise highly repetitive sequences (Murray, 1998). Long introns also distinguished a recent genome sequence of the Chinese conifer (Niu et al., 2022). In the marine realm, the large 4.85 Gb genome of the great white shark (*Carcharodon carcharias*) holds 8.5% repetitive DNA, and in particular LINE-3/CR1 elements (18.75%) (Marra et al., 2019), while for benthic organisms, the mangrove horseshoe crab *Carcinoscorpius rotundicauda* has 35.01% total repeats in its 1.7 Gb genome (Nong et al., 2021). Other genomes are discussed below. SINES and LINES can also move and transpose within a genome, which helps them expand in number and the proportion of a genome. Because of their mostly autonomous replication beyond the control of host genome and no noticeable contribution to the individual organism's fitness, Dawkins and others dubbed these types of sequences as "selfish" (Dawkins, 1976; Charlesworth and Charlesworth, 1983; Hua-Van et al., 2011).

Short tandem repeats (STRs, also known as microsatellite DNA) range from 1 to 10 nucleotides long while minisatellites appear >10 nucleotides and are dispersed randomly throughout a nuclear genome. Unlike mtDNA, nuclear genome-based STRs appear as multiple distinct loci, segregating in Mendelian fashion, and when multiple loci are combined in a study, the STRs have robustly defined hundreds of species populations on land and in the sea (Culver et al., 2000; Chaves-Fonnegra et al., 2015; Vieira et al., 2016). As one example to show the different types of population structures, which can arise due to different reproductive strategies, STRs were analyzed from 316 brooding coral *Seriatopora hystrix* colonies and broadcast 142 spawning *Acropora millepora* in the Pulau Seribu and Spermonde Archipelago of Indonesia in the Coral Triangle in 2012 and 2013 by Van der Ven et al. (2021). The analyses yielded expected results that *S. hystrix* showed greater population differentiation likely due to lower larval dispersal and local water current patterns. Heterozygosities were normal for each species but varied based on locales—e.g., *S. hystrix* showed Ho of 0.238–0.330 in Pulau Seribu and from 0.381 to 0.604 in Spermonde Archipelago. The STR data can estimate the scale of genetic connectivity among populations and with the computer program STRUCTURE indicated two likely distinct populations in the case broadcasting *A. millepora*. This can be contrasted with three populations of *S. hystrix*, which showed greatest biogeographic differences based on the greater distance of positions of the mid- and outer shelves further offshore, also influenced by oceanographic conditions. STR/microsatellite methods still have utility and can be applied to screening the genets of coral genotypes slated for outplanting in reef restoration programs (Baums et al., 2019).

Developing STR markers is often labor-intensive, and therefore, the method has been slightly eclipsed by the use of single-nucleotide polymorphism (SNP) analyses (Allendorf et al., 2022). An SNP represents a variant single nucleotide within a DNA sequence that appears on average about every 1000 bp among individuals. These variations can arise from genetic mutations or be inherited from parents. SNPs can now be easily generated by current massively parallel sequencing methods, and often only a fraction of a genome is required from multiple individuals to assess sufficient numbers of SNPs. The most common technique now used to generate thousands of SNPs is restriction site-associated DNA (RadSeq) or double-digest RadSeq (ddRadSeq) (both considered genotype by sequencing) (Miller et al., 2007; Davey et al., 2011; Peterson et al., 2012; Vendrami et al., 2017). Similar to STRs and mtDNA, SNP data provide the estimation of standard population genetics parameters such as population genetic and nucleotide diversity, compliance with Hardy Weinberg Equilibrium proportions, heterozygosity, F-statistics, and compliance with Hardy-Weinberg Equilibrium expectations. These will characterize genetic differences between individuals and populations at the fundamental level of the nucleotide, providing profiles of populations and individuals, as well as estimating probabilities of association with genetic risk factors for diseases through genome-wide association studies or GWAS (Tam et al., 2019). Furthermore, the application of whole genome sequencing has also been reviewed (Fuentes-Pardo and Ruzzante, 2017).

Because DNA sequences precisely identify individuals and species, they can also be used for forensic taxonomy. Genetic barcoding refers to the use of relatively short informative segments of universal genes such as the ubiquitous 18S rRNA or mitochondrial cytochrome oxidase I (COI) on eukaryotic tissue samples to assist in taxonomic identification of individual organisms (Sampieri et al., 2021). The barcoding method has been effectively applied across multiple taxa since about 2003 (Hebert et al., 2003). Thus, the application of DNA *metabarcoding* (also known as "environmental DNA or "eDNA") can be seen as another useful technological wedge that can explore various aspects of biodiversity sources and sinks, as well as benthic-pelagic coupling (Ficetola et al., 2008; Vrijenhoek, 2009; Ruppert et al., 2019). Metabarcoding derives directly from barcoding, has a more recent origin, and is applied to bulk environmental samples, which hypothetically hold multiple organisms or whole communities of diverse species. Metabarcoding relies more on high-throughput deep DNA sequencing to identify the resident organisms. The type and size of samples will affect how many species can be sequenced and identified. Compared with single sample/genome barcoding, metabarcoding methods allow for the simultaneous identification of multiple taxa within the same sample (Lopez, 2019).

Similar to the application used in a grocery store to quickly scan items, a barcoding method is meant to be facile, inexpensive yet as precise as the chosen genetic barcode to be applied. With respect to coral reef assessments, autonomous reef monitoring structures (ARMSs) combined metabarcoding technology with ecological elegance to produce a tool that has been used at multiple locations (Ransome et al., 2017; Pearman et al., 2019). Metabarcoding combined with ARMS is a powerful tool now commonly used in biodiversity research and ecology, with some examples described throughout this volume (Brainard et al., 2009; Lopez et al., 2019). Perhaps not originally planned to be on the scale that the Tara expeditions carried out for mid-water communities, ARMs nonetheless convey detailed census information about benthic species (Ransome et al., 2017). ARMS has allowed assessment of many organisms that may be present in the particular benthic habitat. These were first deployed in 2009 to monitor biodiversity on coral reefs (Knowlton et al., 2010; Plaisance et al., 2011). ARMS resembles a beehive underwater, made from PVC parts that cost less than $300. According to Brainard et al. (2009), ARMSs are intended to perform the following: "ARMS provide a tool for systematic and consistent (repeatable) observation of spatial patterns and temporal changes of coral reef invertebrate diversity ARMS also provide a potential tool for early detection of cryptic alien species. The relative simplicity of ARMS design allows cost effective assessment and monitoring at local, regional, and global scales. Use of advanced genetic techniques will significantly reduce the time and cost of biodiversity assessments." The structural simplicity of ARMS belies the complexity of their utility and actual functions. When combined with metabarcoding, ARMS can apply high-throughput sequencing techniques to simultaneously identify genetically and quantify multiple species within a sample, an important tool for the benthic ecologist. They find utility for metabarcoding both eukaryotes and prokaryotes. For example, Pearman et al. (2019) deployed 56 ARMS at 10 m depth for 2–3 years at 19 coral reefs of the Red Sea spanning a latitudinal gradient of 16° and a length of approximately 2000 km to survey bacteria community and function. Among variables monitored during ARMS deployment (monthly averages of Chl*a*, SST, particulate organic carbon (POC), and photosynthetically active radiation (PAR), SST appeared as a key driver for various bacterial families in the reef community. Planctomycetaceae, SAR116, and Phyllobacteriaceae increased concomitantly with SST, while Flammeovirgaceae decreased. The presence of specific bacterial taxa also correlated with proximal macrobenthic representatives (hard corals, soft corals, turf, and macroalgae) on the reef. Pseudoalteromonadaceae responded positively to the percentage of macroalgal cover.

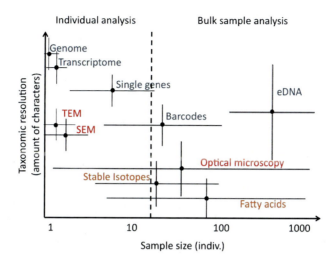

FIG. 2.3

Conceptual diagram of the trade-offs between number of taxonomic characters and sample size in terms of number of individuals for the morphological *(red)*, molecular *(blue)*, and ecological *(orange)* techniques normally used in meiofaunal studies.

Diagram obtained with permission from Fonseca, G., Fontaneto, D., Di Domenico, M. 2018. Addressing biodiversity shortfalls in meiofauna. J. Exp. Mar. Biol. Ecol. 502, 26–38.

Like many methods, however, ARMS and metabarcoding are not perfect tools and have a few drawbacks (Casey et al., 2019): some metabarcoding markers such as the cytochrome oxidase I (COI) do not amplify all taxa, perhaps due to higher substitution rates in some lineages preventing genuine "universal" PCR amplification; the alternative 18S rRNA gene may not show enough variation to distinguish taxa in metabarcodes. The application of traditional microscopy, eDNA, and other molecular methods has been analyzed in the context of meiofaunal characterizations (Fonseca et al., 2018) (Fig. 2.3). The tools could help address seven major gaps in meiofaunal biodiversity knowledge: species identity, species distribution, species abundance, biological traits, evolutionary history, biological interactions, and environmental requirements.

Moreover, the primary sequence of DNA is not the sole predictor or regulator of expressed phenotypic functions. Another regulator is the patterns of modifications made to an organism's hereditary material such as methylation at CpG dinucleotides (cytosine 5mC), affecting gene expression (Suzuki and Bird, 2008). This involves epigenetics, epigenomics, or the changes in gene expression that occur without alterations in the underlying DNA sequence. The modifications can be induced by various environmental factors, such as stress, diet, and toxins, and can influence a range of biological processes, including development, aging, and disease susceptibility (Jones, 2012). Epigenetic modifications include DNA methylation, histone modifications, and non-coding RNA-mediated regulation, among others. In eukaryotes, any changes that tend to "open" up 5′ regulatory sequences on the DNA, by preventing any protein binding, tend to promote gene expression. Conversely, more condensed or bound chromatin will be less active. These changes can be reversible or stable and considered "genetic imprinting" when patterns are transmitted from one generation to the next, contributing to the heritability of traits. In this case, the epigenetic changes can also be viewed as Lamarckian in that the

114 **Chapter 2** Multiple approaches to understanding the benthos

modifications occur in somatic tissue and not the organism's germline. The characterization of epigenetic modifications has yielded to the determination of an epigenome or "methylome" alongside standard genomes, transcriptomes, and metabolomes (Brown, 2018; de Mendoza et al., 2020; Beal et al., 2020). Various studies have shown the impact of epigenetic modifications across various benthic taxa (Bogan et al., 2020; Downey-Wall et al., 2020; Lee et al., 2022). Invertebrates appear to have less methylation than vertebrates (Suzuki and Bird, 2008). Using the technique of Methylation-Sensitive Amplified Polymorphism (MSAP) in 200 colonies *Acropora cervicornis*, 83 restriction fragments displayed variations in DNA methylation that were significantly correlated to seasonal variation (e.g., sea water temperature fluctuations) (Rodríguez-Casariego et al., 2020). Epigenetic patterns may also be connected to a host's microbiome (Eirin-Lopez and Putnam, 2019).

Unfortunately, molecular biologists remain far from designing and customizing epigenetic patterns, i.e., picking sites to methylate, to control expression in vivo for whole organisms. We are still in the midst of sequencing and trying to understand the standard DNA sequence of four nucleotides (A, C, G, and T). We can use CRISPR tools to edit DNA sequences but mostly in small segments. Overall, subtle differences in transmission of the epigenetic signals, which can be either intergenerational (epigenetic change triggered by the environment can be lost) or transgenerational, exist (modification is maintained in the absence of the original stimulus) (Fitz-James and Cavalli, 2022). Less evidence has been found for the latter type of inheritance, indicating more work is required for this leading wedge. Methylome modification and their full expression may also be farther in model laboratory organisms, which do not include many marine species. With current technologies and understanding, combining both methylation and sequencing editing would be akin to playing three-dimensional chess.

Another recent large-scale example in this vein is the California Conservation Genomics Project (CCGP) (Shaffer et al., 2022). Still ongoing at the time of writing, CCGP recognized that single species conservation projects have commendable intentions but may fall short when the whole ecosystem or habitat is lost under the individual species feet. Thus, a primary goal of $12 million CCGP is to "generate a comprehensive database of genomic variation and associated geo-referenced environmental data, and to use that database to help guide the protection of species and ecosystems that may be vulnerable to climate change and other anthropogenic threats" (https://www.californiabiodiversityinitiative.org/pdf/california-biodiversity-action-plan.pdf; https://www.ccgproject.org). In the first 2 years, the CCGP has formed a consortium of scientists (68 laboratories and 114 investigators drawn from all 10 University of California campuses) to address this goal, working on diverse taxa within the targeted geographic locale of California, and all of the diverse habitats within 19 terrestrial ecoregions and inshore marine habitats. The State of California provided a basis of a phase I, which supported 153 species projects encompassing 235 species, with a degree of high conservation concerns. Large-scale genomics data and interpretations will be delivered to the state's agencies to slow or prevent species' declines. The lessons and scale of CCGP could easily be applied to similar problems and scale that we are discussing in the marine realm (Grummer et al., 2019). Challenges to using this model for benthic studies would encompass the proper identification of stakeholders and those willing to support and fund a large-scale project when technically "owns the high seas" (See later discussion of the Tragedy of the Commons in Chapter 4).

One final biotechnological approach for preserving marine biodiversity to consider is establishing viable *cell cultures* of benthic species that are threatened, on the verge of extinction or losing their habitats before they disappear. Unfortunately, establishing continuous cell lines of marine invertebrates is notoriously difficult (Frank et al., 1994; Ventura et al., 2018; Kawamura et al., 2021). This strategy goes beyond saving seeds or tissue samples in frozen zoos, seed banks, or repositories, which are also effective means (Liu et al., 2020). Suggestions and guideposts for carrying out cell culture approaches have been proscribed for several years and with success for many vertebrates and other terrestrial fauna (Friedrich Ben-Nun et al., 2011). For example, Oliver Ryder has for several years resolutely advocated for a "global network of facilities establishing and cryobanking collections of viable cells" of endangered species with the following tangible benefits (Ryder and Onuma, 2018):

- Frozen early-passage cell cultures constitute an expandable resource; cell cultures may be thawed, grown, and a portion banked again, repeatedly.
- Cell freezing provides access to chromosomes and karyotypes. Chromosomes prepared by collecting metaphase cells from dividing cultures allow identification of chromosomal constitution and its variation, genetic mapping, and purification of individual chromosomes or chromosomal regions and facilitate genome assembly by identification of centromere location.
- Frozen cells provide sufficient DNA and RNA from small organisms for multiple sequencing libraries and different platforms.
- The highest-quality DNA extracts enable improved genome assemblies. Cells may be lysed to obtain chromosome-length molecules. Un-degraded in vivo chromatin can be prepared.
- Cell freezing provides a resource for functional studies, including receptor biology, cell signaling, and disease investigation.
- Cell freezing provides a source for reprogramming to produce iPSCs, and iPSCs allow for transcriptome studies from cells of all developmental stages.
- Cell freezing may be used for genetic rescue through in vitro gametogenesis or production of individuals that produce gametes following development of chimeric embryos.
- Cell freezing reduces animal welfare concerns. Biopsies may be obtained opportunistically or postmortem, and in vitro studies lessen the need for whole-animal investigations.
- Future interests may be better served. Banking viable cells in concert with conservation breeding affords the most extensive options for ex situ species conservation.

In the marine realm, freshwater *Hydra* and *Anemonia viridis* cells have been cultured, and until recently, few viable cell lines of marine species have been created (Rinkevich, 2005; Cai and Zhang, 2014; Conkling et al., 2019; Hesp et al., 2023). However, Nowotny et al. (2021) appear to have established some primary cell cultures from *Nematostella vectensis* and *Pocillopora damicornis* (Nowotny et al., 2021), and recent breakthroughs have been made with sponge cell cultures (Conkling et al., 2019; Hesp et al., 2023). Therefore, this research space represents a wide-open door for intensive investigation and exploration for the sake of biodiversity (Rosner et al., 2021).

Overall, this chapter has tried to convey different approaches for effectively addressing large problems of an enormous and diverse habitat, which is the world's seafloor. We all agree that benthic habitats are not monolithic, because they vary across many scales, oceans, and latitudes. But this review should also reveal the dangers and risk of our indifference. We are barely paying attention to the oceans at the surface, so it follows that less attention will be paid (by the general public) to even deeper layers of the ocean. The in-depth tour to a few rich benthic habitats in the next chapter should further convince us that more actions are needed for the conservation of benthic life immediately.

116   **Chapter 2** Multiple approaches to understanding the benthos

References

Abbasi, M.N., Fu, J., Bian, X., Wang, H., Zhang, Y., Li, A., 2020. Recombineering for genetic engineering of natural product biosynthetic pathways. Trends Biotechnol. 38 (7), 715–728.

Achenbach, J., 2012. A Hole at the Bottom of the Sea: The Race to Kill the BP Oil Gusher. Simon and Schuster.

Aguzzi, J., Chatzievangelou, D., Francescangeli, M., Marini, S., Bonofiglio, F., Del Rio, J., Danovaro, R., 2020. The hierarchic treatment of marine ecological information from spatial networks of benthic platforms. Sensors 20 (6), 1751.

Albertin, C.B., Simakov, O., Mitros, T., Wang, Z.Y., Pungor, J.R., Edsinger-Gonzales, E., et al., 2015. The octopus genome and the evolution of cephalopod neural and morphological novelties. Nature 524 (7564), 220–224.

Allen, E.E., Facciotti, D., Bartlett, D.H., 1999. Monounsaturated but not polyunsaturated fatty acids are required for growth of the deep-sea bacterium *Photobacterium profundum* SS9 at high pressure and low temperature. Appl. Environ. Microbiol. 65 (4), 1710–1720.

Allendorf, F.W., Hohenlohe, P.A., Luikart, G., 2010. Genomics and the future of conservation genetics. Nat. Rev. Genet. 11 (10), 697–709.

Allendorf, F.W., Funk, W.C., Aitken, S.N., 2022. Conservation and the Genomics of Populations. Oxford University Press.

Altschul, S.F., Madden, T.L., Schäffer, A.A., Zhang, J., Zhang, Z., Miller, W., Lipman, D.J., 1997. Gapped BLAST and PSI-BLAST: a new generation of protein database search programs. Nucleic Acids Res. 25 (17), 3389–3402.

Armstrong, R.A., Pizarro, O., Roman, C., 2019. Underwater robotic technology for imaging mesophotic coral ecosystems. In: Mesophotic Coral Ecosystems. Springer, pp. 973–988.

Australian Institute of Marine Sciences (AIMS), 2022. Annual Summary Report of Coral Reef Condition 2021/2022. https://www.aims.gov.au/sites/default/files/2022-08/AIMS_LTMP_Report_on%20GBR_coral_status_2021_2022_040822F3.pdf.

Baden, D.G., Bourdelais, A.J., Jackocks, H., Michelliza, S., Naar, J., 2005. Natural and derivative brevetoxins: historical background, multiplicity, and effects. Environ. Health Perspect. 11 (5), 621–625.

Baker, E., Thygesen, K., Harris, P., Andradi-Brown, D., et al., 2016. Mesophotic Coral Ecosystems: A Lifeboat for Coral Reefs? UN Environment, GRID-Arendal.

Ballard, R.D.R.D., 2000. The history of woods hole's deep submergence program. In: 50 Years of Ocean Discovery: National Science Foundation 1950–2000. National Academies Press, pp. 67–84.

Barrangou, R., Doudna, J.A., 2016. Applications of CRISPR technologies in research and beyond. Nat. Biotechnol. 34 (9), 933–941.

Baums, I.B., Baker, A.C., Davies, S.W., Grottoli, A.G., Kenkel, C.D., Kitchen, S.A., et al., 2019. Considerations for maximizing the adaptive potential of restored coral populations in the western Atlantic. Ecol. Appl. 29 (8), e01978.

Bayley, D.T., Mogg, A.O., 2019. New advances in benthic monitoring technology and methodology. In: World Seas: An Environmental Evaluation. Academic Press, pp. 121–132.

Beal, A., Rodriguez-Casariego, J., Rivera-Casas, C., Suarez-Ulloa, V., Eirin-Lopez, J.M., 2020. Environmental epigenomics and its applications in marine organisms. In: Population Genomics: Marine Organisms. Springer, pp. 325–359.

Beijbom, O., Edmunds, P.J., Roelfsema, C., Smith, J., Kline, D.I., Neal, B.P., Dunlap, M.J., Moriarty, V., Fan, T.-Y., Tan, C.-J., 2015. Towards automated annotation of benthic survey images: variability of human experts and operational modes of automation. PLoS One 10, e0130312.

Beukes, L., Robinson, K.S., Liu, K., Rajaniemi, H., Reynolds, A., De Bodard, A., 2017. Science fiction when the future is now. Nature 552 (7685), 329–333.

Bewley, C.A., Faulkner, D.J., 1998. Lithistid sponges: star performers or hosts to the stars. Angew. Chem., Int. Ed. 37 (16), 2162–2178.

Blaxter, M., Archibald, J.M., Childers, A.K., Coddington, J.A., Crandall, K.A., Di Palma, F., et al., 2022. Why sequence all eukaryotes? Proc. Natl. Acad. Sci. 119 (4). https://doi.org/10.1073/pnas.2115636118.

Bluhm, H., 2001. Re-establishment of an abyssal megabenthic community after experimental physical disturbance of the seafloor. Deep Sea Res. Part II 48, 3841–3868. https://doi.org/10.1016/s0967-0645(01)00070-4.

Bogan, S.N., Johnson, K.M., Hofmann, G.E., 2020. Changes in genome-wide methylation and gene expression in response to future p CO_2 extremes in the Antarctic Pteropod *Limacina helicina* antarctica. Front. Mar. Sci. 6, 788.

Bräger, S., Rodriguez, G.Q.R., Mulsow, S., 2020. The current status of environmental requirements for deep seabed mining issued by the International Seabed Authority. Mar. Policy 114, 103258.

Brainard, R., Moffitt, R., Timmers, M., Paulay, G., Plaisance, L., Knowlton, N., et al., 2009. Autonomous reef monitoring structures (ARMS): a tool for monitoring indices of biodiversity in the Pacific Islands. In: 11th Pacific Science Inter-Congress, Papeete, Tahiti, March.

Britten, R.J., Kohne, D.E., 1968. Repeated sequences in DNA: hundreds of thousands of copies of DNA sequences have been incorporated into the genomes of higher organisms. Science 161 (3841), 529–540.

Brockett, T., Richards, C.Z., 1994. Deepsea mining simulator for environmental impact studies. Sea Technol. 35, 77–82.

Brown, T.A., 2018. Genomes 4. Wiley-Liss.

Brown, C.J., Smith, S.J., Lawton, P., Anderson, J.T., 2011. Benthic habitat mapping: a review of progress towards improved understanding of the spatial ecology of the seafloor using acoustic techniques. Estuar. Coast. Shelf Sci. 92 (3), 502–520.

Bull, A.T., Ward, A.C., Goodfellow, M., 2000. Search and discovery strategies for biotechnology: the paradigm shift. Microbiol. Mol. Biol. 64 (3), 573–606.

Burrough, P.A., McDonnell, R.A., Lloyd, C.D., 2015. Principles of Geographical Information Systems. Oxford University Press.

Cai, X., Zhang, Y., 2014. Marine invertebrate cell culture: a decade of development. J. Oceanogr. 70, 405–414.

Callaway, E., Irwin, A., Kozlov, M., Ledford, H., Makri, A., Masood, E., et al., 2022. Nature's 10: ten people who helped shape science in 2022. Nature 612 (7941), 611–625.

Carroll, A.R., Copp, B.R., Davis, R.A., Keyzers, R.A., Prinsep, M.R., 2023. Marine natural products. Nat. Prod. Rep. 40, 275.

Casey, J.M., Meyer, C.P., Morat, F., Brandl, S.J., Planes, S., Parravicini, V., 2019. Reconstructing hyperdiverse food webs: gut content metabarcoding as a tool to disentangle trophic interactions on coral reefs. Methods Ecol. Evol. 10 (8), 1157–1170.

Chadwick, D.J., Whelan, J., 2008. In: Secondary Metabolites: Their Function and Evolution. Wiley and Sons.

Charlesworth, B., Charlesworth, D., 1983. The population dynamics of transposable elements. Genet. Res. 42 (1), 1–27.

Chaves-Fonnegra, A., Feldheim, K.A., Secord, J., Lopez, J.V., 2015. Population structure and dispersal of the coral excavating sponge *Cliona delitrix*. Mol. Ecol. 24, 1447–1466. https://doi.org/10.1111/mec.13134. https://onlinelibrary.wiley.com/doi/10.1111/mec.13134/abstract;jsessionid=33AA0B3E3978DB90BAF76%2009C538D0869.f03t03.

Chekan, J.R., Fallon, T.R., Moore, B.S., 2020. Biosynthesis of marine toxins. Curr. Opin. Chem. Biol. 59, 119–129.

Cho, H., Hwang, C.Y., Kim, J.G., Kang, S., Knittel, K., Choi, A., et al., 2020. A unique benthic microbial community underlying the *Phaeocystis antarctica*-dominated Amundsen Sea polynya, Antarctica: a proxy for assessing the impact of global changes. Front. Mar. Sci. 6, 797.

Cleves, P.A., Tinoco, A.I., Bradford, J., Perrin, D., Bay, L., Pringle, J.R., 2020. Reduced thermal tolerance in a coral carrying CRISPR-induced mutations in the gene for a heat-shock transcription factor. Proc. Natl. Acad. Sci. U.S.A. 117 (46), 28899–28905.

Conkling, M., Hesp, K., Munroe, S., Sandoval, K., Martens, D.E., Sipkema, D., et al., 2019. Breakthrough in marine invertebrate cell culture: sponge cells divide rapidly in improved nutrient medium. Sci. Rep. 9 (1), 17321.

Costantini, M., 2020. Genome mining and synthetic biology in marine natural product discovery. Mar. Drugs 18 (12), 615. https://doi.org/10.3390/md18120615.

Costanza, R., Mageau, M., 1999. What is a healthy ecosystem? Aquat. Ecol. 33, 105–115.

Cowan, D.A., 2000. Microbial genomes—the untapped resource. Trends Biotechnol. 18, 14–16.

Cragg, G.M., Newman, D.J., Snader, K.M., 1997. Natural products in drug discovery and development. J. Nat. Prod. 60 (1), 52–60.

Cuevas, C., Francesch, A., 2009. Development of Yondelis®(trabectedin, ET-743). A semisynthetic process solves the supply problem. Nat. Prod. Rep. 26 (3), 322–337.

Culver, M., Johnson, W.E., Pecon-Slattery, J., O'Brien, S.J., 2000. Genomic ancestry of the American puma (*Puma concolor*). J. Hered. 91 (3), 186–197.

Da Ros, Z., Dell'Anno, A., Morato, T., Sweetman, A.K., Carreiro-Silva, M., Smith, C.J., et al., 2019. The deep sea: the new frontier for ecological restoration. Mar. Policy 108, 103642.

Davey, J.W., Hohenlohe, P.A., Etter, P.D., Boone, J.Q., Catchen, J.M., Blaxter, M.L., 2011. Genome-wide genetic marker discovery and genotyping using next-generation sequencing. Nat. Rev. Genet. 12 (7), 499–510.

Davidson, S.K., Allen, S.W., Lim, G.E., Anderson, C.M., Haygood, M., 2001. Evidence for the biosynthesis of bryostatins by the bacterial symbiont "Candidatus Endobugula sertula" of the bryozoan *Bugula neritina*. Appl. Environ. Microbiol. 67 (10), 4531–4537.

Dawkins, R., 1976. The Selfish Gene. Oxford University Press.

Dayton, P.K., 1971. Competition, disturbance, and community organization: the provision and subsequent utilization of space in a rocky intertidal community. Ecol. Monogr. 41 (4), 351–389.

de Mendoza, A., Lister, R., Bogdanovic, O., 2020. Evolution of DNA methylome diversity in eukaryotes. J. Mol. Biol. 432 (6), 1687–1705.

Dekker, A.G., Phinn, S.R., Anstee, J., Bissett, P., Brando, V.E., Casey, B., et al., 2011. Intercomparison of shallow water bathymetry, hydro-optics, and benthos mapping techniques in Australian and Caribbean coastal environments. Limnol. Oceanogr. Methods 9 (9), 396–425.

Desplat, Y., Warner, J., Blake, E., Vijayan, N, Cuvelier, M., Blackwelder, P., Lopez, J.V., 2023. Morphological and transcriptional effects of crude oil and dispersant exposure on the marine sponge *Cinachyrella alloclada*. Sci. Total Environ. 878, 162832. https://doi.org/10.1016/j.scitotenv.2023.162832.

Dharmaraj, S., 2010. Marine Streptomyces as a novel source of bioactive substances. World J. Microbiol. Biotechnol. 26 (12), 2123–2139.

Doudna, J.A., Charpentier, E., 2014. The new frontier of genome engineering with CRISPR-Cas9. Science 346 (6213), 1258096.

Downey-Wall, A.M., Cameron, L.P., Ford, B.M., McNally, E.M., Venkataraman, Y.R., Roberts, S.B., et al., 2020. Ocean acidification induces subtle shifts in gene expression and DNA methylation in mantle tissue of the Eastern oyster (*Crassostrea virginica*). Front. Mar. Sci. 7, 566419.

Duarte, C.M., Agusti, S., Barbier, E., Britten, G.L., Castilla, J.C., Gattuso, J.P., Fulweiler, R.W., Hughes, T.P., Knowlton, N., Lovelock, C.E., Lotze, H.K., Predragovic, M., Poloczanska, E., Roberts, C., Worm, B., 2020. Rebuilding marine life. Nature 580 (7801), 39–51. https://doi.org/10.1038/s41586-020-2146-7.

Dyshlovoy, S.A., Honecker, F., 2019. Marine compounds and cancer: the first two decades of XXI century. Mar. Drugs 18 (1), 20.

Eddy, S.R., 2013. The ENCODE project: missteps overshadowing a success. Curr. Biol. 23 (7), R259–R261.

Eirin-Lopez, J.M., Putnam, H.M., 2019. Marine environmental epigenetics. Annu. Rev. Mar. Sci. 11, 335–368.

Eiseley, L., 2011. The Immense Journey: An Imaginative Naturalist Explores the Mysteries of Man and Nature. Vintage.

El-Hawary, S.S., Sayed, A.M., Mohammed, R., Hassan, H.M., Zaki, M.A., Rateb, M.E., et al., 2018. Epigenetic modifiers induce bioactive phenolic metabolites in the marine-derived fungus *Penicillium brevicompactum*. Mar. Drugs 16 (8), 253.

Engel, S., Jensen, P.R., Fenical, W., 2002. Chemical ecology of marine microbial defense. J. Chem. Ecol. 28, 1971–1985.

Erwin, T.L., 1982. Tropical forests: their richness in Coleoptera and other arthropod species. Coleopt. Bull. 36, 74–75.

Erwin, T.L., 1997. Biodiversity at its utmost: tropical forest beetles. In: Biodiversity II: Understanding and Protecting Our Biological Resources. Joseph Henry Press, pp. 27–40.

Erwin, T.L., Scott, J.C., 1980. Seasonal and size patterns, trophic structure, and richness of Coleoptera in the tropical arboreal ecosystem: the fauna of the tree Luehea seemannii Triana and Planch in the Canal Zone of Panama. Coleopt. Bull. 34, 305–322.

Faulkner, D.J., 2001. Marine natural products. Nat. Prod. Rep. 18 (1), 1R–49R.

Fennel, W., Neumann, T., 2014. Introduction to the Modelling of Marine Ecosystems. Elsevier.

Ficetola, G.F., Miaud, C., Pompanon, F., Taberlet, P., 2008. Species detection using environmental DNA from water samples. Biol. Lett. 4 (4), 423–425.

Figuerola, B., Avila, C., 2019. The phylum bryozoa as a promising source of anticancer drugs. Mar. Drugs 17 (8), 477.

Fitz-James, M.H., Cavalli, G., 2022. Molecular mechanisms of transgenerational epigenetic inheritance. Nat. Rev. Genet. 23 (6), 325–341.

Fonseca, G., Fontaneto, D., Di Domenico, M., 2018. Addressing biodiversity shortfalls in meiofauna. J. Exp. Mar. Biol. Ecol. 502, 26–38.

Formenti, G., Theissinger, K., Fernandes, C., Bista, I., Bombarely, A., Bleidorn, C., et al., 2022. The era of reference genomes in conservation genomics. Trends Ecol. Evol. 37, 197–202.

Frank, U., Rabinowitz, C., Rinkevich, B., 1994. In vitro establishment of continuous cell cultures and cell lines from ten colonial cnidarians. Mar. Biol. 120, 491–499.

Friedrich Ben-Nun, I., Montague, S.C., Houck, M.L., Tran, H.T., Garitaonandia, I., Leonardo, T.R., et al., 2011. Induced pluripotent stem cells from highly endangered species. Nat. Methods 8 (10), 829–831.

Fuentes-Pardo, A.P., Ruzzante, D.E., 2017. Whole-genome sequencing approaches for conservation biology: advantages, limitations and practical recommendations. Mol. Ecol. 26 (20), 5369–5406.

Fukami, H., Knowlton, N., 2005. Analysis of complete mitochondrial DNA sequences of three members of the *Montastraea annularis* coral species complex (Cnidaria, Anthozoa, Scleractinia). Coral Reefs 24, 410–417.

G10KCOS, 2009. Genome 10K: a proposal to obtain whole-genome sequence for 10,000 vertebrate species. J. Hered. 100, 659–674.

Garson, M.J., 1994. The biosynthesis of sponge secondary metabolites: why is it imporatint. In: Sponges in Time and Space. AA Balkema, pp. 427–440.

Gerwick, W.H., Moore, B.S., 2012. Lessons from the past and charting the future of marine natural products drug discovery and chemical biology. Chem. Biol. 19, 85–98. https://doi.org/10.1016/j.chembiol.2011.12.014.

Ghaffary, S., 2023. Silicon Valley's AI frenzy isn't just another crypto craze. Vox. https://www.vox.com/technology/2023/3/6/23624015/silicon-valley-generative-ai-chat-gpt-crypto-hype-trend).

Ghareeb, M.A., Tammam, M.A., El-Demerdash, A., Atanasov, A.G., 2020. Insights about clinically approved and preclinically investigated marine natural products. Curr. Res. Biotechnol. 2, 88–102.

GIGA Community of Scientists, 2014. The global invertebrate genome alliance (GIGA): developing community resources to study diverse invertebrates. J. Hered. 105, 1–18. https://doi.org/10.1093/jhered/est084 (Supplementary material – https://tinyurl.com/2zwbky5h).

GIGA Community of Scientists (Voolstra, C as lead author), 2017. Advancing genomics through the global invertebrate genomics alliance (GIGA). Invertebr. Syst. 31 (1), 1–7. http://www.publish.csiro.au/is/Fulltext/IS16059.

Giordano, D., Costantini, M., Coppola, D., Lauritano, C., Núñez Pons, L., Ruocco, N., et al., 2018. Biotechnological applications of bioactive peptides from marine sources. Adv. Microb. Physiol. 73, 171–220.

Gonzalez-Rivero, M., Beijbom, O., Rodriguez-Ramirez, A., Bryant, D.E., Ganase, A., Gonzalez-Marrero, Y., et al., 2020. Monitoring of coral reefs using artificial intelligence: a feasible and cost-effective approach. Remote Sens. 12 (3), 489.

Goodwin, S., McPherson, J.D., McCombie, W.R., 2016. Coming of age: ten years of next-generation sequencing technologies. Nat. Rev. Genet. 17 (6), 333–351.

120 **Chapter 2** Multiple approaches to understanding the benthos

Graur, D., Li, W.H., 2000. Fundamentals of Molecular Evolution. Sinauer Associates.

Griffith, F., 1928. The significance of pneumococcal types. Epidemiol. Infect. 27 (2), 113–159.

Grummer, J.A., Beheregaray, L.B., Bernatchez, L., Hand, B.K., Luikart, G., Narum, S.R., Taylor, E.B., 2019. Aquatic landscape genomics and environmental effects on genetic variation. Trends Ecol. Evol. 34 (7), 641–654.

Gunasekera, S.P., Gunasekera, M., Longley, R.E., Schulte, G.K., 1990. Discodermolide: a new bioactive polyhydroxylated lactone from the marine sponge *Discodermia dissoluta*. J. Org. Chem. 55 (16), 4912–4915.

Hamdan, L.J., Coffin, R.B., Sikaroodi, M., Greinert, J., Treude, T., Gillevet, P.M., 2013. Ocean currents shape the microbiome of Arctic marine sediments. ISME J. 7 (4), 685–696.

Hand, K.P., German, C.R., 2018. Exploring ocean worlds on Earth and beyond. Nat. Geosci. 11, 2–4.

Handelsman, J., Rondon, M.R., Brady, S.F., Clardy, J., Goodman, R.M., 1998. Molecular biologicalaccess to the chemistry of unknown soil microbes: a new frontier for natural products. Chem. Biol. 5 (10), R245–R249.

Harari, G.M., Müller, S.R., Stachl, C., Wang, R., Wang, W., Bühner, M., et al., 2019. Sensing sociability: individual differences in young adults' conversation, calling, texting, and app use behaviors in daily life. J. Pers. Soc. Psychol. 119 (1), 204.

Hastie, T., Tibshirani, R., Friedman, J., 2008. The Elements of Statistical Learning, second ed. Springer, ISBN: 0-387-95284-5.

Haygood, M.G., Davidson, S.K., 1997. Small-subunit rRNA genes and in situ hybridization with oligonucleotides specific for the bacterial symbionts in the larvae of the bryozoan *Bugula neritina* and proposal of "Candidatus endobugula sertula". Appl. Environ. Microbiol. 63 (11), 4612–4616.

Hebert, P.D.N., Cywinska, A., Ball, S.L., DeWaard, J.R., 2003. Biological identifications through DNA barcodes. Proc. R. Soc. Lond. B Biol. Sci. 270, 313–321.

Hellwarth, B., 2012. Sealab: America's Forgotten Quest to Live and Work on the Ocean Floor. Simon & Schuster, New York, p. 259 (ISBN 978-0-7432-4745-0. LCCN 2011015725.).

Helm, R.R., Clark, N., Harden-Davies, H., Amon, D., Girguis, P., Bordehore, C., et al., 2021. Protect high seas biodiversity. Science 372 (6546), 1048–1049.

Hesp, K., van der Heijden, J.M., Munroe, S., Sipkema, D., Martens, D.E., Wijffels, R.H., Pomponi, S.A., 2023. First continuous marine sponge cell line established. Sci. Rep. 13 (1), 5766.

Hill, R.T., 2004. Microbes from marine sponges: a treasure trove of biodiversity for natural products discovery. In: Bull (Ed.), Microbial Diversity and Bioprospecting. ASM Press, Washington DC.

Hiraoka, S., Yang, C.C., Iwasaki, W., 2016. Metagenomics and bioinformatics in microbial ecology: current status and beyond. Microbes Environ. 31, 204–212.

Hopwood, D.A., 1997. Genetic contributions to understanding polyketide synthases. Chem. Rev. 97 (7), 2465–2498.

Hotchkiss, R.D., 1965. OSWALD T. AVERY: 1877-1955. Genetics 51, 1–10.

Hua-Van, A., Le Rouzic, A., Boutin, T.S., Filée, J., Capy, P., 2011. The struggle for life of the genome's selfish architects. Biol. Direct 6 (1), 1–29.

Hubby, J.L., Lewontin, R.C., 1966. A molecular approach to the study of genic heterozygosity in natural populations. I. The number of alleles at different loci in *Drosophila pseudoobscura*. Genetics 54 (2), 577.

Hughes, T.P., Rodrigues, M.J., Bellwood, D.R., Ceccarelli, D., Hoegh-Guldberg, O., McCook, L., et al., 2007. Phase shifts, herbivory, and the resilience of coral reefs to climate change. Curr. Biol. 17 (4), 360–365.

Hughes, T.P., Kerry, J.T., Baird, A.H., Connolly, S.R., Dietzel, A., Eakin, C.M., et al., 2018. Global warming transforms coral reef assemblages. Nature 556 (7702), 492–496.

Human Microbiome Project Consortium (HMP)-248 Collaborators, 2012. Structure, function and diversity of the healthy human microbiome. Nature 486, 207–214.

Huo, L., Hug, J.J., Fu, C., Bian, X., Zhang, Y., Müller, R., 2019. Heterologous expression of bacterial natural product biosynthetic pathways. Nat. Prod. Rep. 36 (10), 1412–1436.

Hutchins, E., 1983. Understanding micronesian navigation. In: Mental Models. Lawrence Erlbaum, pp. 191–225.

Hutchison, C.R., 1998. Combinatorial biosynthesis for new drug discovery. Curr. Opin. Microbiol. 1 (3), 319–329.

Imamura, N., Nishijima, M., Takadera, T., Adachi, K., Saiki, M., Sano, H., 1997. New anticancer antibiotics pelagiomicins, produced by a new marine bacterium Pelagiobacter variabilis. J. Antibiot. 50 (1), 8–12.

Ingole, B.S., Ansari, Z.A., Matondkar, S.G.P., Rodriges, N., 1999. Immediate response of meio and macrobenthos to disturbance caused by a benthic disturber. In: Proceedings of the III Ocean Mining Symposium, Goa, India, November 8–10, pp. 191–194.

Ingole, B.S., Goltekar, R., Gonsalves, S., Ansari, Z., 2005. Recovery of deep-sea meiofauna after artificial disturbance in the Central Indian Basin. Mar. Georesour. Geotechnol. 23, 253–266.

Isakov, N., Galron, D., Mustelin, T., Pettit, G.R., Altman, A., 1993. Inhibition of phorbol ester-induced T cell proliferation by bryostatin is associated with rapid degradation of protein kinase C. J. Immunol. 150 (4), 1195–1204.

Jaspars, M., De Pascale, D., Andersen, J.H., Reyes, F., Crawford, A.D., Ianora, A., 2016. The marine biodiscovery pipeline and ocean medicines of tomorrow. J. Mar. Biol. Assoc. U.K. 96, 151–158. https://doi.org/10.1017/s0025315415002106.

Jensen, P.R., Mincer, T.J., Williams, P.G., Fenical, W., 2005. Marine actinomycete diversity and natural product discovery. Anton. Leeuw. 87, 43–48.

Jokiel, P.L., Rodgers, K.S., Kuffner, I.B., Andersson, A.J., Cox, E.F., Mackenzie, F.T., 2008. Ocean acidification and calcifying reef organisms: a mesocosm investigation. Coral Reefs 27, 473–483.

Jones, P.A., 2012. Functions of DNA methylation: islands, start sites, gene bodies and beyond. Nat. Rev. Genet. 13 (7), 484–492.

Jones, N.P., Kabay, L., Semon Lunz, K., Gilliam, D.S., 2021. Temperature stress and disease drives the extirpation of the threatened pillar coral, *Dendrogyra cylindrus*, in Southeast Florida. Sci. Rep. 11 (1), 1–10.

Jordan, G.F., Malloy, R.J., Kofoed, J.W., 1964. Bathymetry and geology of Pourtales Terrace, Florida. Mar. Geol. 1 (3), 259–287. https://doi.org/10.1016/0025-3227(64)90063-5.

Judson, H.F., 1979. The Eighth Day of Creation. Cold Spring Harbor, New York, p. 550.

Kapli, P., Telford, M.J., 2020. Topology-dependent asymmetry in systematic errors affects phylogenetic placement of Ctenophora and Xenacoelomorpha. Sci. Adv. 6 (50), eabc5162.

Kate, K., Laird, S.A., 2019. The Commercial Use of Biodiversity: Access to Genetic Resources and Benefit-Sharing. Routledge.

Kawamura, K., Sekida, S., Nishitsuji, K., et al., 2021. In vitro symbiosis of reef-building coral cells with photosynthetic dinoflagellates. Front. Mar. Sci. 900.

Keeling, P.J., Burki, F., Wilcox, H.M., Allam, B., Allen, E.E., Amaral-Zettler, L.A., et al., 2014. The Marine Microbial Eukaryote Transcriptome Sequencing Project (MMETSP): illuminating the functional diversity of eukaryotic life in the oceans through transcriptome sequencing. PLoS Biol. 12 (6), e1001889.

Keller, N.P., 2019. Fungal secondary metabolism: regulation, function and drug discovery. Nat. Rev. Microbiol. 17 (3), 167–180.

Kennedy, B.R., Cantwell, K., Malik, M., Kelley, C., Potter, J., Elliott, K., et al., 2019. The unknown and the unexplored: insights into the Pacific deep-sea following NOAA CAPSTONE expeditions. Front. Mar. Sci. 6, 480.

Kennicutt, M.C., Brooks, J.M., Bidigare, R.R., Fay, R.R., Wade, T.L., McDonald, T.J., 1985. Vent-type taxa in a hydrocarbon seep region on the Louisiana slope. Nature 317 (6035), 351–353.

Khalifa, S.A., Elias, N., Farag, M.A., Chen, L., Saeed, A., Hegazy, M.E.F., et al., 2019. Marine natural products: a source of novel anticancer drugs. Mar. Drugs 17 (9), 491.

Kilfoyle, K., Walker, B.K., Gregg, K., Fiscoi, D.P., Spieler, D.E., 2018. Southeast Florida Coral Reef Fishery-Independent Baseline Assessment: 2012-2016 Summary Report. National Oceanic and Atmospheric Administration Coral Reef Conservation Program. https://floridadep.gov/rcp/coral/documents/southeast-florida-coral-reef-fishery-independent-baseline-assessment-2012-2016. (Accessed 3 September 2023).

Kim, B.M., Kang, S., Ahn, D.H., Jung, S.H., Rhee, H., Yoo, J.S., et al., 2018. The genome of common long-arm octopus *Octopus minor*. Gigascience 7 (11), giy119.

Kimura, M., 1983. The Neutral Theory of Molecular Evolution. Cambridge University Press.

Kirch, P.V., 2017. On the Road of the Winds: An Archaeological History of the Pacific Islands before European Contact. University of California Press (2017).

Knight, R., Jansson, J., Field, D., Fierer, N., Desai, N., Fuhrman, J.A., et al., 2012. Unlocking the potential of metagenomics through replicated experimental design. Nat. Biotechnol. 30, 513–520.

Knowlton, N., 1993. Sibling species in the sea. Annu. Rev. Ecol. Syst. 24 (1), 189–216.

Knowlton, N., Mate, J.L., Guzman, H.M., Rowan, R., Jara, J., 1997. Direct evidence for reproductive isolation among the three species of the *Montastraea annularis* complex in Central America (Panama and Honduras). Mar. Biol. 127, 705–711.

Knowlton, N., Brainard, R.E., Fisher, R., Moews, M., Plaisance, L., Caley, M.J., 2010. Coral reef biodiversity. In: Life in the World's Oceans: Diversity Distribution and Abundance. John Wiley & Sons, pp. 65–74.

Koepfli, K.-P., Paten, B., Antunes, A., Belov, K., Bustamante, C., Castoe, T., Clawson, H., Crawford, A., Distel, D., Durbin, R., Earl, D., Fujita, M., Gamble, T., Georges, A., Gemmell, N., Marshall Graves, J., Green, R.E., Hickey, G., Jarvis, E., Johnson, W., Korf, I., Kuhn, R., Larkin, D., Lewin, H., Lopez, J.V., Ma, J., Bonet, T.M., Miller, W., Murphy, R., Pevzner, P., Shapiro, B., Steiner, C., Tamazian, G., Venkatesh, B., Wang, J., Wiley, E., Yang, H., Zhang, G., Haussler, D., Ryder, O., O'Brien, S.J., for the G10KCOS, 2015. The genome 10K project and the state of vertebrate genomics: a way forward. Rev. Anim. Genet. 3, 57–111.

Kohler, K.E., Gill, S.M., 2006. Coral point count with excel extensions (CPCe): a visual basic program for the determination of coral and substrate coverage using random point count methodology. Comput. Geosci. 32 (9), 1259–1269.

Kornfuehrer, T., Eustáquio, A.S., 2019. Diversification of polyketide structures via synthase engineering. MedChemComm 10 (8), 1256–1272.

Kursar, T.A., Caballero-George, C.C., Capson, T.L., Cubilla-Rios, L., Gerwick, W.H., Heller, M.V., et al., 2007. Linking bioprospecting with sustainable development and conservation: the Panama case. Biodivers. Conserv. 16, 2789–2800.

Leal, M.C., Anaya-Rojas, J.M., Munro, M.H., Blunt, J.W., Melian, C.J., Calado, R., Lürig, M.D., 2020. Fifty years of capacity building in the search for new marine natural products. Proc. Natl. Acad. Sci. 117 (39), 24165–24172.

Lee, B., Ko, Y., Kwon, G., Lee, S., Ku, K., Kim, J., Kang, K., 2018. Exploiting biological systems: toward eco-friendly and high-efficiency rechargeable batteries. Joule 2 (1), 61–75.

Lee, Y.H., Kim, M.S., Wang, M., Bhandari, R.K., Park, H.G., Wu, R.S.S., Lee, J.S., 2022. Epigenetic plasticity enables copepods to cope with ocean acidification. Nat. Clim. Chang. 12 (10), 918–927.

Lejzerowicz, F., Gooday, A.J., Barrenechea Angeles, I., Cordier, T., Morard, R., et al., 2021. Eukaryotic biodiversity and spatial patterns in the Clarion-Clipperton Zone and other abyssal regions: insights from sediment DNA and RNA metabarcoding. Front. Mar. Sci. 8, 671033.

Leray, M., Knowlton, N., Ho, S., Nguyen, B., Machida, R., 2019. GenBank is a reliable resource for 21st century biodiversity research. Proc. Natl. Acad. Sci. U.S.A. 116 (45), 22651–22656.

Lester, S.E., Rassweiler, A., McCoy, S.J., Dubel, A.K., Donovan, M.K., Miller, M.W., et al., 2020. Caribbean reefs of the anthropocene: variance in ecosystem metrics indicates bright spots on coral depauperate reefs. Glob. Chang. Biol. 26 (9), 4785–4799.

Levin, R.A., Voolstra, C.R., Agrawal, S., Steinberg, P.D., Suggett, D.J., Van Oppen, M.J., 2017. Engineering strategies to decode and enhance the genomes of coral symbionts. Front. Microbiol. 8, 1220.

Levin, L.A., Bett, B.J., Gates, A.R., Heimbach, P., Howe, B.M., Janssen, F., et al., 2019. Global observing needs in the deep ocean. Front. Mar. Sci. 6, 241.

Levitan, D.R., Fukami, H., Jara, J., Kline, D., McGovern, T.M., McGhee, K.E., et al., 2004. Mechanisms of reproductive isolation among sympatric broadcast-spawning corals of the *Montastraea annularis* species complex. Evolution 58 (2), 308–323.

Lewin, H.A., Larkin, D.M., Pontius, J., O'Brien, S.J., 2009. Every genome sequence needs a good map. Genome Res. 19 (11), 1925–1928.

Lewin, H.A., Robinson, G.E., Kress, W.J., Baker, W.J., Coddington, J., Crandall, K.A., et al., 2018. Earth BioGenome project: sequencing life for the future of life. Proc. Natl. Acad. Sci. 115 (17), 4325–4333.

Lewin, H.A., Richards, S., Zhang, G., 2022. The earth BioGenome project 2020: starting the clock. Proc. Natl. Acad. Sci. U. S. A. 119 (4), e2115635118. 18 January 2022 https://doi.org/10.1073/pnas.2115635118.

Lewis, J.P., Tarnecki, J.H., Garner, S.B., et al., 2020. Changes in reef fish community structure following the deepwater horizon oil spill. Sci. Rep. 10, 5621. https://doi.org/10.1038/s41598-020-62574-y.

Lewontin, R.C., Hubby, J.L., 1966. A molecular approach to the study of genic heterozygosity in natural populations. II. Amount of variation and degree of heterozygosity in natural populations of *Drosophila pseudoobscura*. Genetics 54 (2), 595.

Lhee, D., Lee, J., Ettahi, K., Cho, C.H., Ha, J.S., Chan, Y.F., et al., 2021. Amoeba genome reveals dominant host contribution to plastid endosymbiosis. Mol. Biol. Evol. 38 (2), 344–357.

Li, A.J., Leung, P.T., Bao, V.W., Yi, A.X., Leung, K.M., 2014. Temperature-dependent toxicities of four common chemical pollutants to the marine medaka fish, copepod and rotifer. Ecotoxicology 23, 1564–1573.

Li, J., Mara, P., Schubotz, F., et al., 2020. Recycling and metabolic flexibility dictate life in the lower oceanic crust. Nature 579, 250–255. https://doi.org/10.1038/s41586-020-2075-5.

Li, G., Chen, X., Zhou, F., Liang, Y., Xiao, Y., Cao, X., et al., 2021. Self-powered soft robot in the Mariana trench. Nature 591 (7848), 66–71.

Li, Y., Sun, X.M., Dang, Y.R., Liu, N.H., Qin, Q.L., Zhang, Y.Q., Zhang, X.Y., 2022. Genomic analysis of Marinomonas profundi M1K-6T reveals its adaptation to deep-sea environment of the Mariana trench. Mar. Genomics 62, 100935.

Liang, X., Luo, D., Luesch, H., 2019. Advances in exploring the therapeutic potential of marine natural products. Pharmacol. Res. 147, 104373.

Lim-Fong, G.E., Regali, L.A., Haygood, M.G., 2008. Evolutionary relationships of "Candidatus Endobugula" bacterial symbionts and their Bugula bryozoan hosts. Appl. Environ. Microbiol. 74 (11), 3605–3609.

Lin, T.H., Chen, C., Watanabe, H.K., Kawagucci, S., Yamamoto, H., Akamatsu, T., 2019. Using soundscapes to assess deep-sea benthic ecosystems. Trends Ecol. Evol. 34 (12), 1066–1069.

Link, M.C., 1973. Windows in the Sea. Smithsonian Institution Press.

Liu, U., Cossu, T.A., Davies, R.M., Forest, F., Dickie, J.B., Breman, E., 2020. Conserving orthodox seeds of globally threatened plants ex situ in the Millennium Seed Bank, Royal Botanic Gardens, Kew, UK: the status of seed collections. Biodivers. Conserv. 29 (9–10), 2901–2949.

Longley, R.E., Caddigan, D., Harmody, D., Gunasekera, M., Gunasekera, S.P., 1991. Discodermolide—a new, marine-derived immunosuppressive compound. II. In vivo studies. Transplantation 52 (4), 656–661.

Lopez, J.V., 2019. After the taxonomic identification phase: addressing the functions of symbiotic communities within marine invertebrates. In: Li, Z. (Ed.), Symbionts of Marine Sponges and Corals. Springer-Verlag, Dordrecht, The Netherlands, pp. 105–144, https://doi.org/10.1007/978-94-024-1612-1_8.

Lopez, J.V., Culver, M., Stephens, J.C., Johnson, W.E., O'Brien, S.J., 1997. Rates of nuclear and cytoplasmic mitochondrial sequence divergence in mammals. Mol. Biol. Evol. 14 (3), 277–286.

Lopez, J.V., Peixoto, R., Rosado, A.S., 2019. Inevitable future: space colonization beyond earth with microbes first. FEMS Microbiol. Ecol. 95 (10), fiz127. https://doi.org/10.1093/femsec/fiz127. https://academic.oup.com/femsec/article/95/10/fiz127/5553461.

Mahatma, R., 2009. Meiofauna Communities of the Pacific Nodule Province: Abundance, Diversity and Community Structure (Ph.D. thesis). Carl von Ossietzky Universitat, Oldenburg (142 pp.).

Malde, K., Handegard, N.O., Eikvil, L., Salberg, A.B., 2020. Machine intelligence and the data-driven future of marine science. ICES J. Mar. Sci. 77 (4), 1274–1285.

Mallet, D., Pelletier, D., 2014. Underwater video techniques for observing coastal marine biodiversity: a review of sixty years of publications (1952–2012). Fish. Res. 154, 44–62.

124   **Chapter 2** Multiple approaches to understanding the benthos

Mardis, E.R., 2011. A decade's perspective on DNA sequencing technology. Nature 470 (7333), 198–203.

Margulies, M., Egholm, M., Altman, W.E., Attiya, S., Bader, J.S., et al., 2005. enome sequencing in microfabricated high-density picolitre reactors. Nature 437 (7057), 376–380.

Marlétaz, F., Peijnenburg, K.T., Goto, T., Satoh, N., Rokhsar, D.S., 2019. A new spiralian phylogeny places the enigmatic arrow worms among gnathiferans. Curr. Biol. 29 (2), 312–318.

Marra, N.J., Stanhope, M.J., Jue, N.K., Wang, M., Sun, Q., Pavinski Bitar, P., et al., 2019. White shark genome reveals ancient elasmobranch adaptations associated with wound healing and the maintenance of genome stability. Proc. Natl. Acad. Sci. 116 (10), 4446–4455.

Martinez, C.M., 2009. NOAA Ocean Explorer: Okeanos Explorer: A New Paradigm for Exploration: Telepresence. Retrieved from https://oceanexplorer.noaa.gov/. (Accessed 9 August 2009).

Mayer, L., Jakobsson, M., Allen, G., Dorschel, B., Falconer, R., Ferrini, V., et al., 2018. The Nippon Foundation—GEBCO seabed 2030 project: the quest to see the world's oceans completely mapped by 2030. Geosciences 8 (2), 63.

McKenzie, L.J., Nordlund, L.M., Jones, B.L., Cullen-Unsworth, L.C., Roelfsema, C., Unsworth, R.K., 2020. The global distribution of seagrass meadows. Environ. Res. Lett. 15 (7), 074041.

McNutt, M., Alexander, V., Ausubel, J., Ballard, R., Chance, T., Douglas, P., Earle, S., Estes, J., Fornari, D., Gordon, A., et al., 2000. Discovering the Earth's Final Frontier: A US Strategy for Ocean Exploration. US Department of Commerce, National Oceanic and Atmospheric Administration.

Medawar, P., 1967. The Art of the Soluble. Methuen & Co./Barnes and Noble, London/New York, NY.

Medema, M.H., Fischbach, M.A., 2015. Computational approaches to natural product discovery. Nat. Chem. Biol. 11, 639–648.

Miljutin, D.M., Miljutina, M.A., Arbizu, P.M., Galéron, J., 2011. Deep-sea nematode assemblage has not recovered 26 years after experimental mining of polymetallic nodules (Clarion-Clipperton Fracture Zone, Tropical Eastern Pacific). Deep-Sea Res. I Oceanogr. Res. Pap. 58 (8), 885–897.

Miller, M.R., Dunham, J.P., Amores, A., Cresko, W.A., Johnson, E.A., 2007. Rapid and cost-effective polymorphism identification and genotyping using restriction site associated DNA (RAD) markers. Genome Res. 17 (2), 240–248.

Miller, I.J., Vanee, N., Fong, S.S., Lim-Fong, G.E., Kwan, J.C., 2016. Lack of overt genome reduction in the bryostatin-producing bryozoan symbiont "Candidatus Endobugula sertula". Appl. Environ. Microbiol. 82 (22), 6573–6583.

Mita, A., Lockhart, C., Chen, T.L., Boshinski, K., Curtright, J., Cooper, W., Hammond, L., Rothenberg, M., Rowinsky, E., Sharma, S., 2004. A phase I pharmacokinetic (PK) trial of XAA296A (discodermolide) administered every 3 wks to adult patients with advanced solid malignancies. J. Clin. Oncol. 10, 2025.

Mooney, T.A., Di Iorio, L., Lammers, M., Lin, T.H., Nedelec, S.L., Parsons, M., et al., 2020. Listening forward: approaching marine biodiversity assessments using acoustic methods. R. Soc. Open Sci. 7 (8), 201287.

Mukherjee, S., 2011. The Emperor of all Maladies: A Biography of Cancer. Simon and Schuster.

Mukherjee, S., 2017. The Gene: An Intimate History. Scribner.

Muller-Karger, F.E., Miloslavich, P., Bax, N.J., Simmons, S., Costello, M.J., Sousa Pinto, I., et al., 2018. Advancing marine biological observations and data requirements of the complementary essential ocean variables (EOVs) and essential biodiversity variables (EBVs) frameworks. Front. Mar. Sci. 5, 211.

Murray, B.G., 1998. Nuclear DNA amounts in gymnosperms. Ann. Bot. 82 (Suppl 1), 3–15.

Nakamura, M., Srinivasan, P., Chavez, M., Carter, M.A., Dominguez, A.A., La Russa, M., et al., 2019. Anti-CRISPR-mediated control of gene editing and synthetic circuits in eukaryotic cells. Nat. Commun. 10 (1), 194.

Nakanishi, M., Hashimoto, J., 2011. A precise bathymetric map of the world's deepest seafloor, Challenger Deep in the Mariana Trench. Mar. Geophys. Res. 32, 455–463.

Narum, S.R., Buerkle, C.A., Davey, J.W., Miller, M.R., Hohenlohe, P.A., 2013. Genotyping-by-sequencing in ecological and conservation genomics. Mol. Ecol. 22 (11), 2841.

Nath, R., Manna, R., 2023. From posthumanism to ethics of artificial intelligence. AI Soc. 38 (1), 185–196.

Neumann, A.C., Kofoed, J.W., Keller, G.H., 1977. Lithoherms in the straits of Florida. Geology 5 (1), 4–10.

Newman, D.J., Hill, R.T., 2006. New drugs from marine microbes: the tide is turning. J. Ind. Microbiol. Biotechnol. 33 (7), 539–544.

Niu, S., Li, J., Bo, W., Yang, W., Zuccolo, A., Giacomello, S., et al., 2022. The Chinese pine genome and methylome unveil key features of conifer evolution. Cell 185 (1), 204–217.

Nong, W., Qu, Z., Li, Y., Barton-Owen, T., Wong, A.Y., Yip, H.Y., et al., 2021. Horseshoe crab genomes reveal the evolution of genes and microRNAs after three rounds of whole genome duplication. Commun. Biol. 4 (1), 83.

Nowotny, J.D., Connelly, M.T., Traylor-Knowles, N., 2021. Novel methods to establish whole-body primary cell cultures for the cnidarians *Nematostella vectensis* and *Pocillopora damicornis*. Sci. Rep. 11 (1), 4086.

Nuttall, G.H.F., 1904. Blood Immunity and Blood Relationship: A Demonstration of Certain Blood-Relationships amongst Animals by Means of the Precipitation Test for Blood. Cambridge University Press, Cambridge, p. 462.

Nyankson, E., Rodene, D., Gupta, R.B., 2016. Advancements in crude oil spill remediation research after the deepwater horizon oil spill. Water Air Soil Pollut. 227, 1–22.

Obura, D.O., Aeby, G., Amornthammarong, N., Appeltans, W., et al., 2019. Coral reef monitoring, reef assessment technologies, and ecosystem-based management. Front. Mar. Sci. 6, 580.

Ohta, T., 1992. The nearly neutral theory of molecular evolution. Annu. Rev. Ecol. Evol. Syst. 23 (1), 263–286.

Paoli, L., Ruscheweyh, H.J., Forneris, C.C., Hubrich, F., Kautsar, S., Bhushan, A., et al., 2022. Biosynthetic potential of the global ocean microbiome. Nature 607 (7917), 111–118.

Park, J.W., Nam, S.J., Yoon, Y.J., 2017. Enabling techniques in the search for new antibiotics: combinatorial biosynthesis of sugar-containing antibiotics. Biochem. Pharmacol. 134, 56–73.

Parsekar, A.S., Jobby, R., 2022. Deep-sea extremophiles and their diversity in the Indian Ocean. In: Extremophiles: A Paradox of Nature with Biotechnological Implications. vol. 1. De Gruyter, p. 65.

Paul, V.J. (Ed.), 2020. Ecological Roles of Marine Natural Products. Cornell University Press.

Pearman, J.K., Aylagas, E., Voolstra, C.R., Anlauf, H., Villalobos, R., Carvalho, S., 2019. Disentangling the complex microbial community of coral reefs using standardized autonomous reef monitoring structures (ARMS). Mol. Ecol. 28 (15), 3496–3507.

Pearman, T.R.R., Robert, K., Callaway, A., Hall, R., Iacono, C.L., Huvenne, V.A., 2020. Improving the predictive capability of benthic species distribution models by incorporating oceanographic data—towards holistic ecological modelling of a submarine canyon. Prog. Oceanogr. 184, 102338.

Penn, K., Wang, J., Fernando, S.C., Thompson, J.R., 2014. Secondary metabolite gene expression and interplay of bacterial functions in a tropical freshwater cyanobacterial bloom. ISME J. 8, 1866–1878.

Peterson, B.K., Weber, J.N., Kay, E.H., Fisher, H.S., Hoekstra, H.E., 2012. Double digest RADseq: an inexpensive method for de novo SNP discovery and genotyping in model and non-model species. PLoS One 7 (5), e37135.

Pevsner, J., 2015. Bioinformatics and Functional Genomics. John Wiley & Sons.

Pfeifer, B.A., Khosla, C., 2001. Biosynthesis of polyketides in heterologous hosts. Microbiol. Mol. Biol. Rev. 65 (1), 106–118.

Piel, J., Hui, D., Wen, G., Butzke, D., Platzer, M., Fusetani, N., Matsunaga, S., 2004. Antitumor polyketide biosynthesis by an uncultivated bacterial symbiont of the marine sponge Theonella swinhoei. Proc. Natl. Acad. Sci. U.S.A. 101 (46), 16222–16227.

Plaisance, L., Caley, M.J., Brainard, R.E., Knowlton, N., 2011. The diversity of coral reefs: what are we missing? PLoS One 6 (10), e25026.

Pomponi, S.A., 1999. The bioprocess—technological potential of the sea. J. Biotechnol. 70, 5–13.

Popper, K., 1959. The Logic of Scientific Discovery. Abingdon-on-Thames, Routledge (2002) https://archive.org/details/The_Logic_Of_Scientific_Discovery_Karl_Popper.pdf/page/n3/mode/2up.

Prampolini, M., Angeletti, L., Castellan, G., Grande, V., Le Bas, T., Taviani, M., Foglini, F., 2021. Benthic habitat map of the southern adriatic sea (Mediterranean Sea) from object-based image analysis of multi-source acoustic backscatter data. Remote Sens. 13 (15), 2913.

Radziejewska, T., 2002. Responses of deep-sea meiobenthic communities to sediment disturbance simulating effects of polymetallic nodule mining. Int. Rev. Hydrobiol. 87 (4), 457–477.

Radziejewska, T., Rokicka-Praxmajer, J., Stoyanova, V., 2001. IOM BIE revisited: meiobenthos at the IOM BIE site 5 years after the experimental disturbance. In: Proceedings of the Fourth Ocean Mining Symposium, Szczecin, Poland, September 23–27, pp. 63–68.

Ramirez-Llodra, E., Tyler, P.A., Baker, M.C., Bergstad, O.A., Escobar, E., Van Dover, C.L., 2011. Man and the last great wilderness: human impact on the deep sea. PLoS One 6 (8), e22588.

Ransome, E., Geller, J.B., Timmers, M., Leray, M., Mahardini, A., Sembiring, A., 2017. The importance of standardization for biodiversity comparisons: a case study using autonomous reef monitoring structures (ARMS) and metabarcoding to measure cryptic diversity on Mo'orea coral reefs, French Polynesia. PLoS One 12 (4), e0175066.

Rayko, M., Komissarov, A., Lim-Fong, G., Rhodes, A.C., Kwan, J., Kliver, C., Chesnokova, P.S., O'Brien, S.J., Lopez, J.V., 2020. Draft genome of Bryozoan *Bugula neritina*—a colonial animal packing powerful symbionts and potential medicines. Sci. Data 7, 356. https://doi.org/10.1038/s41597-020-00684-y.

Renegar, D.A., Turner, N.R., Riegl, B.M., Dodge, R.E., Knap, A.H., Schuler, P.A., 2017. Acute and subacute toxicity of the polycyclic aromatic hydrocarbon 1-methylnaphthalene to the shallow-water coral *Porites divaricata*: application of a novel exposure protocol. Environ. Toxicol. Chem. 36 (1), 212–219.

Rhie, A., McCarthy, S.A., Fedrigo, O., Damas, J., Formenti, G., Koren, S., et al., 2021. Towards complete and error-free genome assemblies of all vertebrate species. Nature 592 (7856), 737–746.

Riesgo, A., Farrar, N., Windsor, P.J., Giribet, G., Leys, S.P., 2014. The analysis of eight transcriptomes from all poriferan classes reveals surprising genetic complexity in sponges. Mol. Biol. Evol. 31 (5), 1102–1120. https://doi.org/10.1093/molbev/msu057.

Rigali, S., Schlicht, M., Hoskisson, P., Nothaft, H., Merzbacher, M., Joris, B., Titgemeyer, F., 2004. Extending the classification of bacterial transcription factors beyond the helix–turn–helix motif as an alternative approach to discover new cis/trans relationships. Nucleic Acids Res. 32 (11), 3418–3426.

Rinkevich, B., 2005. Marine invertebrate cell cultures: new millennium trends. Mar. Biotechnol. 7, 429–439.

Robertson, B.E., 2021. Galaxy formation and reionization: key unknowns and expected breakthroughs by the James Webb space telescope. Annu. Rev. Astron. Astrophys. 60, 121–158.

Robinson, G.E., Hackett, K.J., Purcell-Miramontes, M., Brown, S.J., Evans, J.D., Goldsmith, M.R., Lawson, D., Okamuro, J., Robertson, H.M., Schneider, D.J., 2011. Creating a buzz about insect genomes. Science 331, 1386.

Rodríguez-Casariego, J.A., Mercado-Molina, A.E., Garcia-Souto, D., Ortiz-Rivera, I.M., Lopes, C., Baums, I.B., et al., 2020. Genome-wide DNA methylation analysis reveals a conserved epigenetic response to seasonal environmental variation in the staghorn coral *Acropora cervicornis*. Front. Mar. Sci. 7, 560424.

Rodríguez-Romero, A., Jiménez-Tenorio, N., Basallote, M.D., Orte, M.R.D., Blasco, J., Riba, I., 2014. Predicting the impacts of CO_2 leakage from subseabed storage: effects of metal accumulation and toxicity on the model benthic organism Ruditapes philippinarum. Environ. Sci. Technol. 48 (20), 12292–12301.

Ronaghi, M., 2001. Pyrosequencing sheds light on DNA sequencing. Genome Res. 11 (1), 3–11.

Roose, K., 2023. A.I. Poses 'Risk of Extinction,' Industry Leaders Warn. New York Times. https://www.nytimes.com/2023/05/30/technology/ai-threat-warning.html?smid=url-share. (30 May 2023).

Rosenthal, J.P., Katz, F.N., 2003. Natural products research partnerships with multiple objectives in global biodiversity hot spots: nine years of the international cooperative biodiversity groups program. In: Microbial Diversity and Bioprospecting. John Wiley & Sons, Inc, pp. 458–466.

Rosner, A., Armengaud, J., Ballarin, L., Barnay-Verdier, S., Cima, F., Coelho, A.V., et al., 2021. Stem cells of aquatic invertebrates as an advanced tool for assessing ecotoxicological impacts. Sci. Total Environ. 771, 144565.

Rowe, G.T., Wei, C., Nunnally, C., Haedrich, R., Montagna, P., Baguley, J.G., et al., 2008. Comparative biomass structure and estimated carbon flow in food webs in the deep Gulf of Mexico. Deep-Sea Res. II Top. Stud. Oceanogr. 55 (24–26), 2699–2711.

Ruijne, F., Kuipers, O.P., 2021. Combinatorial biosynthesis for the generation of new-to-nature peptide antimicrobials. Biochem. Soc. Trans. 49 (1), 203–215.

Ruppert, K.M., Kline, R.J., Rahman, M.S., 2019. Past, present, and future perspectives of environmental DNA (eDNA) metabarcoding: a systematic review in methods, monitoring, and applications of global eDNA. Glob. Ecol. Conserv. 17, e00547.

Russell, S.J., Norvig, P., 2009. Artificial Intelligence: A Modern Approach, third ed. Prentice Hall, Upper Saddle River, NJ, USA.

Ryder, O.A., Onuma, M., 2018. Viable cell culture banking for biodiversity characterization and conservation. Annu. Rev. Anim. Biosci. 6, 83–98.

Rydning, J., Reinsel, D., Gantz, J., 2018. The Digitization of the World from Edge to Core. International Data Corporation, 16, Framingham. IDC White Paper – #US44413318. Seagate.

Salvaris, M., Dean, D., Tok, W.H., 2018. Deep learning with azure. In: Building and Deploying Artificial Intelligence Solutions on Microsoft AI Platform. Apress.

Sampieri, B.R., Vieira, P.E., Teixeira, M.A., Seixas, V.C., Pagliosa, P.R., Amaral, A.C.Z., Costa, F.O., 2021. Molecular diversity within the genus Laeonereis (Annelida, Nereididae) along the West Atlantic coast: paving the way for integrative taxonomy. PeerJ 9, e11364.

Sanger, F., Nicklen, S., Coulson, A.R., 1977. DNA sequencing with chain-terminating inhibitors. Proc. Natl. Acad. Sci. 74 (12), 5463–5467.

Savini, A., Vertino, A., Marchese, F., Beuck, L., Freiwald, A., 2014. Mapping cold-water coral habitats at different scales within the Northern Ionian Sea (Central Mediterranean): an assessment of coral coverage and associated vulnerability. PLoS One 9 (1), e87108.

Schmidt, E.W., 2008. Trading molecules and tracking targets in symbiotic interactions. Nat. Chem. Biol. 4 (8), 466–473.

Schorn, M.A., Alanjary, M.M., Aguinaldo, K., Korobeynikov, A., Podell, S., Patin, N., et al., 2016. Sequencing rare marine actinomycete genomes reveals high density of unique natural product biosynthetic gene clusters. Microbiology 162 (12), 2075.

Secretariat of the Convention on Biological Diversity, 2011. The Strategic Plan for Biodiversity 2011-2020 and the Aichi Biodiversity Targets. https://www.cbd.int/decision/cop/?id=12268. (Accessed 31 August 2023).

Shaffer, H.B., Toffelmier, E., Corbett-Detig, R.B., Escalona, M., Erickson, B., Fiedler, P., et al., 2022. Landscape genomics to enable conservation actions: the California conservation genomics project. J. Hered. 113 (6), 577–588.

Shakya, M., Lo, C.C., Chain, P.S., 2019. Advances and challenges in metatranscriptomic analysis. Front. Genet. 10, 904.

Sharma, R., 2001. Indian deep-sea environmental experiment (INDEX): an appraisal. Deep-Sea Res. II 48, 3295–3307.

Shih, P.J., Saadat, H., Parameswaran, S., Gamaarachchi, H., 2023. Efficient real-time selective genome sequencing on resource-constrained devices. GigaScience 12, giad046.

Shirayama, Y., 1999. Biological results of the JET project: an overview. In: Proceedings of the III Ocean Mining Symposium, Goa, India, November 8–10, pp. 185–188.

Shirayama, Y., Fukushima, T., Matsui, T., Kuboki, E., 2001. The responses of deepsea benthic organisms to experimental removal of the surface sediment. In: Proceedings of the IV Ocean Mining Symposium, Szczecin, Poland, September 23–27, pp. 77–81.

Shiva, V., 2007. Bioprospecting as sophisticated biopiracy. Signs J. Women Cult. Soc. 32 (2), 307–313.

Shivram, H., Cress, B.F., Knott, G.J., Doudna, J.A., 2021. Controlling and enhancing CRISPR systems. Nat. Chem. Biol. 17 (1), 10–19.

Shwab, E.K., Bok, J.W., Tribus, M., Galehr, J., Graessle, S., Keller, N.P., 2007. Histone deacetylase activity regulates chemical diversity in *Aspergillus*. Eukaryot. Cell 6, 1656–1664.

Sigwart, J.D., Wicksten, M.K., Jackson, M.G., Herrera, S., 2019. Deep-sea video technology tracks a monoplacophoran to the end of its trail (Mollusca, Tryblidia). Mar. Biodivers. 49 (2), 825–832.

Sigwart, J.D., Blasiak, R., Jaspars, M., Jouffray, J.B., Tasdemir, D., 2021. Unlocking the potential of marine biodiscovery. Nat. Prod. Rep. 38 (7), 1235–1242.

Simon-Lledó, E., Bett, B.J., Huvenne, V.A., Köser, K., Schoening, T., Greinert, J., Jones, D.O., 2019. Biological effects 26 years after simulated deep-sea mining. Sci. Rep. 9 (1), 1–13.

Smith, T.E., Pond, C.D., Pierce, E., Harmer, Z.P., Kwan, J., Zachariah, M.M., et al., 2018. Accessing chemical diversity from the uncultivated symbionts of small marine animals. Nat. Chem. Biol. 14 (2), 179–185.

Sodergren, E., Weinstock, G.M., Davidson, E.H., Cameron, R.A., Gibbs, R.A., Angerer, R.C., et al., 2006. The genome of the sea urchin *Strongylocentrotus purpuratus*. Science 314 (5801), 941–952.

Sporn, M.B., 1996. The war on cancer. Lancet 347 (9012), 1377–1381.

Staunton, J., Weissman, K.J., 2001. Polyketide biosynthesis: a millennium review. Nat. Prod. Rep. 18 (4), 380–416.

Steiner, C.C., Putnam, A.S., Hoeck, P.E., Ryder, O.A., 2013. Conservation genomics of threatened animal species. Annu. Rev. Anim. Biosci. 1 (1), 261–281.

Stephens, Z.D., Lee, S.Y., Faghri, F., Campbell, R.H., Zhai, C., Efron, M.J., et al., 2015. Big data: astronomical or genomical? PLoS Biol. 13 (7), e1002195.

Stincone, P., Brandelli, A., 2020. Marine bacteria as source of antimicrobial compounds. Crit. Rev. Biotechnol. 40 (3), 306–319.

Stone, M.J., Williams, D.H., 1992. On the evolution of functional secondary metabolites (natural products). Mol. Microbiol. 6 (1), 29–34.

Sueur, J., Farina, A., 2015. Ecoacoustics: the ecological investigation and interpretation of environmental sound. Biosemiotics 8, 493–502.

Sturtevant, A.H., 1913. The linear arrangement of six sex? Linked factors in Drosophila, as shown by their mode of association. J. Exp. Zool. 14 (1), 43–59.

Sunagawa, S., Coelho, L.P., Chaffron, S., Kultima, J.R., et al., 2015. Ocean plankton. Structure and function of the global ocean microbiome. Science 348, 1261359.

Suzuki, M.M., Bird, A., 2008. DNA methylation landscapes: provocative insights from epigenomics. Nat. Rev. Genet. 9 (6), 465–476.

Tam, V., Patel, N., Turcotte, M., Bossé, Y., Paré, G., Meyre, D., 2019. Benefits and limitations of genome-wide association studies. Nat. Rev. Genet. 20 (8), 467–484.

Taylor, M.L., Roterman, C.N., 2017. Invertebrate population genetics across Earth's largest habitat: The deep-sea floor. Mol. Ecol. 26 (19), 4872–4896.

Thaler, A.D., Amon, D., 2019. 262 voyages beneath the sea: a global assessment of macro-and megafaunal biodiversity and research effort at deep-sea hydrothermal vents. PeerJ 7, e7397.

Thiel, H., Schriever, G., 1989. Cruise Report DISCOL 1, SONNEFcruise 61, Balboa/PanamaFCallao/Peru with Contribution by C. Borowski, C. Bussau, D. Hansen, J. Melles, J. Post, K. Steinkamp, K. Watson. Berichte Aus Dem Zentrum f.Ur Meeres- Und Klimaforschungder. vol. 3 Universität Hamburg. 75 pp.

Thiel, H., 1991. From MESEDA to DISCOL: a new approach to deep-sea mining risk assessments. Mar. Min. 10, 369–386.

Thompson, L.R., Sanders, J.G., McDonald, D., Amir, A., Ladau, J., Locey, K.J., et al., 2017. A communal catalogue reveals Earth's multiscale microbial diversity. Nature 551 (7681), 457–463.

Thomson, C.W., Murray, J., Nares, G.S., Thomson, F.T., 1889. Report on the Scientific Results of the Voyage of HMS Challenger During the Years 1873-76 Under the Command of Captain George S. Nares... and the Late Captain Frank Tourle Thomson, R.N., vol. 2. HM Stationery Office.

Thosteson, E.D., Widder, E.A., Cimaglia, C.A., Taylor, J.W., Burns, B.C., Paglen, K.J., 2009. New technology for ecosystem-based management: marine monitoring with the ORCA Kilroy network. In: OCEANS 2009-EUROPE 1–7, IEEE, 2009. https://ieeexplore.ieee.org/document/5278229.

Thrush, S.F., Lohrer, A.M., 2012. Why bother going outside: the role of observational studies in understanding biodiversity-ecosystem function relationships. In: Biodiversity, Ecosystem Functioning, & Human Wellbeing an Ecological and Econimic Perspective. Oxford University Press, Oxford, pp. 200–214.

Tanhua, T., McCurdy, A., Fischer, A., Appeltans, W., Bax, N., Currie, K., et al., 2019. What we have learned from the framework for ocean observing: evolution of the global ocean observing system. Front. Mar. Sci. 6, 471.

Tollefson, J., Gibney, E., 2022. Nuclear-fusion lab achieves 'ignition': what does it mean? Nature 612 (7941), 597–598.

Treibergs, K.A., Giribet, G., 2020. Differential gene expression between polymorphic zooids of the marine bryozoan Bugulina stolonifera. Genes Genomes Genet. 10 (10), 3843–3857.

Turner, N.R., Parkerton, T.F., Renegar, D.A., 2021. Toxicity of two representative petroleum hydrocarbons, toluene and phenanthrene, to five Atlantic coral species. Mar. Pollut. Bull. 169, 112560.

Valiela, I., Valiela, I., 1995. Marine ecological processes. vol. 686 Springer, New York.

van der Ven, R.M., Heynderickx, H., Kochzius, M., 2021. Differences in genetic diversity and divergence between brooding and broadcast spawning corals across two spatial scales in the Coral Triangle region. Mar. Biol. 168, 1–15.

van Oppen, M.J., Coleman, M.A., 2022. Advancing the protection of marine life through genomics. PLoS Biol. 20 (10), e3001801.

van Oppen, M.J., Oakeshott, J.G., 2020. A breakthrough in understanding the molecular basis of coral heat tolerance. Proc. Natl. Acad. Sci. 117 (46), 28546–28548.

Vaughan, D.E. (Ed.), 2021. Active Coral Restoration: Techniques for a Changing Planet. J. Ross Publishing.

Vendrami, D.L., Telesca, L., Weigand, H., Weiss, M., Fawcett, K., Lehman, K., et al., 2017. RAD sequencing resolves fine-scale population structure in a benthic invertebrate: implications for understanding phenotypic plasticity. R. Soc. Open Sci. 4 (2), 160548.

Ventura, P., et al., 2018. Cnidarian primary cell culture as a tool to investigate the effect of thermal stress at cellular level. Mar. Biotechnol. 20, 144–154.

Verne, J., 1871. Vingt Mille Lieues Sous les Mers Twenty Thousand Leagues Under the Sea. SeaWolf Press (February 16, 2019).

Vieira, M.L.C., Santini, L., Diniz, A.L., Munhoz, C.D.F., 2016. Microsatellite markers: what they mean and why they are so useful. Genet. Mol. Biol. 39, 312–328.

Vieira, H., Leal, M.C., Calado, R., 2020. Fifty shades of blue: how blue biotechnology is shaping the bioeconomy. Trends Biotechnol. 38 (9), 940–943.

Vonnahme, T.R., Molari, M., Janssen, F., Wenzhöfer, F., Haeckel, M., Titschack, J., Boetius, A., 2020. Effects of a deep-sea mining experiment on seafloor microbial communities and functions after 26 years. Sci. Adv. 6 (18), eaaz5922.

Vrijenhoek, R.C., 2009. Cryptic species, phenotypic plasticity, and complex life histories: assessing deep-sea faunal diversity with molecular markers. Deep-Sea Res. II Top. Stud. Oceanogr. 56 (19–20), 1713–1723.

Walker, B.K., 2012. Spatial analyses of benthic habitats to define coral reef ecosystem regions and potential biogeographic boundaries along a latitudinal gradient. PLoS One 7 (1), e30466.

Walker, B.K., Riegl, B., Dodge, R.E., 2008. Mapping coral reef habitats in Southeast Florida using a combined technique approach. J. Coast. Res. 24 (5), 1138–1150.

Walker, B.K., Jordan, L.K., Spieler, R.E., 2009. Relationship of reef fish assemblages and topographic complexity on southeastern Florida coral reef habitats. J. Coast. Res. 10053, 39–48.

Walker, B.K., Messing, C., Ash, J., Brooke, S., Reed, J.K., Farrington, S., 2021. Regionalization of benthic hard-bottom communities across the Pourtalès Terrace, Florida. Deep Sea Res. Part I Oceanogr. Res. Pap. 172, 103514. https://doi.org/10.1016/j.dsr.2021.103514.

Wang, K., Shen, Y., Yang, Y., Gan, X., Liu, G., Hu, K., et al., 2019. Morphology and genome of a snailfish from the Mariana Trench provide insights into deep-sea adaptation. Nat. Ecol. Evol. 3 (5), 823–833.

Watson, J.D., Bell, T., Baker, S., Gann, A., Levine, M., Losick, R., 2013. The Molecular Biology of the Gene, seventh ed. Pearson.

130 **Chapter 2** Multiple approaches to understanding the benthos

Widder, E.A., 1992. Mixed light imaging system for recording bioluminescence behaviours. J. Mar. Biol. Assoc. U. K. 72 (1), 131–138.

Widder, E., 2021. Below the Edge of Darkness: A Memoir of Exploring Light and Life in the Deep Sea. Random House.

Widder, E.A., 2007. Sly eye for the shy guy: peering into the depths with new sensors. Oceanography 20 (4), 46–51.

Williams, S.L., Ambo-Rappe, R., Sur, C., Abbott, J.M., Limbong, S.R., 2017. Species richness accelerates marine ecosystem restoration in the coral triangle. Proc. Natl. Acad. Sci. 114 (45), 11986–11991.

Wilson, E.O., 1987. The little things that run the world (the importance and conservation of invertebrates). Conserv. Biol. 1 (4), 344–456.

Winnikoff, J.R., Haddock, S.H., Budin, I., 2021. Depth-and temperature-specific fatty acid adaptations in ctenophores from extreme habitats. J. Exp. Biol. 224 (21), jeb242800.

Yancey, P.H., 2020. Cellular responses in marine animals to hydrostatic pressure. J. Exp. Zool. A Ecol. Integr. Physiol. 333 (6), 398–420.

Yang, Z., He, J., Wei, X., Ju, J., Ma, J., 2020. Exploration and genome mining of natural products from marine Streptomyces. Appl. Microbiol. Biotechnol. 104 (1), 67–76.

Yoshioka, Y., Yamashita, H., Suzuki, G., Zayasu, Y., Tada, I., Kanda, M., et al., 2021. Whole-genome transcriptome analyses of native symbionts reveal host coral genomic novelties for establishing coral–algae symbioses. Genome Biol. Evol. 13 (1), evaa240.

Zarrella, I., Herten, K., Maes, G.E., Tai, S., Yang, M., Seuntjens, E., et al., 2019. The survey and reference assisted assembly of the *Octopus vulgaris* genome. Sci. Data 6 (1), 1–8.

Zhang, L., Hashimoto, T., Qin, B., Hashimoto, J., Kozone, I., Kawahara, T., et al., 2017. Characterization of giant modular PKSs provides insight into genetic mechanism for structural diversification of aminopolyol polyketides. Angew. Chem. Int. Ed. 56 (7), 1740–1745.

Zuckerkandl, E., Pauling, L., 1965. Evolutionary divergence and convergence in proteins. In: Evolving genes and proteins. Academic Press, pp. 97–166.

CHAPTER

Diversity hotspots on the benthos— Case studies highlight hidden treasures

3

For those of us focused on the seabed, we have probably heard the whimsical idea, "if we could only drain the oceans for a day," e.g., to see what treasures may lie on the seafloor, some from antiquity, sunken ships, raw materials and biodiversity. More realistically, some people can dive and visit the seabed in person via SCUBA or by ROVs discussed in Chapter 2. There are few experiences that match being at the bottom of the ocean, hovering above a healthy coral reef, and looking up through the clear water to the surface, which can appear like a ceiling when visibility reaches over 100 ft. I have experienced unobstructed views like these a few times, but opine it applies to several reef sites that have yet escaped some of the threats described in Chapter 4. This is also why it can be a challenge to elevate any one habitat here—e.g., San Andreas, Curacao, Maldives, or the Red Sea.

This chapter puts a spotlight on a few well-known biodiversity treasures from benthic habitats. To choose a biodiversity "hotspot," it is necesssary to parse the criteria. We can always turn to the literature, which also partially entails the various efforts to classify oceanic regions for their diversity, but may never be completed. Moreover, although a biodiversity "hotspot" concept may appear straightforward, several definitions do exist, and so we may consider which ones to apply (Myers, 1990; Hughes et al., 2002; Reid, 1998). In this chapter, I plan to view biodiversity on multiple wide-ranging scales, although the primary guidelines for prioritizing are richness, endemism, and threat levels.

Because of our basic ignorance of fully understanding many benthic habitats, picking hotspots for this book turns out to be a nearly random exercise. The closer we investigate a basin or region for its biodiversity, benthic hotspots appear nearly everywhere. This could be akin to "over-splitting" a phylogenetic reconstruction such that every difference becomes a rationale for bifurcation. Nonetheless, multiple studies of patchiness can also point us toward specific locales with high biodiversity density based on geomorphological features, which then lead to habitat variation: the Coral Triangle area in the Western Pacific Ocean, the Caribbean, the Mediterranean, the Red Sea, areas in the Arctic Ocean (Greenland Sea), and areas of the western South Pacific, including the northeast Australian shelf (Fischer et al., 2019). However, these are by no means exclusive. A comprehensive and relatively recent overview of the many marine habitats we omit here is included in three volumes of *World Seas: An Environmental Evaluation* (Hamel, 2018; Sheppard, 2018). Many shallow biodiverse reefs have been characterized at least cursorily. The biggest discoveries of biodiversity await us at greater oceanic depths.

Another basis for focusing on specific areas or biodiversity hotspots will rely on the relative high endemism of the site, defined as holding species or taxa that are unique and *exclusively* found in that particular place. Endemism for mobile, pelagic taxa is more difficult to establish than that for sessile taxa.

Assessments and Conservation of Biological Diversity From Coral Reefs to the Deep Sea. https://doi.org/10.1016/B978-0-12-824112-7.00003-0
Copyright © 2024 Elsevier Inc. All rights reserved.

131

132   **Chapter 3** Diversity hotspots on the benthos

We have mentioned a few long-standing efforts to create biogeographical classifications of the oceans. For example, UNESCO proposed a classification with 38 provinces (14 bathyal, 14 abyssal, and 10 hadal) with over 10 hydrothermal vent provinces (UNESCO, 2009). Principles governing the classification were as follows: (1) consider the pelagic and benthic environments separately; (2) biogeographic classifications would not be based upon distinctive areas or focal species; (3) taxonomic identities would drive the classification scheme and could be separated from ecological biomes; (4) emphasis would be on "generally recognizable communities of species"; (5) differentiate the various ecological structures (pelagic and benthic), which could define the province; and 6) the classification should be hierarchical. This effort was followed by Watling et al. (2013), who continued with the idea of the classification and examined "detailed hydrographic and organic-matter flux data to hypothesize the position and extent of deep-sea provinces and some of the proposed biogeographic provinces using studies of selected benthic species." It makes sense that species distributions of benthic species best delineate biogeographic regions (Briggs and Bowen, 2013). Indeed, a seminal analyses of biogeographic realms across the globe concluded that "species-rich benthic taxa (as opposed to pelagic) such as arthropods and mollusks contributed most to endemicity" (Costello et al., 2017). The latter analysis counted 18 continental-shelf and 12 offshore deep-sea realms with many faunal dsitributions following depth and latitude.

Some organismal populations can be so replete, with the number of individuals appearing overwhelmingly numerous that *infinity* seems perceptible (Aczel, 2001). Then, we realize that biology has its limits. Nonetheless, can envision iconic examples such as the seemingly endless herds of wildebeest on Serengeti savannahs, monarch butterflies converging on their final mating grounds, fiddler crabs on a pristine intertidal, or box jellyfish floating during a full bloom. Moreover, if we could train a video camera to view most of the activity occurring at microscopic levels ($<1\,mm$), we would again be amazed and easily comprehend that unicellular life on this planet is effectively limitless (and where most of the planet's genetic diversity resides). Such novel in situ views with a yet-to-be invented field microscope would provide topics for nature documentaries (see Chapter 5). When considering the meiofauna mentioned earlier, some benthic habitats may be viewed in the same light. In some bathyal locales, collecting at any random coordinate will likely yield at least one or two new species. This could be a comforting thought if the predictions of the Anthropocene continue to unfold in shallower regions. For now, we can likely acknowledge that certain habitats appear to have a higher number of total species (high species richness) than others. Similarly, the possible combinations and sequences of the genetic code, derived from only four basic biochemical letters, can also become limitless when 4^n represents a possible gene sequence of length n.

For this chapter, we define a "diversity hotspot" as a geographic area, which boasts a rich and varied assortment of species, particularly those that are endemic to the region. These hotspots often occur in tropical locations and display high species richness. Unfortunately, many diversity hotspots face peril due to human activities such as habitat destruction and climate change and are therefore viewed as crucial areas for conservation efforts. With this definition in mind, we can now view benthic habitats and their inherent biodiversity from this hotspot perspective.

The choice of hotspots made in this book may be unabashedly subjective. It is partially based on my own personal and limited experience and knowledge of benthic habitats. Furthermore, marking these sites can be itself an exercise in preordained randomness. For most likely, based on both our collective lack of a true consensus and knowledge of the largest habitat, one could throw a dart at an oceanic map, hit any marine locale, outfit an expedition, and find several novel species. This is the nature of the beast.

Another concept that helps mark these diversity hotspots are specific "keystone" (sometimes referred to as "foundation") species that are present. Keystones are defined as species or taxa that perform many vital functions for the community and have a disproportionately large effect on its native environment relative to its abundance and other resident species (Dayton, 1972; Slattery and Lesser, 2021). Quantitative and semiqualitative multispecies trophic network analysis can help determine the ecological roles of specific species (Ortiz et al., 2013). We keep in mind that the origin and maturation of habitats and ecosystems require a community of diverse interactions and are not built upon just one or two species, even if they provide a primary function (Ulanowicz, 2012).

We can contrast species diversity between shallow and deep benthic sites, or compare habitats found in tropical, subtropical, and arctic latitudes. Many apply either latitudinal or bathymetric parameters for measuring species richness gradients (Rex, 1973; Danovaro et al., 2010; Valentine and Jablonski, 2015; Chaudhary et al., 2017). Generally, species richness declines from the Equator to higher latitudes, peaking at mid latitudes at the edge of tropics due to increased temperature gradient stability and bathymetrically, species richness declines from shallow water to deep seas, with a small peak near the bathyal zone (e.g., gastropods and nematodes) due to increased environmental stability (Brown, 2014; Saeedi et al., 2019). Moreover, the regular monitoring and comparative analyses of reefs have also been documented by experts, such as the Global Coral Reef Monitoring Network (Wilkinson et al., 2016; Obura et al., 2019). For this volume, I have taken into account space and time limitations for this topic. That is, we cannot provide full or equal coverage to every deserving or unique species–rich benthic habitat possible. However, I can direct attention to locales that have accumulated a large number of peer-reviewed studies. Diverse shallow coral reefs occur around the world. For example, the habitats in the Red Sea, Maldives, Chagas, Brazil, Okinawa, Belize, Honduras, Colombia, and more locales could easily qualify for spotlighting here. Many hotspots are also in varying states of health (Riegl, 2003; Fine et al., 2019; Schleyer et al., 2018). I also refer to some other hotspots in the context of other special topics in this volume.

Continuing a Census of Marine Life

As a context for choosing hotspots, the Census of Marine Life (CML) could be used and represent an ambitious project designed to connect a global network of researchers to assess and explain the diversity, distribution, and abundance of life in the oceans. The CML began in 2000 and ran until 2010 and was funded with $78 million from the Sloan Foundation and $550 million from non-Sloan sources around the world, including national governments, international organizations, and maritime industries. There were 14 specific benchmark projects, and researchers from 80 nations used tools such as deep-sea submersibles and DNA analysis. The project resulted in the discovery of thousands of new species and a greater understanding of ocean ecosystems. A legacy of this census was the development of the Ocean Biogeographical Information System (OBIS) containing tens of millions of records on hundreds of thousands of marine species. The 14 projects and areas encompassed by the CML were as follows:

- Arctic Ocean: ArcOD (Arctic Ocean Diversity)
- Antarctic Ocean: CAML (Census of Antarctic Marine Life)
- Mid-Ocean Ridges: MAR-ECO (Mid-Atlantic Ridge Ecosystem Project)

134 **Chapter 3** Diversity hotspots on the benthos

- Vents and Seeps: ChEss (Biogeography of Deep-Water Chemosynthetic Ecosystems)
- Abyssal Plains: CeDAMar (Census of Diversity of Abyssal Marine Life)
- Seamounts: CenSeam (Global Census of Marine Life on Seamounts)
- Continental Margins: COMARGE (Continental Margin Ecosystems)
- Continental Shelves: POST (Pacific Ocean Shelf Tracking Project)
- Near Shore: NaGISA (Natural Geography in Shore Areas)
- Coral Reefs: CReefs (Census of Coral Reefs)
- Regional Ecosystems: GoMA (Gulf of Maine Program)
- Microbes: ICoMM (International Census of Marine Microbes)
- Zooplankton: CMarZ (Census of Marine Zooplankton)
- Top Predators: TOPP (Tagging of Pacific Predators)

The CML has published multiple reports on their findings, including the "Census of Marine Life: A Decade of Discovery," which summarizes this project's overall results (https://sloan.org/programs/completed-programs/census-of-marine-life) (Brasier et al., 2016; Poore et al., 2015; Ransome et al., 2017; Stępień et al., 2019). However valuable this information is, it probably has a shelf life, and numbers will be updated after each publication. Many biologists realize that the figures of ~1.8 to 2.0 million eukaryotic species only cover those formally classified, while the likely number of extant eukaryotic species could reach up to 12–15 million, including 8.1 million plants and animals (Roskov et al., 2013; Lewin et al., 2022). We typically take a census every decade for our own species, country by country. However, since we are concerned about many aspects of biodiversity, perhaps society should consider having a regular CML for the benthos and our natural habitats.

Cold water extremes at Arctic, Antarctic, and deep sea habitats

One nearly universal constant about the benthos is the low temperature at abyssal depths and lower found across most oceans. (High pressure is another one). Except for the very rare submarine volcano or hot thermal vents dotting the ocean bottom, temperatures in the aphotic zone average about 6°C. Therefore, the relatively constant temperatures at deeper waters would seem to be one primary bridge between habitats, allowing organisms the potential to connect or at least converge across long distances. And yet, as discussed earlier regarding genetic connectivity, benthic organisms do run into barriers. We can find distinct structures of populations that are species-dependent, and this can lead to the eventual divergence of species. This situation can be applied to the polar extremes as well. The Arctic and Antarctic benthic habitats differ significantly in topographic structure, if not temperature and pressure. Dayton et al. (1994) summarized the differences of the two poles: "The Arctic has broad shallow continental shelves with seasonally fluctuating physical conditions and a massive fresh water impact in the northern coastal zones, low seasonality of pack ice and little vertical mixing. In contrast, the Antarctic has over twice the oceanic surface area, deep narrow shelves, and, except for ice cover, a relatively stable physical environment with very little terrestrial input." Yet, the fauna of the two extremes appears very different. This could be due to the semi-isolation of the southernmost continent by the Antarctic Circumpolar Current.

To start off with the top of the list of the CML, and one of the least known of benthic habitats of major oceans is the Arctic seabed. This zone has a general accessibility, which could be likened to extraterrestrial solar system destinations. In other words, traveling to this destination is dangerous and costly. As is well known, much of the Arctic typically has large ice coverage most of the year, making accessibility most facile only during a few summer months. Although global warming has likely helped shrink the extent of arctic sea ice sheets over the last several years (Stroeve and Notz, 2018), most expeditions require costly voyages in harsh environmental conditions, far from support networks. Therefore, expeditions are not common or trivial. In 2004, a joint Russian-American program called RUSALCA (Russian-American Long Term Census of the Arctic) carried out 40-day voyage into the Bering Strait and northward to the Pacific side of the Arctic Ocean. The Russian Academy of Sciences and the Russian Federal Service for Hydrometeorology and Environmental Monitoring haul carried out benthic trawls (https://oceanexplorer.noaa.gov/explorations/09arctic/welcome.html).

Exploring the biodiversity of the polar regions would however almost guarantee positive gains on the initial investment. Fischer et al. (2019) indicated that the Arctic was likely to hold very high levels of biodiversity due to concomitant high diversity in geomorphology and benthic habitats and following the "habitat heterogeneity" hypothesis that would generate relatively greater biodiversity. At the other pole, Antarctic benthic habitats also host interesting biodiversity and high species richness. This would include the symbionts of resident invertebrates (Giudice et al., 2019). Aronson and Blake (2001) were puzzled by and likened Antarctic shallow communities to Paleozoic marine communities and modern deep-sea communities, which are depauperate in large predators. They found dense Antarctic ophiuroid and crinoid populations. Multiple surveys of the area indicate the impact of important abiotic factors such as ice coverage and moving iceberg disturbance (Thrush et al., 2006) and seasonal changes (Caputi et al., 2020). Studies of benthic areas in the Ross and Weddell seas are more recent and fewer than shallow, lower latitude coastal areas but have revealed significant biodiversity nonetheless (Linse et al., 2006; Brandt et al., 2007; Gutt, 1988, 2008, 2013; Teschke et al., 2019; Teschke and Brey, 2019a, b, 2020). For example, Teschke et al. (2020) have compiled a large amount of environmental and ecological data from various expeditions in the Weddell sea that describe and support high biodiversity counts. Marini et al. (2022) have also recently described cutting-edge approaches for noninvasive long-term monitoring of the Antarctic benthos.

Multiple entities from Canada, NOAA, or USGS have carried out expeditions, mostly to map the continental margins, and generated several benthic surveys (https://oer.hpc.msstate.edu/ECS/arctic/HLY1202_cruise_report.pdf). Thus, overall arctic habitats easily remain a prime focus area for future biodiversity assessments. To fill in some gaps, a new database based on trait-based approaches—life history, morphological, physiological and behavioral characteristics (i.e., traits) of species found in the Arctic (Degen and Faulwetter, 2019).

To geographically move further south from the Arctic sea, through the Bering strait, we can find ourselves in slightly more hospitable waters and regions. Indeed, in 2023, NOAA plans to send the *Okeanos Explorer* out to sea for 160 days to explore and map the Exclusive Economic Zone (EEZ) waters off the US west coast and Alaska. As the largest state of the United States, Alaska holds 70% of the nation's continental shelf habitat, while its EEZ encompasses about 1 million square nautical miles, of which 69% remains unmapped. Seascape Alaska project intends to fill in the gaps (Ocean Science and Technology Subcommittee Policy 2020). An interesting cold water region worth mentioning for its benthic biodiversity is the Aleutian Islands. Multiple surveys of shallow and deep water benthos have been carried out over the years (Krieger and Wing, 2002). For example, a 2006 and 2007 benthic survey as part of the Alaska Monitoring and Assessment Program showed communities (<20) as dependent on crustose corraline algae (Chenelot et al., 2011). Stone and Shotwell (2007)

136 **Chapter 3** Diversity hotspots on the benthos

characterized deep coral ecosystems and found high diversity of soft and stony corals: One hundred and forty-one unique coral taxa include 11 species of stony corals, 14 species of black corals, 15 species of true soft corals (including six species of stoloniferans), 63 species of gorgonians, 10 species of sea pens, and 28 species of stylasterids. Saeedi and Brandt (2020) have compiled a more recent survey of the Pacific southwest benthos. As seen above, exploring the benthos at higher latitudes of the arctic brings challenges that have yet to be fully overcome. Moreover, abyssal and hadal zones have similar temperatures and accessibility issues. Yet, more than likely many abyssal and hadal zone benthic habitats will also provide diversity hotspots (Jamieson, 2015).

Profiles of multiple coral reefs

Coral reefs are one of most diverse and productive oligotrophic ecosystems, largely due to the complex habitat provided by scleractinian (stony or hard) corals, ostensibly the key ecosystem engineers of the habitat. The hard corals can only build permanent reefs through calcification and a symbiotic partnership with photosynthetic dinoflagellate algae, mostly in the family Symbiodiniaceae. Yet, like many hyper-diverse tropical ecosystems, stony coral and their reefs are threatened by a broad range of anthropogenic stressors (e.g., overfishing, eutrophication, and sedimentation; Carpenter et al., 2008; Barlow et al., 2018). Of greatest concern is the increasing frequency of marine heatwaves that cause mass coral bleaching events, threatening coral reefs globally (Hughes et al., 2018). Impacts from coral bleaching events can be long-lasting (e.g., Burt et al., 2011) and may alter species composition (e.g., Hughes et al., 2018), recruitment (Burt and Bauman, 2020), and calcification rates (Perry et al., 2017) resulting in the loss of key ecological functions (Alvarez-Filip et al., 2013). Notably, climate-induced bleaching events coupled with local anthropogenic stressors are already altering the structure and function of coral reefs (Bellwood et al., 2019). Sustained and ongoing habitat degradation on coral reef ecosystems is leading to the overall loss of biodiversity (e.g., reef fishes, invertebrates) and erosion of important ecosystem services. Although defining keystone species can be complex and debatable among biologists for specific habitats, reef-building scleractinian (stony or hard) corals represent obvious foundations for shallow reefs, since their calcium carbonate skeletons provide most of the primary structure. However, a coral reef remains a *community* of diverse organisms with stony corals providing the substrate for other species to anchor upon. Other species can still play foundation roles. For example, damselfish live on reef habitats and can provide nitrogen for sessile organisms. Sometimes, keystone species' major functions are not fully recognized until they are nearly lost, such as the long spined sea urchin *Diadema antillarum* (Lessios, 1988). Of greatest concern is whether ongoing coral or comparable keystone species diminution will eventually lead to the extinction of other species or the whole ecosystem altogether.

In spite of many years of research and advances in coral reef biology, relatively few models can forecast important factors that influence reef biodiversity (Warwick and Clarke, 2001). One relevant model was elaborated by Van Woesik (2000) and can be contrasted with the Littler's relative dominance model based on top-down predation factors (Littler et al., 2006). Van Woesik raised provocative issues regarding biodiversity: "Why are there so many coral species in some places? We can posit that this question should be the crux of many future studies focused on conserving and understanding the basis of reef biodiversity. The answer may become more apparent if we revise the question as 'Why do some places support so few coral species?' In benign sites a coral reef may support a high diversity of coral species by 'default', however where recruitment is not limiting the selection against some species in harsh environments may be the process regulating local diversity." Van Woesik's model also considered the extinction of species in the ecosystem, which provides a guidepost for further research. Another model based on a state transition Markov chain focused on the dynamics of algal and coral excavating sponge *Cliona delitrix* overgrowth followed by successive mass bleaching events (Chaves-Fonnegra et al., 2017). The transition states were defined as healthy coral (HC), deteriorated coral (IC), dead coral covered with macroalgae and other organisms (AL), coral with sponge (CS), and sponge (S). Using 217 coral colonies monitored over 10 years (2000–2010) from the Florida reef tract (FRT—see more below) off of Broward county Florida, predictions over 100 years indicated a shift to algae-coral-dominated reefs into algae-sponge-dominated reefs.

With this background, we can now highlight and visit a few specific diversity hotspots, which will include reefs of various types and latitudes. Two types of reef hotspot will be compared below. The reefs of the Western Atlantic (WA) have been characterized as highly stressed and threatened over the last few decades (Gardner et al., 2003; Perry et al., 2013). Many of these WA reefs lie proximal to large urban areas and have fewer endemic coral scleractinian species than western Pacific reefs, such as those on the GBR and coral triangle. Because of multiple environmental stresses, which will be discussed in the next chapter, such as rising sea surface temperatures, the 1983 mass die-off of *D. antillarum*, hurricanes, and ocean acidification, WA and many Caribbean reefs have suffered a > 50% loss of coral coverage since the 1970s (Alves et al., 2022). Both WA and Pacific reef areas have however commanded a great deal of research focus, and so a comparison of their relative standing at this time could be informative.

A resident's view of the Florida Reef Tract

Live stony coral growth occurs off the mainland coast of Southeast Florida from Miami-Dade County north to central Palm Beach County. This is also a highly urbanized area, which is home to more than 6 million residents. The reef system in this area is known as the Florida Reef Tract (FRT) and encompasses the Southeast Florida Coral Reef Ecosystem Conservation Area (Coral ECA), which ranges from the northern boundary of Biscayne National Park in Miami-Dade County to the St. Lucie Inlet in Martin County (approximately 170 km) and within 3 km off the mainland Atlantic coast of Florida. The Coral ECA reefs continues further south into the Florida Keys National Marine Sanctuary. Unlike other Caribbean reefs, the FRT reef communities attenuate with latitude as they transition into a temperate climate, making this system unique and valuable for the continental United States.

The geomorphology, climatology, oceanographic process, biogeography, and benthic biodiversity of the FRT was comprehensively reviewed by Banks et al. (2008). The FRT arose in the late Pleistocene epoch with the outer reef consisting mostly of relic acroporids (Macintyre and Milliman, 1970; Lighty, 1977; Lighty et al., 1978; Macintyre, 1988). Unfortunately, Caribbean reefs have been known to be in decline for the last two decades (Perry et al., 2013), and patterns of decline appear unprecedented in the fossil record within the past few millennia (Aronson et al., 2002). From Palm Beach County to Martin County, the reef system comprises limestone ridges and terraces colonized by reef biota, which help delineate ecologically relevant regions (Walker and Gilliam, 2013). In 2014, the FRT began experiencing widespread coral mortality from stony coral tissue loss disease (SCTLD). This will be discussed more in detail in the next chapter. Nonetheless, compared with other extensive barrier reef systems in

138 Chapter 3 Diversity hotspots on the benthos

the WA, such as Belize or in the Bahamas (Alves et al., 2022), sustained existence of the FRT requires the vigilance of a large cadre of dedicated scientists and citizens concerned about coral reef conservation and protection.

Typical of several western Atlantic reefs, the South Florida Coral ECA consists of a series of linear reef complexes (referred to as reefs, reef tracts, or reef terraces) running parallel to the shore (Moyer et al., 2003; Banks et al., 2008; Walker et al., 2008) (Fig. 3.1). Coral reef ecosystem biogeographic subregions were defined extensively by Walker (2012). For example, the near reef complex (NRC) occurs at 3–5 m depth, which is relatively flat, low-relief, solid carbonate rock. The Inner Reef (also referred to as the "First Reef") crests in 3–8 m depths. The Middle Reef ("Second Reef") crests at 15 m depths. A large sand area separates the Outer and Middle Reef complexes. The Outer Reef ("Third Reef") crests in 15–21 m depths. The Outer Reef is the most continuous reef complex, extending from Miami-Dade County to northern Palm Beach County. Inshore of these reef complexes, there are extensive nearshore ridges and colonized pavement areas. NRC has been shown to have statistically different benthic and fish populations than the Middle and Outer reefs (Moyer et al., 2003; Walker et al., 2009). Up to 43 species of scleractinian corals have been documented in the *S. Florida* FRT. Caribbean reefs host a smaller number of hard corals compared with the Pacific. Before the recent SCTLD outbreak, *Montastraea cavernosa* had the highest coral cover with *Siderastrea siderea* being the most abundant numerically. Stony coral species on the FRT include the following: *Acropora cervicornis*, *Agaricia* spp., *Colpophyllia natans*, *Dichocoenia stokesii*, *Diploria labyrinthiformis*, *Eusmilia fastigiata*, *Helioseris cucullata*, *Isophyllia* spp., *Madracis arenterna*, *Madracis decactis*, *Meandrina meandrites*, *Montastraea cavernosa*, *Mycetophyllia* spp., *Oculina* spp., *Orbicella* spp., *Phyllangia americana*, *Porites astreoides*, *Pseudodiploria* spp., *Scolymia* spp., *Siderastrea* spp., and *Solenastrea bournoni*, *Stephanocoenia intersepta* (Gilliam et al., 2021).

The dominant Coral ECA sponges are the basket sponge, *Xestospongia muta, Anthosigmella varians,* and *Spheciospongia vespariuum*. In the Southeast Florida Coral Reef Evaluation and Monitoring (SECREMP) Project 2020 Year 18 Final Report, the status of iconic or indicator taxa (stony corals, sponges, octocorals, macroalgae) for the FRT is reviewed from the inception year of 2003 (Jones et al., 2020; Hayes et al., 2022). The survey covers the four major S. Florida counties of Martin, Palm Beach, Broward, and Miami-Dade as shown in Fig. 3.1. A major finding was that "The Southeast Florida Coral Reef Ecosystem Conservation Area (Coral ECA) experienced significant stony coral assemblage declines across the study period, with significant losses determined for all stony coral metrics examined (cover, Live Tissue Area (LTA), and density)." As the FRT continues to observe declines in stony coral, more attention should probably turn to the surviving taxa. For example, *Antillogorgia americana* (formerly *Pseudopterogorgia americana*), *Eunicea flexuosa* (formerly *Plexaura flexuosa*), and *Gorgonia ventalina* represent indicator octocoral species, but many other soft coral species exist on the FRT (Moyer et al., 2003). Broward county reefs in particular appear "dominated by soft corals, possibly in response to wave action." Although octocoral cover generally declines during the whole period, regionally, year-to-year analysis of octocoral cover showed a significant increase from 2019 to 2020. For example, *Antillogorgia americana* mean colony height reached its minimum in 2020 and was significantly smaller than in 2013–2018; however colony density was significantly higher in 2020 than all other years. These results indicate that the region experienced a decline in colony size and/or an increase in partial mortality in the larger-size classes. More updated species counts are available through NCRMP (See https://www.coris.noaa.gov/monitoring/biological.html).

In contrast to the FRT, the Pacific "Coral Triangle" (CT) encompasses about 5.7 million square kilometers (2,200,000 sq. mi) of waters bounded by Indonesia, the Philippines, Papua New Guinea,

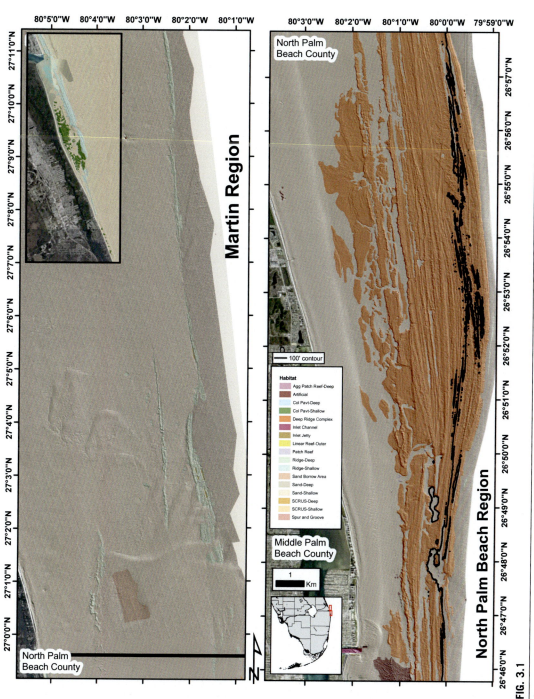

FIG. 3.1

Overview maps of the Florida Reef Tract of Martin, Broward, and Miami-Dade counties in S. Florida. Habitats are color-coded and shown in the legend. An URL to a flyover video of this region can be seen at https://www.youtube.com/watch?v=z1QLmi2lsdg.

Figures were prepared by Dr. Brian Walker.

(Continued)

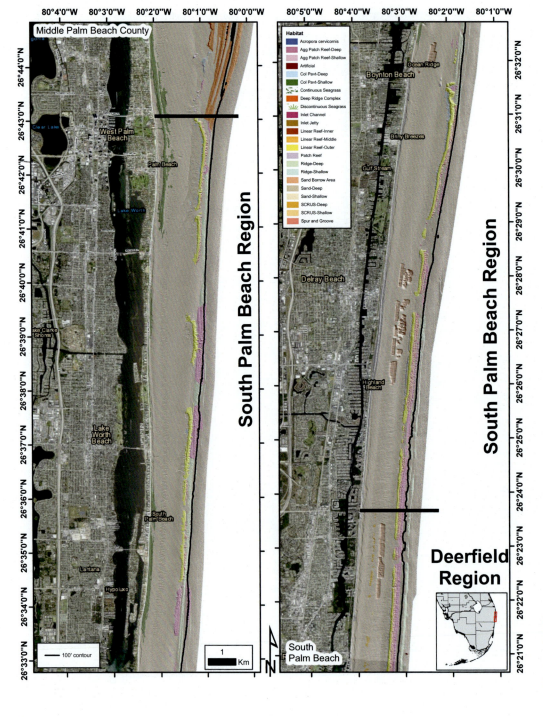

FIG. 3.1, CONT'D *(Continued)*

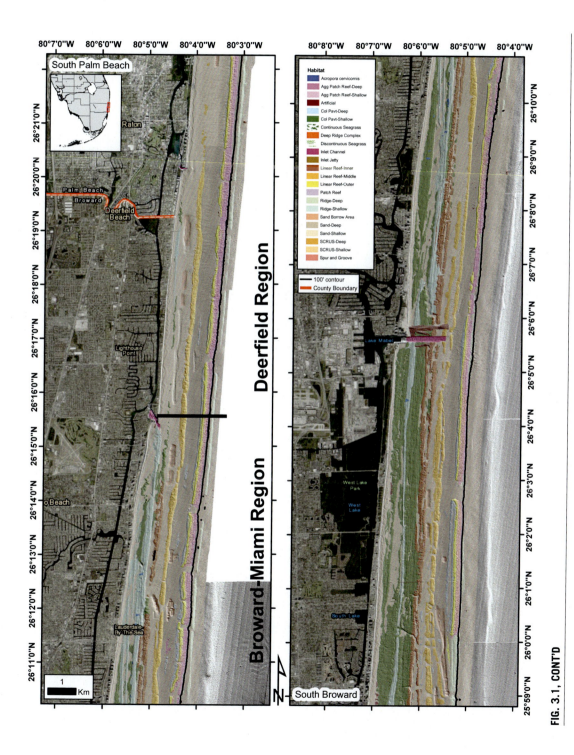

FIG. 3.1, CONT'D (Continued)

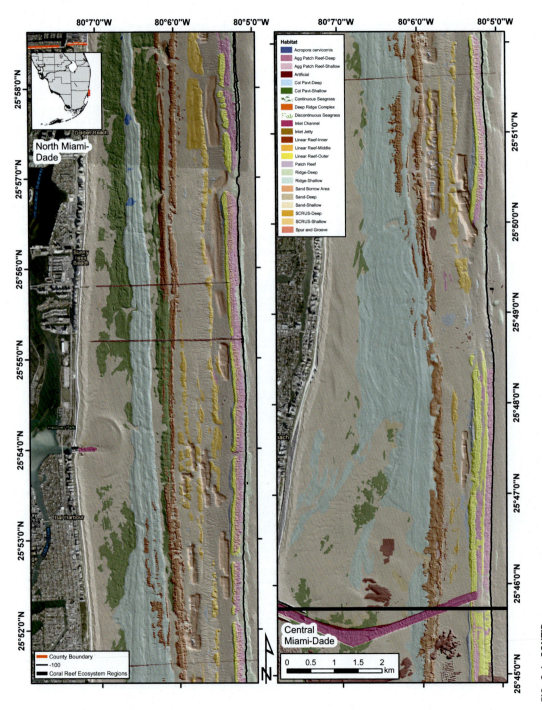

FIG. 3.1, CONT'D

Solomon Islands, and East Timor/Timor-Leste and boasts the highest biodiversity of coral and fish species in the world. The oceanic regions span portions of the Indian Ocean and the Western and Central Pacific, sometimes together called the IWP. The Triangle spans approximately 29 million hectares and contains over 500 species of coral and more than 3000 species of fish. Estimates have been made that "this area holds 76% of the world's shallow-water reef-building coral species, 37% of its reef fish species, 50% of its razor clam species, six out of seven of the world's sea turtle species, and the world's largest mangrove forest" (Hutomo and Moosa, 2005; Asaad et al., 2018). Various reasons have been proposed for how this large volume of diversity arose, with the CT being either a: (1) center of origin, (2) center of accumulation, (3) center of survival, or (4) center of overlap (Ekman, 1953; Gaither et al., 2011; Gaither and Rocha, 2013). Most of these center on the idea of convergence of favorable ecological and evolutionary processes in a tropical locale (Veron et al., 2009, 2011). Another possible factor stems from the centre of overlap hypothesis which involves the Indo-Malay-Philippine (IMP) region. The Triangle occurs in the tropics, but even coral diversity is not confined to shallow seas, since diverse species have been found down to 4000 m. Mesophotic reef connectivity is discussed more below.

Regarding the Coral Triangle, multiple studies have discussed the ecological and conservation parameters that have led to the ascendance of the Triangle (Briggs, 1999; Green et al., 2014; Beger et al., 2015). A major conclusion from Hoeksema (2007) was that "sufficient habitat heterogeneity occurred on coral reefs in this area which combined favorably with gradients in nutrient input, salinity, siltation, wave exposure, and turbidity" could have produced the high biodiversity viewed today. This study particularly focused on the Fungiidae, the Mushroom Coral family (Scleractinia, Fungiidae) to exemplify patterns of biodiversity. However, the treatise goes on to address debates about whether centers of biodiversity can also be centers of origin (speciation) or accumulation.

This special biogeographic region also supports the livelihoods of more than 120 million people who depend on the marine resources provided by the coral reef ecosystems for their food security, income, and cultural heritage. Fisheries of multiple organisms reach into the millions of dollars (Teh et al., 2017). Similar to other reefs, the Coral Triangle also plays a crucial role in protecting coastal communities from storms and wave damage. The location of the CT in southeast Asia also helps contribute to the realization that 64% of the world's coral reefs occur in developing countries with nearshore dense populations (Pascal et al., 2016). This underscores the need for continued stewardship of these high biodiversity zones. Asaad et al. (2018) have mapped priority areas and found that 13% of the Coral Triangle was clustered into hotspots of high biodiversity importance (Fig. 3.2).

Because the Coral Triangle covers a wide swath of individual reefs and national jurisdictions, I will only touch upon a few examples within the area itself. For example, Tubbataha Reefs Natural Park (TRNP), the largest and best-enforced no-take marine protected area (MPA), represents a planned MPA and relatively pristine reef within the Philippines (Dygico et al., 2013). Covering 970 km^2 and established in 1988, a survey of TRNP should provide an estimate of best-case scenarios for the region. At the time of this survey, Dygico indicated a stable fish count and coral coverage: in 2011, demersal fish density appeared to be 465 individuals/100 m^2, hard coral coverage at both shallow and deep reefs was around 50%, and soft coral coverage was about 10%.

More recently, the general state of Philippines reefs has been reviewed by Licuanan et al. (2019a,b). With a total of 25,000 km^2, Philippine reefs represent the third largest in the world (Gomez et al., 1994). The archipelago has over 7000, mostly small, islands, with 2000 of those inhabited by people, while also including various benthic habitats such as seagrasses, mangroves, and deep reefs. Coral reef

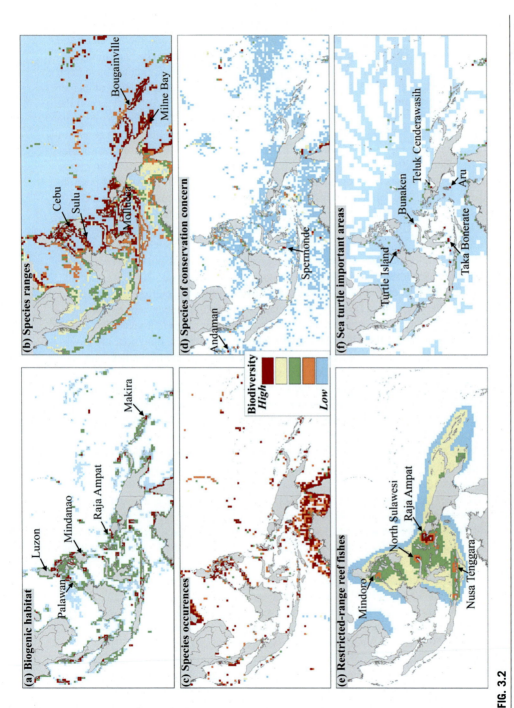

FIG. 3.2

Representative biodiversity within the Coral Triangle. Areas of biodiversity importance in the Coral Triangle based on each criterion in 0.5° cells: (A) Coverage of coral reefs, mangroves, and seagrasses. (B) Species richness (occurrences) based on ES_{50} of 19,251 species. (C) Species richness based on the overlapped ranges of 10,672 species. (D) Richness of species of conservation concern based on ES_{35} of 834 species. (E) Restricted-range species based on the distribution of 373 CT endemic reef fishes. (F) Areas important for sea turtles based on the distribution of six sea turtle species. White cells = no data.

Reproduced with permission from Asaad, I., Lundquist, C.J., Erdmann, M.V., Costello, M.J. 2018. Delineating priority areas for marine biodiversity conservation in the Coral Triangle. Biol. Conserv. 222, 198–211.

FIG. 3.3

Example of shallow coral reef habitats in the Philippines. In situ photographs of Philippine reefs in the Coral Triangle. *Upper left* and bottom pictures are from Sombrero Island, Batangas. *Upper right* image is from South Miniloc, El Nido, Palawan.

Photographs courtesy of Caitlin Crisostomo.

assessments indicate that the islands have an average of 22% stony coral cover, with some sites reaching up to 40% at some sites across the archipelago (Licuanan et al., 2017). Fig. 3.3 shows the high coral coverage and biodiversity of a few representative Philippine reefs.

Moreover, Ablan-Lagman (2018) provided a deep review of the contentious yet biodiversity-rich Spratly Islands, which falls within the bounds of the Coral Triangle. To reiterate just a few of her points here, these islands hold a significant amount of reef biodiversity and should be treated with extensive care. Covering an area of 160,000–500,000 km^2 of ocean, the Spratlys dot the South China Seas, which can reach to an average depth of 4700 m (and maximum depth of 5559 m). The Islands emerge from slope depths of <200 m and are composed of a "biogenic carbonate platform that lie on the crest of uplifted fault blocks known as horsts which arose in the Late Cretaceous and Early Oligocene." (Sun et al., 2011; Ablan-Lagman, 2018). From this base, the following organismal counts have been made for the Spratly Islands: "a total of 2927 marine species, including 776 benthic species, 382 species of hard coral, 524 species of marine fish, 262 species of algae and seagrass, 35 species of seabirds, and 20 species of marine mammals and sea turtles." Similar to other Coral Triangle habitats, the reefs can

146 **Chapter 3** Diversity hotspots on the benthos

sustain endangered giant clam species (*Tridacna*) (DeBoer et al., 2014), whose genomes are being sequenced for the Moore/Sanger Aquatic Symbiosis Project (McKenna et al., 2021; JC Li in preparation). Also, the Spratly Islands may also be a source of organismal propagules for other reef areas (McManus and Menez, 1998). With regard to geopolitics, the Spratly Islands occur within an uncomfortable placement of the Triangle. Multiple countries, such as China, Malaysia, the Philippines, Taiwan, Vietnam, and Brunei, claim the islands as their own, which has led to the installation of military bases on several islands. Unfortunately, part of the interest in the islands stems from oil and gas prospecting and not ecosystem protections or biodiversity valuation per se (Hutchison and Vijayan, 2010).

Just on the border of the Coral Triangle, the 340+ islands of Palau sit between the Philippines and Guam. The marine biodiversity of these islands has been documented, such as in octocorals (McFadden et al., 2014). Analyzing the effects of latitude and the richness of the regional species, Witman et al. (2004) found that Palau had some the highest-level epifaunal invertebrate-richness communities compared with higher-latitude sites, which may be based on the following latitudinal differences: (i) the process of colonization with lower dispersal of larvae or propagules at low latitudes or lower predation on competitive dominants; (ii) habitat disturbances, resulting in greater amounts of open substrate at high latitudes; (iii) enhanced competitive resource partitioning in the tropics resulting in fewer vacant niches, greater saturation, and less regional pool influence; and (iv) productivity, because the regional pool effect might be greater at intermediate levels.

A recent literature review has provided another interesting assessment of Coral Triangle research, which underscores the potentially large biodiversity assessment gaps alluded to above (Reimer et al., 2022):

In this literature review, we collected papers from the Web of Science (1995–2021) focused specifically on coral reef restoration from six countries and regions around the Coral Triangle (Japan, Taiwan, mainland China, the Philippines, Malaysia, Indonesia) to examine how much coral reef restoration research has been performed in each area, when it was performed, what methodologies were used, what organisms were targeted, and whether any assessment of biodiversity was included. Our results show great disparity in the research efforts of each area, with the Philippines clearly leading research in the region with almost half of the literature examined, followed by Japan and Indonesia, with nascent efforts in mainland China, Taiwan, and Malaysia. Overall, for the region, research appears to be increasing with time. Research in most areas was concentrated in one or two locations, and almost exclusively focused only on corals. Only approximately 38% of papers mentioned biodiversity in any manner, and only 14% included organisms other than scleractinian corals in their results.

The review makes recommendations such as increasing coordination among NGO's and diverse governments, prioritizing the publication peer-reviewed scientific papers on the Triangle, and also increasing transparency of results to stakeholders. Thus overall, it seems that there is only a latent awareness for a large treasure chest of biodiversity and potential, but a clear deficiency to protect and assess its entirety.

The twilight zones

Most mesophotic coral ecosystems (MCEs) occur in warm water, but remain light-dependent starting at 30–40 m and reach to the bottom of the photic zone, which varies by location (Hinderstein et al., 2010; Reaka et al., 2016; Kahng et al., 2019) (Fig. 3.4). Mesophotic coral ecosystems remain relatively

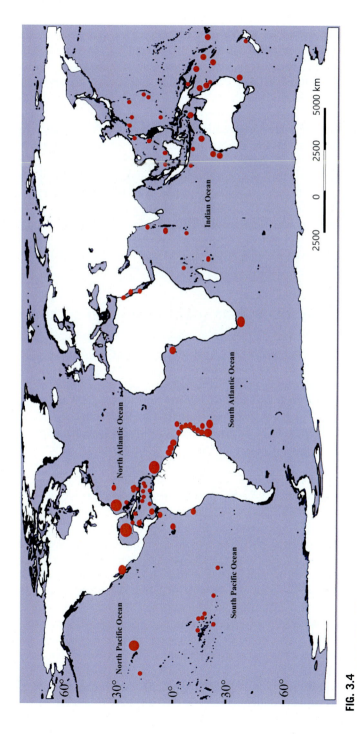

FIG. 3.4

Mesophotic coral ecosystems (MCEs) worldwide. *Circle size* indicates number of MCEs in the region.

Reproduced with permission from de Oliveira Soares, M., de Araújo, J.T., Ferreira, S.M.C., Santos, B.A., Boavida, J.R.H., Costantini, F., Rossi, S. 2020. Why do mesophotic coral ecosystems have to be protected? Sci. Total Environ. 726, 138456.

148 **Chapter 3** Diversity hotspots on the benthos

understudied but may offer the best hope for the survival of some reef benthos in general. One factor for this is that MCEs are sometimes continuous with shallower reefs, which can help with local recruitment and affect connectivity across depth gradients (Andradi-Brown et al., 2016). Direction of gene flow is not always predictable, however. For example, after studying the *Montastraea cavernosa* across the Caribbean, Serrano et al. (2014) found that migration appeared asymmetric, predominantly from shallow to intermediate/deep habitat migration as shown by using DNA microsatellite loci to track genetic variation.

MCEs have been characterized at the Great Barrier Reef, Australia, north eastern Brazil, and the Hawaiian Archipelago among other locales (Kahng et al., 2010; Bridge et al., 2011; Rooney et al., 2010). The MCE's of Hawaii encompass about 13,180 km^2, which is 62% larger than the potential habitat of shallower Hawaiian reefs. The Hawaiian Archipelago has eight main Hawaiian Islands and nine Northwestern Hawaiian Islands with their MCEs having a patchy distribution throughout the archipelago. The better developed and deeper MCEs appear near the southern end.

MCE's have been recently reviewed (Rocha et al., 2018; de Oliveira Soares et al., 2020). However, Pyle and Copus (2019) also indicated that 70% of papers have been published only since 2012, indicating that much work remains to be done on this relatively understudied twilight habitat.

The Clarion-Clipperton Fracture Zone (CCZ): One jewel of the deep sea

Deep sea—bathyal, abyssal, and hadal—zones are considered diversity hotspots (Danovaro et al., 2017; Da Ros et al., 2019). Several studies now expand our background of deep ocean habitats (Jamieson, 2015). For example, using metabarcoding methods, Cordier et al. (2022) show that diversity in deep ocean sediments appears at least threefold more than pelagic realms. Studies of the Clipperton fracture zone, also known as the Clarion Clipperton Fracture Zone (CCZ), also help answer this question. The CCZ occurs as a geological submarine fracture zone in the Pacific Ocean around. Depths can range from 800 to 4000 m. The CCZ has gained attention for a number of reasons, including the abundant presence of manganese nodules. This transition metal is crucial in metallurgy, including the forging of steel alloys, and as an oxidizing agent. The International Seabed Authority (ISA) estimates that the total amount of nodules in the Clarion Clipperton Zone exceeds 21 billion tons (Bt), containing about 5.95 Bt of manganese, 0.27 Bt of nickel, 0.23 Bt of copper, and 0.05 Bt of cobalt (International Seabed Authority, 2010). This fracture zone has a length of around 7240 km (4500 miles) and spans approximately 4,500,000–6,000,000 km^2 (1,700,000 sq. mi).

How often do we see sudden reversal of fortune scenarios, when a seemingly predictable disaster is flipped to produce the opposite outcome? Unfortunately, too rarely. However, there may be some hope that the CCZ may resemble more of the improbable second situation. This is because since 2016, an unprecedented avalanche of scientific attention has now been paid to the CCZ since the discovery of the nodules. With the likely foray of mineral prospecting, such as metallic nodules, the pace of exploration and the concomitant need for baseline biodiversity data on targeted areas will accelerate. Significant spatial patterns and heterogeneity occur in the CCZ, with some areas having higher biodiversity and others having lower biodiversity. These patterns are likely influenced by physical, chemical, and biological factors unique to the CCZ.

The concept of the slope-abyss source-sink (SASS) hypothesis was proposed by Rex et al. (2005) to explain that abyssal fauna population maintenance and health mostly depend on larval dispersal from more productive areas on the surrounding continental slopes. Rather organic nutrient inputs tend to be minimal. SASS may, however, cause problems for the CCZ and other areas ripe for extraction and commercialization. Since the hypothesis deems the abyssal seafloor as a biodiversity sink, rather than a high-diversity region with centers of evolutionary radiation, then the CCZ does not have to be protected as a biodiversity hotspot. Any degradation of biodiversity will be repopulated eventually from other areas.

After the posing of an SASS hypothesis, various studies have discovered a high number of unique and rare species inhabiting the CCZ, including many that appear endemic. For example, in a study, the CCZ's megafauna, Remotely Operated Vehicle (ROV) *Remora* III, collected specimens from UK eastern portion of the CCZ, 13°49′N, 116°36′W, at 3900–4400 m depth (Amon et al., 2016). The results found that "seven of the 12 species were new to science, including three new genera: *Abyssoprimnoa gemina* Cairns, 2015, a new genus and species; two undescribed genera of the coral family Isididae, a new species of *Phelliactis anemone;* a new species of primnoid *Calyptrophora persephone* Cairns, 2015, and two new species of *Caulophacus* sponges." Also, agencies such as the ISA have set up policies to help protect the CCZ (Bräger et al., 2020). These are the spotlights that should be turned into virtual floodlights that could help dispel our ignorance of the seafloor. According to accumulating studies (Gooday et al., 2021), the CCZ harbors an unexpectedly high level of biodiversity and is home to at least 1215 species of animals that live in the deep sea. The consensus notes that many new species have yet to be described. Another study published by Washburn et al. (2021) also asserted that the CCZ is vital for benthic biodiversity conservation, due to the high number of endemic species living there. The study estimates that nearly half of the megafauna species found in the CCZ are endemic to the region. For comparison, metabarcoding efforts were performed in the Clarion-Clipperton Zone (CCZ) in the eastern Pacific Ocean, which is an area of intense study and activity because of rich mineral deposits (Hauquier et al., 2019; Lejzerowicz et al., 2021).

Lejzerowicz et al. (2021) conducted an extensive metabarcoding study to investigate the biodiversity and spatial patterns of eukaryotic organisms in the Clarion-Clipperton Zone (CCZ) and other abyssal regions. Using 18S, both foraminiferans and metazoans were targeted for analyses. They found that the diversity of eukaryotes was high, and distribution patterns varied depending on the depth and location of the samples. Over 60% of benthic foraminiferal and almost a third of eukaryotic operational taxonomic units (OTUs) were not assignable.

Comparison with other abyssal regions: The study compared the biodiversity and spatial patterns of eukaryotic organisms in the CCZ with other abyssal regions, such as the East Pacific Rise and the Angola Basin. The authors found that the diversity of eukaryotes was generally higher in the CCZ compared with other abyssal regions, highlighting the importance of this region for assessing and fully understanding the dynamics of deep-sea biodiversity with habitat heterogeneity.

Abundance patterns in the CCZ were as follows: Nematoda as highest (present in 97.9% of the samples), followed by Platyhelminthes/Xenacoelomorpha (77.9%), and Annelida (63.3%). The Echinodermata, Gastrotricha, Mollusca, Nemertea, and Porifera occurred in, at most, only 23.3% of the CCZ samples. The nemertea abundances were consistent with Hauquier et al. (2019). Prior to 2004, only one kinorhynch species had been identified in the vicinity (Central American East Pacific area) to the species level, *Campyloderes* cf. *vanhoeffeni* Zelinka, 1913 (Neuhaus, 2004). However, recently Sánchez et al. (2022) have characterized multiple rare Kinorhyncha species (also known as mud dragons) in the CCZ. Recent characterizations of the community structure of abyssal benthos with a box corer from 12 stations within the OMS areas of the CCZ yielded a total of 42 families of

Polychaeta, 15 families of Copepoda, 10 families of Tanaidacea, and 8 families of Isopoda were reported. To conclude, ongoing research suggests that the Clarion-Clipperton Zone to be a unique important area for biodiversity conservation because of high species richness and endemism, which will likely resolve with further characterization.

Hydrothermal vents and seeps

Descriptions of the first discovery of hot thermal vents at deep spreading on the ocean floor have been likened to discovering new worlds or Apollo's landing on the moon (https://www.whoi.edu/feature/history-hydrothermal-vents/discovery/1977.html) (Corliss et al., 1979). The difference is that millions of people did not witness most of the seafloor discoveries in real time (or anytime), although the habitats occur in our own backyard and are worthy of awe. Thus, no ticker tape parades will likely be planned for those who operate the AUVs or submersibles, searching and often finding new biological habitats and organisms.

The unique biologicals at spreading sea floor rifts have been fairly well documented. Life there is almost completely driven by chemosynthetic symbioses (Van Dover et al., 2001; Dubilier et al., 2008). With regard to pelagic benthic coupling, the effects are obvious. The black smokers spew potential energy in the form of heat (hot water up to 400°C), acidity (pH 2), sulfides, high concentrations of methane, carbon dioxide, as well as multiples types of heavy metals and minerals (and arsenic-containing compounds) into the immediate vicinity (Von Damm, 1990; Sarrazin et al., 1999). Lastly, various studies have asserted that vent community influence can spread farther beyond the immediate vent than previously realized (Levin et al., 2016).

The continuum of biodiversity to their genetic sources

Besides the descriptions of geographic locales as biodiversity, we can also view the meaning of physical location or "address" at the molecular level, a few levels downward. Once again from a genetic perspective, we can characterize biodiversity and its basis in the genetic code, and that specific genes directly related to benthic adaptations store this code at specific addresses otherwise known as *loci* on a specific chromosome. In its earliest days, genetics was not an academic discipline and had no data on the physical location of where these loci of heredity were or belonged. . The Austrian Augustinian friar Gregor Mendel, the father of genetics, carried out experiments on pea plants (*Pisum sativum*) and other organisms, which elegantly characterized and predicted the inheritance of easily observable traits based on single genetic loci (then two and three, etc.) (Punnett, 1922). Further along the history of genetics without going into too much detail (but see Futuyma, 2017), we encounter the population geneticists, such as Ronald Fisher, Sewall Wright, JBS Haldane, George Gaylord Simpson, and Ernst Mayr and others brilliantly formulated the new synthesis of population genetics, using terms and mathematical formulas to describe concepts such as the behavior of abstract genetic loci within mathematical landscapes and valleys (Fisher, 1930; Haldane, 1932; Huxley, 1942; Mayr, 1944; Simpson, 1944; Wright, 1986). These insights were generated without reading or viewing any data or nucleic acid and protein sequences. Now, whole genome sequencing brings much of the genome into physical relief for multiple organisms. Benthic habitats and their resident diverse fauna and flora will likely soon yield the diverse array of novel genes that have been evolving for millennia (Wolf et al., 2002; Shendure and Ji, 2008; Koonin, 2009; Lewin et al., 2022; Blaxter et al., 2022). Characterizing and subsequently conserving these gene sequences follow the paradigm shift of focusing on more genes to sequence and analyze, rather than many individuals and few, limited number of their genes (McMahon et al., 2014):

> Applying genomics to conservation will thus mean going from 'one gene, many individuals' to a 'few individuals, several genes' approach. When conservation biologists and stakeholders realize this, sampling and monitoring programs need to be changed accordingly. While the lab costs for a genomic may study still supersede the lab costs for traditional conservation genetics, a less labour intensive sampling may cut the costs of the conservation project on the nonlab part of the project.

This strategy could revert back to the many individuals plus all of their genes (i.e., whole genomes), but this in the distant future when genomes can be sequenced and interpreted at very low costs. For now, we can also view this current discourse as trying to map a longstanding goal—understanding the connection and details of "genotype to phenotype" (G2P) (Dowell et al., 2010). We now understand a little better that a single gene does not work in a vacuum, but rather interacts with a large suite of genes within any cell or organism. At the same time, it can also be noted that just one or two pivotal mutations in a protein can alter its shape, function, and possible fitness of the host organism. Thus, the continuum of G2P stems from single gene variation to genotypic spectra, which can influence individuals and their populations. G2P concepts also have a basis in evolution-developmental biology, since the expression of specific genes with concomitant effects on organismal development can be traced simultaneously (Stracke and Hejnol, 2023).

When we assert that nucleic acid (DNA/RNA) codes and their expression are the basis of the phenotype, the physical manifestation into organismal development, morphology, and function (Dawkins, 1976), then we can also posit that specific sequence haplotypes (a set of alleles inherited together) differentiate by evolutionary processes (natural selection, drift, mutation) sometimes provide adaptations, become the basis and markers of speciation (Nei and Kumar, 2000). The latter criteria could be labeled as a *synapomorphy*, evolutionary, and phylogenetic term for a "shared, derived trait" coined by Willi Hennig (Graur and Li, 2000). ADAR (adenosine deaminase acting on RNA) proteins, which convert adenosine into inosine in double-stranded RNAs, could provide a good example of a synapomorphy. They appear to be in all metazoans except ctenophores (Grice and Degnan, 2015). These derived sequences can also be called taxon-restricted genes (TRG), if they are limited to only a few lineages (Wilson et al., 2005; Khalturin et al., 2009; Johnson, 2018). In spite of the jargon, the distinction here effectively recapitulates our major goal to assess benthic biodiversity.

Conversely, we can also keep in mind that homologous genes or DNA sequences also occur across multiple lineages, similar to the genes identified for specific animal clades by Dunn and Ryan (2015). These sequences will also likely vary through their differential expression as they become subjected to different environments or selective forces. Moreover, we can keep in mind that non-coding ultra-conserved elements (UCEs) are maintained within genomes and also vary across lineages, making them useful as new phylogenetic markers (Faircloth et al., 2012; Petersen et al., 2022). These sequences can encode regulatory, promoter-like DNA motifs, which control gene expression and other important cellular activities. With our capacity to carry out recombinant or gene editing experiments in the future, characterizing diverse UCEs and determining their function represent "treasure chests" of DNA sequences that could possibly be activated in some future environmental context.

152   Chapter 3 Diversity hotspots on the benthos

However, for now, we will mostly focus on unique, lineage-specific coding sequences. In this context, we can address the mechanisms of how novel DNA sequences actually arise within an organismal lineage or taxon to become a genuine gene. In essence, *how are unique coding gene sequences born?*

Duplication of sequences followed by mutations that allow them to fit into new cellular functions has been a long-standing molecular paradigm for novel gene origins and has many supporting examples since its inception (Ohno, 1970). However, determining de novo birth is still not trivial and requires evolutionary inferences and genomics analyses (Van Oss and Carvunis, 2019). Only a relative few experimental attempts to identify de novo gene birth have been made, using model systems such as yeast, mice, and *Arabidopsis thaliana* (Carvunis et al., 2012). In *Drosophila*, several studies confirmed that up to five genes evolved exclusively within a narrow lineage (Begun et al., 2006; Levine et al., 2006). As more non-model animal genomes start to populate databases, we will see more robust patterns and subsequently better models for gene birth crystallize.

For example, besides knowing a DNA sequence, its chromosomal arrangement (synteny) among previously known sequences present in other species should be evaluated for context. Van Oss and Carvunis (2019) goes on to describe an approach called "genomic phylostratigraphy," which incorporates phylogenetics and extensive bioinformatics searches. As expected, the birth of functional gene sequences, which can add to biodiversity, is not trivial and often occurs gradually. The assertions of genetics pioneer, François Jacob, reminds us again of *creation ex nihilo* and its rarity in biology (Jacob, 1977).

Since we are now fully immersed in the era of deep genomic sequencing, below is a limited list of specific genes and their functions that have already been discovered from multiple studies. Similar to biogeographic hotspots, unique gene and protein sequences specifically adapted to unique seabed habitats underscore biodiversity and the value of their conservation. Generating a catalogue of interesting and unique coding genes validates the burgeoning efforts of sequencing of whole reference genomes (Theissinger et al., 2023).

As we go through a brief listing of specialized gene sequences, their function and origins can be viewed as products of ecological specialization, isolation, or the success of lineage sorting. For example, Tautz and Domazet-Lošo (2011) analyzed the presence of *orphan* genes, which are coding sequences, which lack homologs in other lineages and appear to increase during adaptive radiations. This makes sense as phenotypic innovation is required to colonize new niches, and genes are the basis of evolutionary novelties (Oakley, 2017). By contrast, unused gene and pathways will be lost during evolution or degrade over time (Guijarro-Clarke et al., 2020).

Biodiversity includes an array of unique genes and functions

Rho-associated kinase (ROCK) gene homologs have been found in benthic organisms such as Pacific oysters (*Crassostrea gigas*) and whiteleg shrimp (*Penaeus vannamei*) (Mao et al., 2020). These genes play a role in the regulation of cell shape and phagosome formation as an immunological response (Song et al., 2021).

Silicatein enzymes assemble siliceous spicules, the basic component of demosponge skeletal structure, and thus play a pivotal role in biomineralization. Multiple silicatein gene sequences have now been cloned and sequenced from multiple species (Kozhemyako et al., 2010). Analyses of silicatein revealed a close relationship with cathepsins, a gene family of proteases. Engineering the catalytic cysteine for serine and the replacement of adjacent amino acids (S<u>S</u>W → A<u>S</u>Y; the catalytic serine is underlined) helped convert cathepsin to silicatein via the creation of a pocket allowing recognition of a Si(OH)$_4$ molecule (Fairhead et al., 2008). Further genetic engineering of recombinant silicatein has produced heterologous expression in plant cells to display the extracellular synthesis of metal nano-particles, as an example of "Green synthesis" (Shkryl et al., 2018).

<u>Protocadherin</u> genes encode cell-adhesion molecules that play a critical role in the development and maintenance of neural circuits in animals. These genes have been identified in a wide range of marine animals, including cnidarians, echinoderms, and mollusks. In marine animals, protocadherin genes are primarily involved in the development and maintenance of neural circuits. These genes play a crucial role in the formation of synapses between neurons, as well as in the establishment of neuronal connectivity during development. Thus, it was not totally surprising to see active expression and an expansion of protocadherin genes in recently determined octopus (*Octopus vulgaris, Octopus bimaculoides*) and genomes and transcriptomes (Albertin et al., 2015; Styfhals et al., 2019). Long considered to be a unique vertebrate synapomorphy, the new octopus gene sequences indicate that protocadherins display signs of *convergent* evolution of advanced neuronal capacity in two distinct metazoan lineages Lophotrochozoa (i.e., cephalopods) and Vertebrata (Morishita and Yagi, 2007; Styfhals et al., 2019). In addition, protocadherin genes are also involved in the regulation of cell proliferation, differentiation, and apoptosis in various tissues.

In a more general adhesion phenotype, <u>biological cementing substances</u> allow several benthic organisms to glue themselves to hard substrate and withstand wave action, turbidity, and strong currents. The process of cementation varies across the marine taxa that require the stability, have been identified in mollusks, polychaetes, and echinoderms, and has gained attention from industry (Almeida et al., 2020; Davey et al., 2021). Adhesion-type proteins must be secreted into the aqueous environment and also interact with metals, lipids, and polysaccharides. Cement proteins (and their genes) called (AaCPs) AaCP19 and AaCP52 have been identified from the barnacle *Amphibalanus amphitrite* and appear to have no sequence homologs (So et al., 2016). A lysyl oxidase (LOX) appears involved in polymerization. In bivalves, such as *Mytilus*, mussel foot proteins (mfp) have been characterized and contain DOPA, a catecholic amino acid stemming from posttranslational modification of tyrosine (Lee et al., 2011; Davey et al., 2021).

<u>PAX</u> genes have been implicated in eye development in all bilaterians (Kozmik, 2008; Huang et al., 2022). Specifically, after sequencing the genome of the ancient nautilus (*Nautilus pompilius*), *Pax4/6* gene expression appears active and involved in its pinhole eye development, which allows it to adapt to twilight benthic habits (Huang et al., 2022).

<u>Antifreeze Proteins (AFPs)</u> and genes have been found in deep-sea fish and play a role in preventing the formation of ice crystals in their bodies, which would otherwise cause damage in the cold and icy abyssal environments (Chen et al., 1997; Daane and Detrich III, 2022). Some AFP genes appear to have arisen de novo and as example in which the ORF was shown to precede that of the regulatory promoter region (Zhuang et al., 2019). Related to AFPs, fewer examples of ice-binding proteins (IBPs), which have been found across invertebrates, must cope with freezing conditions. IBPs act to protect organisms from ice by suppressing the freezing point relative to the melting point, preventing changes in ice crystal size through ice recrystallization inhibition and allow organisms to control when and where ice forms in the body via nucleation (Bar-Dolev et al., 2020). For example, evidence of IBPs has been found in blue mussels (*Mytilus edulis*), barnacles (*Semibalanus balanoides*), and the psychrophilic diatom (*Fragilariopsis cylindrus*) (Bacillariophyceae) (Theede et al., 1976; Krell et al., 2008).

154 **Chapter 3** Diversity hotspots on the benthos

Box et al. (2022) recently surveyed various invertebrate genome sequences to mine for IBPs and found a higher-than-expected cache of potentially new IBP genes.

Some genes have been found necessary for thermoregulation in some deep-sea organisms and play a role in the regulation of body temperature in response to the extreme pressure and temperature conditions of the abyssal environment. For example, variable expression of genes was found between the mussel *Bathymodiolus thermophilus* and the annelid *Paralvinella pandorae irlandei* when exposed to hot and cold variable regimes (Boutet et al., 2009). This included intracellular hemoglobin, nidogen protein, and Rab7 in *P. pandorae*, and the SPARC protein, cyclophilin, foot protein, and adhesive plaque protein in *B. thermophilus*. By contrast, other proteins such as membrane-spanning SNARE synataxin proteins, which participate in exocystosis, appear to be under positive selection and possible convergent evolution in deep-sea decapod species (Yuan et al., 2020). Samples of *Alvinocaris longirostris* and *Shinkaia crosnieri* were collected from a hydrothermal vent of the Okinawa Trough, and samples of *Rimicaris* sp. and *Austinograea alayseae* collected from a hydrothermal vent in the Manus Basin were contrasted with nine other shallow water or model decapods for gene expression and expansion differences. Yuan et al. (2020) characterized a total of 391 positively selected genes, 109 parallel substituted genes, and 33 significantly expanded gene families in the deep-sea decapods.

Glutathione S-transferase and other transporter genes have been found in some deep-sea invertebrates and play a role in detoxifying harmful substances, such as heavy metals and organic pollutants, in the bathyal and abyssal environments. Novel and lineage-specific C1QDC and transporter proteins were identified in the gills of the bathymodiolin mussels, with possible functions in chemosymbiosis and recognition (Zheng et al., 2017). Gene expression comparisons were made between shallow and deep mussels, *Bathymodiolus manusensis*, *B. platifrons*, *M. kurilensis*, and *P. viridis*, and yielded 88,651, 67,768, 75,614, and 60,036 proteins, respectively. After ortholog analyses, evidence was found for positive selection in that would favor deep sea adaptations. For example, metal-binding (EF-hand domain pair, EF-hand domain, EF-hand for Ca^{2+} binding, and zinc finger-like domains for Zn^{2+} binding), tetratricopeptide repeat in protein kinase domains appeared more prevalent, along with deep sea–specific proteoglycan 4, myeloid differentiation factor 88c, interleukins, interferon regulatory factors, alpha macroglobulin, complement component C3-like protein, low-density lipoprotein receptor-related protein 6, Toll-like receptors, transporters in the class of iron transporters (ferroportin1, copper-transporting ATPase 1, zinc transporter ZIP1, ferrous iron transport protein B, taurine transporter, ammonium transporter Rh type B-A), and many other nutrient, vitamin, and coenzyme transporters in the ATP-binding cassette transporter superfamily (ABCs), the permease family and solute carrier family (SLCs) appeared at higher levels of expression in deep sea mussels.

Green Fluorescent Protein (GFP) gene was first isolated and characterized in the jellyfish *Aequorea victoria* (Shimomura et al., 1962) and also occurs in *Obelia* (a hydroid) and *Renilla* (a sea pansy). Bioluminescence occurs when GFP interacts the jellyfish protein aequorin in the presence of Ca^{2+} ions. In the excited state, the bioluminescence has a blue glow, but upon relaxation energy is transferred back to GFP and green light is produced (Morin and Hastings, 1971). This research allowed the development of useful reporter protein assays for experimental studies and thus led to the Nobel Prize in 2008 (Roda, 2010).

Also, on the topic of submarine illumination, bioluminescent *LuxA* and *luxB* genes code for luciferase, an enzyme that catalyzes the oxidation of flavin mononucleotide (FMN) in a reaction that creates blue-green light for bacteria, mostly within the Gammaproteobacteria class, such as *Vibrio fischeri* (Ruby and Morin, 1979). These bacteria cluster phylogenetically into three families (*Vibrionaceae*,

Shewanellaceae, and *Enterobacteriaceae*), which have been known to encode many functional *Lux* genes. However, a recent sequence survey identified an additional 401 previously unknown luciferase-related genes in public databases (Vannier et al., 2020).

Osmotic Stress Response (OSR) genes have been studied in some marine invertebrates, which have to physiologically regulate responses to salinity changes, which can vary greatly in the marine environment. Transcriptomes (36,344 unigenes annotated with by gene ontology [GO] analysis) were analyzed from the Manila clam, *Ruditapes philippinarum*, an important aquacultured bivalve, which also must adapt to variable intertidal benthic conditions (Nie et al., 2017). Gene expression of mitochondrial matrix enzyme glutamate dehydrogenase 1 (GLUD1) and glutamine synthetase (GLUS) appeared significantly increased under low-salinity conditions. This activity indicated that nitrogen and glutamate metabolism and their homeostasis affect osmolality tolerance in *R. philippinarum*. Moreover, after using the Kyoto Encyclopedia of Genes and Genomes (KEGG) database (https://www.genome.jp/kegg/) to classify the transcript data into likely biochemical pathways and functions, the concomitant increase in intracellular free amino acids (FAAs) in the clam appeared consistent with how intertidal organisms adapt to salinity changes (Meng et al., 2013).

High hydrostatic pressure can damage DNA and tissues integrity in many organisms (Rothschild and Mancinelli (2001). To deal with these extreme conditions in bathyal and deeper zones, Yancey (2020) describe how crustaceans increase their production of *piezolytes*, small organic molecules with pressure-counteracting properties. One of the most important of these osmolytes is trimethylamine N-oxide (TMAO), which produced by oxygenation of trimethylamine via previously mentioned, hepatic enzyme, flavin-containing monooxygenase-3 (FMO3) (Downing et al., 2018; Subramaniam and Fletcher, 2018). In a recent sequencing of the Yap hadal snailfish whole genome (found at 7000 m), five copies of the fmo3 gene were identified (Mu et al., 2021). Moreover, gut bacteria have been found in this snailfish, which have the TMA-lyase gene required to form the TMAO precursor and its biosynthesis. Additionally, for hadal species, expression of genes such as *acyl-CoA oxidase* and desaturases increases because they are involved in lipid metabolism or have the ability to create unsaturated fatty acids (Allen et al., 1999; Lan et al., 2017). This will help maintain the fluidity of cellular membranes at high pressures. Moreover, thermostable patatin-like proteins (PLPs) have been identified from metagenomic libraries derived from the hydrothermal vent communities of the Guaymas Basin (Fu et al., 2015). The PLPs were previously known to exhibit lipid acyl hydrolase activities, and lipases and lipid hydroxylases would be expected to be active at bathyal and abyssal zones. These also appear related to phospholipases.

Heat Shock Proteins (HSPs) have been found in deep-sea animals and play a role in protecting the organisms from stress and damage caused by changes in temperature, pressure, and other environmental factors. Heat shock proteins also have major roles as chaperoning proteins in the cell toward their intended targets, so they have a regulatory function. Heat Shock factor 1 (hsf1) (Ip et al., 2022) *Hsp90* genes in deep sea animals appear to have a positive selection of the same alanine(A)-to-serine(S) substitution. Metatranscriptomics can provide comparative approaches and yield differentially expressed genes under controlled conditions. Using CRISPR methods, Cleves et al. (2020) showed that mutations in the HSF1 gene significantly affected coral thermal tolerance. Other hsf proteins appear to be activated when environmental stresses occur, such as in the *Cinachyrella alloclada* sponge after exposure to oil spill polyaromatic hydrocarbons (Desplat et al., 2023).

Anthozoan Notch ligand-like (*AnNLL*) molecules appear upregulated with several other genes during active symbiosis of *Acropora tenuis* (Yoshioka et al., 2021). The Notch pathway is a highly

156 **Chapter 3** Diversity hotspots on the benthos

conserved cell signaling pathway found in many animals. This study identified several other upregulated gene sequences associated when presented with its native *Symbiodinium* alga symbiont: transcription factor functions (*Hint domain*, *P53 DNA binding domain*, *Transcription factor DP*, *Homeobox domain*, *E2F/DP family winged-helix DNA-binding domain*), prosaposin (*PSAP*) (a glycoprotein involved in antioxidant), and proteins with sugar transporter domains.

Homeobox (HOX) genes represent a group of highly conserved genes that are well known to play crucial roles in the development and patterning of animal body plans (McGinnis and Krumlauf, 1992). First characterized in *Drosophila*, HOX genes have since been identified in a wide spectrum of animals, including marine phyla cnidarians, mollusks, and echinoderms (Finnerty et al., 2003). The origin of HOX genes can be traced back to the common ancestor of all animals, which lived more than 600 million years ago. Over time, these genes underwent extensive duplication and diversification, resulting in the formation of clusters of genes that control the development of specific body regions. The function of HOX genes in marine animals is to regulate the formation and positioning of different body segments during embryonic development. This is achieved through the activation of downstream target genes that control cell differentiation and tissue patterning (Pearson et al., 2005; Mallo and Alonso, 2013). The sea urchin *Stronglyocentrotus purpuratus* genome holds a large 588 kb HOX gene cluster. The expression patterns of its 11 HOX genes are complex, and the precise functions of all except Hox11/13b's cell adhesion activity remain unknown (Ikuta, 2011). Overall, the prototypical Hox gene arrangement of many non-bilateria has not been determined from the genomes currently available.

Mucin, a polymer composed of large glycoproteins and secreted by many marine organisms, may be involved as both defenses against pathogens and mechanisms to capture beneficial symbionts from the environment (Chugh et al., 2015; Zheng et al., 2017). Similar to genes with high utility, its sequences may have undergone positive selection. Despite excessive release upon stressors, mucous production in corals and other cnidarians is not well studied or understood (Jatkar et al., 2010; Wright et al., 2019; Gao et al., 2021).

Like Hox, Sox gene clusters encode transcription factors that regulate a variety of important development pathways whose expression is dependent on the taxon and tissue type (Shinzato et al., 2008). The family of sox genes has actually been a mainstay of genetics studies, as *Sox* members, such as the *Dichaete* gene locus, were discovered in the early days of *Drosophila* genetics research. *Sox* genes have a long history in genetic research, even predating the concept of a mammalian sex-determining factor SRY. One of the first dominant alleles ever discovered in *Drosophila* was the wing and bristle mutation *Dichaete*, recovered by Calvin Bridges in 1915 (Bridges and Morgan, 1923). The *Sox* genes have been identified in the genomes of a wide spectrum of metazoans, and the number appears to increase with the complexity of body plans. For example, placozoan *Trichoplax* and sponges have only about four *Sox* genes, while ctenophores and cnidarians display a noticeable increase (Phochanukul and Russell, 2010): the scarlet sea anemone (*Nematostella vectensis*) has 14 *Sox* genes (Putnam et al., 2008); the freshwater hydra (*Hydra magnipapillata*) has 12 genes; the sea hydra (*C. hemisphaerica*) has 10 genes; the rose coral (*Acropora millepora*) has six genes (Shinzato et al., 2008); and comb jelly (*Pleurobrachia pileus*) has 13 *Sox* genes (Jager et al., 2006, 2008). Also, Echinodermata have six *Sox* genes (Dolmatov et al., 2021), and the lancelet *Branchiostoma floridae*, likely predecessor of the vertebrates, has at least 13 Sox genes. The increase of gene number, most likely by duplication, indicates that *Sox* genes probably play important roles in cnidarian development. In the coral *A. millepora*, Sox plays roles in neural development and early embryogenesis. In *N. vectensis*, the *NvSox2* transcription factors act as a binary switch between the development of two alternative stinging cell fates (Babonis et al., 2023).

By far, most diversity in gene sequences and the metabolic pathways they may encode can be found in marine microbes. For example, microbial sequence diversity positively correlated with their geographic location and biome type (Charlop-Powers et al., 2015). In the more recent survey of microbial metabolic potential, Paoli et al. (2022) identified 40,000 putative mostly new biosynthetic gene clusters from cultures and recently assembled microbial genomes (MAGs), including phospeptin and pythonamide pathways. I have already referred to secondary metabolite potential in the last chapter so will not dwell upon it here, except to foreshadow the utility for driving bioeconomies discussed in Chapter 5.

All of these examples provide a cross-sectional sampling of benthic biodiversity across both space and time. From the above listings, one can appreciate that comprehensive biodiversity assessments can encompass the broad spectrum of molecular, genetic, organismal, population, and ecosystem-level variation for any location and include all of the intrinsic connections.

References

Ablan-Lagman, A., 2018. The Spratly Islands. In: World Seas: An Environmental Evaluation. Academic Press, pp. 583–591.

Aczel, A.D., 2001. The Mystery of the Aleph: Mathematics, the Kabbalah, and the Search for Infinity. Simon and Schuster, New York.

Albertin, C.B., Simakov, O., Mitros, T., Wang, Z.Y., Pungor, J.R., Edsinger-Gonzales, E., et al., 2015. The octopus genome and the evolution of cephalopod neural and morphological novelties. Nature 524 (7564), 220–224.

Allen, E.E., Facciotti, D., Bartlett, D.H., 1999. Monounsaturated but not polyunsaturated fatty acids are required for growth of the deep-sea bacterium *Photobacterium profundum* SS9 at high pressure and low temperature. Appl. Environ. Microbiol. 65 (4), 1710–1720.

Almeida, M., Reis, R.L., Silva, T.H., 2020. Marine invertebrates are a source of bioadhesives with biomimetic interest. Mater. Sci. Eng. C 108, 110467.

Alvarez-Filip, L., Carricart-Ganivet, J.P., Horta-Puga, G., Iglesias-Prieto, R., 2013. Shifts in coral-assemblage composition do not ensure persistence of reef functionality. Sci. Rep. 3 (1), 3486.

Alves, C., Valdivia, A., Aronson, R.B., Bood, N., Castillo, K.D., Cox, C., et al., 2022. Twenty years of change in benthic communities across the Belizean Barrier Reef. PLoS One 17 (1), e0249155.

Amon, D.J., Ziegler, A.F., Dahlgren, T.G., Glover, A.G., Goineau, A., Gooday, A.J., et al., 2016. Insights into the abundance and diversity of abyssal megafauna in a polymetallic-nodule region in the eastern Clarion-Clipperton Zone. Sci. Rep. 6 (1), 1–12.

Andradi-Brown, D., Appeldoorn, R.S., Baker, E., Ballantine, D., Bejarano, I., Bridge, T.L., Colin, P.L., Eyal, G., Harris, P., Holstein, D., Jones, R., Kahng, S.E., Laverick, J., Loya, Y., Pochon, X., Pomponi, S.A., Puglise, K. A., Pyle, R.L., Reaka, M.L., Ruiz, J.H., Sealover, N., Semmler, R.F., Schizas, N., Schmidt, W., Sherman, C., Sinniger, F., Slattery, M., Spalding, H.L., Villalobos, S.G., Weil, E., Wood, E., 2016. Mesophotic Coral Ecosystems: A Lifeboat for Coral Reefs? UN Environment, GRID-Arendal.

Aronson, R.B., Blake, D.B., 2001. Global climate change and the origin of modern benthic communities in Antarctica. Am. Zool. 41 (1), 27–39.

Aronson, R.B., Macintyre, I.G., Precht, W.F., Murdoch, T.J., Wapnick, C.M., 2002. The expanding scale of species turnover events on coral reefs in Belize. Ecol. Monogr. 72 (2), 233–249.

Asaad, I., Lundquist, C.J., Erdmann, M.V., Costello, M.J., 2018. Delineating priority areas for marine biodiversity conservation in the Coral Triangle. Biol. Conserv. 222, 198–211.

Babonis, L.S., Enjolras, C., Reft, A.J., Foster, B.M., Hugosson, F., et al., 2023. Single-cell atavism reveals an ancient mechanism of cell type diversification in a sea anemone. Nat. Commun. 14 (1), 885.

Banks, K.E., Riegl, B.M., Richards, V.P., Walker, B.K., Helmle, K.P., Jordan, L.K.B., Phipps, J., Shivji, M., Spieler, R.E., Dodge, R.E., 2008. The reef tract of continental Southeast Florida (Miami-Dade, Broward,

158 Chapter 3 Diversity hotspots on the benthos

and Palm Beach Counties, USA). In: Riegl, B., Dodge, R.E. (Eds.), Coral Reefs of the USA. Springer-Verlag, Dordrecht, pp. 125–172.

Bar-Dolev, M., Basu, K., Braslavsky, I., Davies, P.L., 2020. Structure–function of IBPs and their interactions with ice. In: Antifreeze Proteins Volume 2: Biochemistry, Molecular Biology and Applications, pp. 69–107.

Barlow, J., França, F., Gardner, T.A., Hicks, C.C., Lennox, G.D., Berenguer, E., et al., 2018. The future of hyperdiverse tropical ecosystems. Nature 559 (7715), 517–526.

Beger, M., McGowan, J., Treml, E.A., Green, A.L., White, A.T., Wolff, N.H., et al., 2015. Integrating regional conservation priorities for multiple objectives into national policy. Nat. Commun. 6 (1), 8208.

Begun, D.J., Lindfors, H.A., Thompson, M.E., Holloway, A.K., 2006. Recently evolved genes identified from Drosophila yakuba and D. erecta accessory gland expressed sequence tags. Genetics 172 (3), 1675–1681.

Bellwood, D.R., Streit, R.P., Brandl, S.J., Tebbett, S.B., 2019. The meaning of the term 'function' in ecology: a coral reef perspective. Funct. Ecol. 33 (6), 948–961.

Blaxter, M., Archibald, J.M., Childers, A.K., Coddington, J.A., Crandall, K.A., Di Palma, F., et al., 2022. Why sequence all eukaryotes? Proc. Natl. Acad. Sci. 119 (4). https://doi.org/10.1073/pnas.2115636118.

Boutet, I., Jollivet, D., Shillito, B., Moraga, D., Tanguy, A., 2009. Molecular identification of differentially regulated genes in the hydrothermal-vent species Bathymodiolus thermophilus and Paralvinella pandorae in response to temperature. BMC Genomics 10 (1), 1–17.

Box, I.C., Matthews, B.J., Marshall, K.E., 2022. Molecular evidence of intertidal habitats selecting for repeated ice-binding protein evolution in invertebrates. J. Exp. Biol. 225 (Suppl_1), jeb243409.

Bräger, S., Rodriguez, G.Q.R., Mulsow, S., 2020. The current status of environmental requirements for deep seabed mining issued by the International Seabed Authority. Mar. Policy 114, 103258.

Brandt, A., Gooday, A.J., Brix, S.B., Brökeland, W., Cedhagen, T., Choudhury, M., Cornelius, N., Danis, B., De Mesel, I., Diaz, R.J., Gillan, D.C., Ebbe, B., Howe, J., Janussen, D., Kaiser, S., Linse, K., Malyutina, M., Brandao, S., Pawlowski, J., Raupach, M., Vanreusel, A., 2007. The Southern Ocean deep sea: first insights into biodiversity and biogeography. Nature 447, 307–311. https://doi.org/10.1038/nature05827.

Brasier, M.J., Wiklund, H., Neal, L., Jeffreys, R., Linse, K., Ruhl, H., Glover, A.G., 2016. DNA barcoding uncovers cryptic diversity in 50% of deep-sea Antarctic polychaetes. R. Soc. Open Sci. 3 (11), 160432.

Bridge, T.C., Done, T.J., Friedman, A., Beaman, R.J., Williams, S.B., Pizarro, O., Webster, J.M., 2011. Variability in mesophotic coral reef communities along the Great Barrier Reef, Australia. Mar. Ecol. Prog. Ser. 428, 63–75.

Bridges, C.B., Morgan, T.H., 1923. The third-chromosome group of mutant characters of Drosophila melanogaster. Publs. Carnegie Inst. 327, 1–251.

Briggs, J.C., 1999. Coincident biogeographic patterns: indo-West Pacific Ocean. Evolution 53 (2), 326–335.

Briggs, J.C., Bowen, B.W., 2013. Marine shelf habitat: biogeography and evolution. J. Biogeogr. 40 (6), 1023–1035.

Brown, J.H., 2014. Why are there so many species in the tropics? J. Biogeogr. 41 (1), 8–22.

Burt, J.A., Bauman, A.G., 2020. Suppressed coral settlement following mass bleaching in the southern Persian/Arabian Gulf. Aquat. Ecosyst. Health Manag. 23 (2), 166–174.

Burt, J., Al-Harthi, S., Al-Cibahy, A., 2011. Long-term impacts of coral bleaching events on the world's warmest reefs. Mar. Environ. Res. 72 (4), 225–229.

Caputi, S.S., Careddu, G., Calizza, E., Fiorentino, F., Maccapan, D., Rossi, L., Costantini, M.L., 2020. Seasonal food web dynamics in the Antarctic benthos of Tethys Bay (Ross Sea): implications for biodiversity persistence under different seasonal sea-ice coverage. Front. Mar. Sci. 7, 594454.

Carpenter, K.E., Abrar, M., Aeby, G., Aronson, R.B., Banks, S., Bruckner, A., et al., 2008. One-third of reef-building corals face elevated extinction risk from climate change and local impacts. Science 321, 560–563.

Carvunis, A.R., Rolland, T., Wapinski, I., Calderwood, M.A., Yildirim, M.A., Simonis, N., et al., 2012. Protogenes and de novo gene birth. Nature 487 (7407), 370–374.

Charlop-Powers, Z., Owen, J.G., Reddy, B.V.B., Ternei, M.A., Guimarães, D.O., de Frias, U.A., et al., 2015. Global biogeographic sampling of bacterial secondary metabolism. elife 4, e05048.

Chaudhary, C., Saeedi, H., Costello, M.J., 2017. Marine species richness is bimodal with latitude: a reply to Fernandez and marques. Trends Ecol. Evol. 32 (4), 234–237.

Chaves-Fonnegra, A., Riegl, B., Zea, Z., Lopez, J.V., Gilliam, D.S., 2017. Bleaching events regulate shifts from corals to excavating sponges in algae dominated reefs. Glob. Chang. Biol. 24 (2), 773–785.

Chen, L., DeVries, A.L., Cheng, C.H.C., 1997. Convergent evolution of antifreeze glycoproteins in Antarctic notothenioid fish and Arctic cod. Proc. Natl. Acad. Sci. 94 (8), 3817–3822.

Chenelot, H., Jewett, S.C., Hoberg, M.K., 2011. Macrobenthos of the nearshore Aleutian Archipelago, with emphasis on invertebrates associated with *Clathromorphum nereostratum* (Rhodophyta, Corallinaceae). Mar. Biodivers. 41, 413–424.

Chugh, S., Gnanapragassam, V.S., Jain, M., Rachagani, S., Ponnusamy, M.P., Batra, S.K., 2015. Pathobiological implications of mucin glycans in cancer: sweet poison and novel targets. Biochim. Biophys. Acta 1856, 211–225.

Cleves, P.A., Tinoco, A.I., Bradford, J., Perrin, D., Bay, L.K., Pringle, J.R., 2020. Reduced thermal tolerance in a coral carrying CRISPR-induced mutations in the gene for a heat-shock transcription factor. Proc. Natl. Acad. Sci. 117 (46), 28899–28905.

Cordier, T., Angeles, I.B., Henry, N., Lejzerowicz, F., Berney, C., Morard, R., Brandt, A., Cambon-Bonavita, M., Guidi, L., Lombard, F., Martínez, P., Arbizu, R.M., Orejas, C., Poulain, J., Smith, C.R., Wincker, P., Arnaud-Haond, S., Gooday, A.J., de Vargas, C., Pawlowski, J., 2022. Patterns of eukaryotic diversity from the surface to the deep-ocean sediment. Sci. Adv. 8, eabj9309. https://doi.org/10.1126/sciadv.abj9309.

Corliss, J.B., Dymond, J., Gordon, L.I., Edmond, J.M., von Herzen, R.P., Ballard, R.D., et al., 1979. Submarine thermal springs on the Galapagos Rift. Science 203 (4385), 1073–1083.

Costello, M.J., Tsai, P., Wong, P.S., Cheung, A.K.L., Basher, Z., Chaudhary, C., 2017. Marine biogeographic realms and species endemicity. Nat. Commun. 8 (1), 1–10.

Da Ros, Z., Dell'Anno, A., Morato, T., Sweetman, A.K., Carreiro-Silva, M., Smith, C.J., et al., 2019. The deep sea: the new frontier for ecological restoration. Mar. Policy 108, 103642.

Daane, J.M., Detrich III, H.W., 2022. Adaptations and diversity of Antarctic fishes: a genomic perspective. Annu. Rev. Anim. Biosci. 10, 39–62.

Danovaro, R., Dell'Anno, A., Pusceddu, A., Gambi, C., Heiner, I., Kristensen, R.M., 2010. The first metazoa living in permanently anoxic conditions. BMC Biol. 8 (1), 30.

Danovaro, R., Corinaldesi, C., Dell'Anno, A., Snelgrove, P.V.R., 2017. The deep sea under global change. Curr. Biol. 27, R461–R465. https://doi.org/10.1016/j.cub.2017.02.046.

Davey, P.A., Power, A.M., Santos, R., Bertemes, P., Ladurner, P., Palmowski, P., et al., 2021. Omics-based molecular analyses of adhesion by aquatic invertebrates. Biol. Rev. 96 (3), 1051–1075.

Dawkins, R., 1976. The Selfish Gene. Oxford University Press.

Dayton, P.K., 1972. Toward an understanding of community resilience and the potential effects of enrichments to the benthos at McMurdo Sound, Antarctica. In: Proceedings of the Colloquium on Conservation Problems in Antarctica. Allen Press, Lawrence, pp. 81–96.

Dayton, P.K., Mordida, B.J., Bacon, F., 1994. Polar marine communities. Am. Zool. 34 (1), 90–99.

de Oliveira Soares, M., de Araújo, J.T., Ferreira, S.M.C., Santos, B.A., Boavida, J.R.H., Costantini, F., Rossi, S., 2020. Why do mesophotic coral ecosystems have to be protected? Sci. Total Environ. 726, 138456.

DeBoer, T.S., Naguit, M.R., Erdmann, M.V., Ablan-Lagman, M.C.A., Ambariyanto, Carpenter, K.E., et al., 2014. Concordance between phylogeographic and biogeographic boundaries in the coral triangle: conservation implications based on comparative analyses of multiple giant clam species. Bull. Mar. Sci. 90 (1), 277–300.

Degen, R., Faulwetter, S., 2019. The Arctic traits database – a repository of Arctic benthic invertebrate traits. Earth Syst. Sci. Data 11 (1), 301–322.

Desplat, Y., Blake, E.J., Cuvelier, M., Warner, J.F., Vijayan, N., Blackwelder, P., Lopez, J.V., 2023. Morphological and transcriptional effects of crude oil and dispersant exposure on the marine sponge *Cinachyrella alloclada*. Sci. Total Environ. 878, 162832. https://doi.org/10.1016/j.scitotenv.2023.162832.

160 **Chapter 3** Diversity hotspots on the benthos

Dolmatov, I.Y., Kalacheva, N.V., Tkacheva, E.S., Shulga, A.P., Zavalnaya, E.G., Shamshurina, E.V., et al., 2021. Expression of Piwi, MMP, TIMP, and Sox during gut regeneration in holothurian *Eupentacta fraudatrix* (Holothuroidea, Dendrochirotida). Genes 12 (8), 1292.

Dowell, R.D., Ryan, O., Jansen, A., Cheung, D., Agarwala, S., Danford, T., et al., 2010. Genotype to phenotype: a complex problem. Science 328 (5977), 469.

Downing, A.B., Wallace, G.T., Yancey, P.H., 2018. Organic osmolytes of amphipods from littoral to hadal zones: increases with depth in trimethylamine N-oxide, scyllo-inositol and other potential pressure counteractants. Deep-Sea Res. I Oceanogr. Res. Pap. 138, 1–10.

Dubilier, N., Bergin, C., Lott, C., 2008. Symbiotic diversity in marine animals: the art of harnessing chemosynthesis. Nat. Rev. Microbiol. 6 (10), 725–740.

Dunn, C.W., Ryan, J.F., 2015. The evolution of animal genomes. Curr. Opin. Genet. Dev. 35, 25–32.

Dygico, M., Songco, A., White, A.T., Green, S.J., 2013. Achieving MPA effectiveness through application of responsive governance incentives in the Tubbataha reefs. Mar. Policy 41, 87–94.

Ekman, S., 1953. Zoogeography of the Sea. Sidgwick and Jackson.

Faircloth, B.C., McCormack, J.E., Crawford, N.G., Harvey, M.G., Brumfield, R.T., Glenn, T.C., 2012. Ultraconserved elements anchor thousands of genetic markers spanning multiple evolutionary timescales. Syst. Biol. 61 (5), 717–726.

Fairhead, M., Johnson, K.A., Kowatz, T., McMahon, S.A., Carter, L.G., Oke, M., et al., 2008. Crystal structure and silica condensing activities of silicatein α–cathepsin L chimeras. Chem. Commun. 15, 1765–1767.

Fine, M., Cinar, M., Voolstra, C.R., Safa, A., Rinkevich, B., Laffoley, D., et al., 2019. Coral reefs of the Red Sea—challenges and potential solutions. Reg. Stud. Mar. Sci. 25, 100498.

Finnerty, J.R., Paulson, D., Burton, P., Pang, K., Martindale, M.Q., 2003. Early evolution of a homeobox gene: the parahox gene Gsx in the Cnidaria and the Bilateria. Evol. Dev. 5 (4), 331–345.

Fischer, A., Bhakta, D., Macmillan-Lawler, M., Harris, P., 2019. Existing global marine protected area network is not representative or comprehensive measured against seafloor geomorphic features and benthic habitats. Ocean Coast. Manag. 167, 176–187.

Fisher, R.A., 1930. The Genetical Theory of Natural Selection. Clarendon Press, Oxford.

Fu, L., He, Y., Xu, F., Ma, Q., Wang, F., Xu, J., 2015. Characterization of a novel thermostable patatin-like protein from a Guaymas basin metagenomic library. Extremophiles 19, 829–840.

Futuyma, D., 2017. Evolution, 4th. Sinauer Associates.

Gaither, M.R., Bowen, B.W., Bordenave, T.R., Rocha, L.A., Gomez, J.A., van Herwerden, L., Craig, M.T., 2011. Phylogeography of the reef fish Cephalopholisargus (Epinephelidae) indicates Pleistocene isolation across the Indo-Pacific Barrier with contemporary overlap in the Coral Triangle. BMC Evol. Biol. 11, 189.

Gaither, M.R., Rocha, L.A., 2013. Origins of species richness in the Indo-Malay-Philippine biodiversity hotspot: evidence for the centre of overlap hypothesis. J. Biogeogr. 40 (9), 1638–1648.

Gao, C., Garren, M., Penn, K., Fernandez, V.I., Seymour, J.R., Thompson, J.R., et al., 2021. Coral mucus rapidly induces chemokinesis and genome-wide transcriptional shifts toward early pathogenesis in a bacterial coral pathogen. ISME J. 15 (12), 3668–3682.

Gardner, T.A., Côté, I.M., Gill, J.A., Grant, A., Watkinson, A.R., 2003. Long-term region-wide declines in Caribbean corals. Science 301 (5635), 958–960.

Gilliam, D.S., Hayes, N.K., Ruzicka, R., Colella, M., 2021. Southeast Florida Coral Reef Evaluation and Monitoring Project 2020 Year 18 Final Report. Miami Beach, FL, Florida DEP & FWC, p. 82. https://floridadep.gov/sites/default/files/Yr%2018%20SECREMP_Comprehensive%20Report_Final.pdf.

Giudice, A.L., Azzaro, M., Schiaparelli, S., 2019. Microbial symbionts of Antarctic marine benthic invertebrates. In: The Ecological Role of Micro-organisms in the Antarctic Environment. Springer, pp. 277–296.

Gomez, E.D., Alino, P.M., Yap, H.T., Licuanan, W.Y., 1994. A review of the status of Philippine reefs. Mar. Pollut. Bull. 29 (1–3), 62–68.

Gooday, A.J., Lejzerowicz, F., Goineau, A., Holzmann, M., Kamenskaya, O., Kitazato, H., et al., 2021. The biodiversity and distribution of abyssal benthic foraminifera and their possible ecological roles: a synthesis across the Clarion-Clipperton Zone. Front. Mar. Sci. 8, 634726.

Graur, D., Li, W.-S., 2000. Fundamentals of Molecular Evolution. Oxford University Press.

Green, A.L., Fernandes, L., Almany, G., Abesamis, R., McLeod, E., Aliño, P.M., et al., 2014. Designing marine reserves for fisheries management, biodiversity conservation, and climate change adaptation. Coast. Manag. 42 (2), 143–159.

Grice, L.F., Degnan, B.M., 2015. The origin of the ADAR gene family and animal RNA editing. BMC Evol. Biol. 15, 1–7.

Guijarro-Clarke, C., Holland, P.W.H., Paps, J., 2020. Widespread patterns of gene loss in the evolution of the animal kingdom. Nat. Ecol. Evol. 4 (4), 519–523.

Gutt, J., 1988. On the Distribution and ecology of sea cucumbers (Holothuroidea, Echinodermata) in the Weddell Sea (Antarctica). Ber. Polarforsch. 41. https://doi.org/10.2312/BzP_0041_1988.

Gutt, J., 2008. The Expedition ANTARKTIS-XXIII/8 of the research vessel "Polarstern" in 2006/2007. Ber. Polarforsch. 569. https://doi.org/10.2312/BzPM_0569_2008.

Gutt, J., 2013. The expedition ANTARKTIS-XXIX/3 of the research vessel "Polarstern" in 2013. Ber. Polarforsch. 665. https://doi.org/10.2312/BzPM_0665_2013.

Haldane, J.B.S., 1932. The Causes of Evolution. Longmans, Green & Co, London.

Hamel, J.F. (Ed.), 2018. World Seas: An Environmental Evaluation: Volume II: The Indian Ocean to the Pacific.

Hauquier, F., Macheriotou, L., Bezerra, T.N., Egho, G., Martínez Arbizu, P., Vanreusel, A., 2019. Distribution of free-living marine nematodes in the Clarion–Clipperton Zone: implications for future deep-sea mining scenarios. Biogeosciences 16 (18), 3475–3489.

Hayes, N.K., Walton, C.J., Gilliam, D.S., 2022. Tissue loss disease outbreak significantly alters the Southeast Florida stony coral assemblage. Front. Mar. Sci. 9. https://doi.org/10.3389/fmars.2022.975894.

Hinderstein, L.M., Marr, J.C.A., Martinez, F.A., Dowgiallo, M.J., Puglise, K.A., Pyle, R.L., et al., 2010. Theme section on "Mesophotic coral ecosystems: characterization, ecology, and management". Coral Reefs 29, 247–251.

Hoeksema, B.W., 2007. Delineation of the Indo-Malayan centre of maximum marine biodiversity: the Coral Triangle. In: Renema, W. (Ed.), Biogeography, Time, and Place: Distributions, Barriers, and Islands. Springer, Dordrecht, pp. 117–178. XII + 416 p.

Huang, Z., Huang, W., Liu, X., Han, Z., Liu, G., Boamah, G.A., et al., 2022. Genomic insights into the adaptation and evolution of the nautilus, an ancient but evolving "living fossil". Mol. Ecol. Resour. 22 (1), 15–27.

Hughes, T.P., Bellwood, D.R., Connolly, S.R., 2002. Biodiversity hotspots, centres of endemicity, and the conservation of coral reefs. Ecol. Lett. 5 (6), 775–784.

Hughes, T.P., Anderson, K.D., Connolly, S.R., Heron, S.F., Kerry, J.T., Lough, J.M., Baird, A.H., Baum, J.K., Berumen, M.L., Bridge, T.C., Claar, D.C., Mark Eakin, C., Gilmour, J.P., Graham, N.A.J., Harrison, H., Hobbs, J.-P.A., Hoey, A.S., Hoogenboom, M., Lowe, R.J., McCulloch, M.T., Pandolfi, J.M., Pratchett, M., Schoepf, V., Torda, G., Wilson, S.K., 2018. Spatial and temporal patterns of massbleaching of corals in the Anthropocene. Science 359 (6371), 80–83. https://doi.org/10.1126/science.aan8048.

Hutchison, C.S., Vijayan, V.R., 2010. What are the Spratly Islands? J. Asian Earth Sci. 39 (5), 371–385.

Hutomo, M., Moosa, M.K., 2005. Indonesian marine and coastal biodiversity: present status. Indian J. Mar. Sci. 34 (1), 88–97.

Huxley, J.S., 1942. Evolution: The Modern Synthesis. Allen and Unwin, London.

Ikuta, T., 2011. Evolution of invertebrate deuterostomes and Hox/ParaHox genes. Genom. Proteom. Bioinform. 9 (3), 77–96.

International Seabed Authority (ISA), 2010. A Geological Model of Polymetallic Nodule Deposits in the Clarion-Clipperton Fracture Zone and Prospector's Guide for Polymetallic Nodule Deposits in the Clarion Clipperton Fracture Zone. Technical Study: No. 6, ISBN: 978-976-95268-2-2.

162 **Chapter 3** Diversity hotspots on the benthos

Ip, J.C.H., Zhang, Y., Xie, J.Y., Yeung, Y.H., Qiu, J.W., 2022. Comparative transcriptomics of two coral holo-bionts collected during the 2017 El Niño heat wave reveal differential stress response mechanisms. Mar. Pollut. Bull. 182, 114017.

Jacob, F., 1977. Evolution and tinkering. Science 196 (4295), 1161–1166.

Jager, M., Queinnec, E., Houliston, E., Manuel, M., 2006. Expansion of the SOX gene family predated the emergence of the Bilateria. Mol. Phylogenet. Evol. 39, 468–477.

Jager, M., Queinnec, E., Chiori, R., Le Guyader, H., Manuel, M., 2008. Insights into the early evolution of SOX genes from expression analyses in a ctenophore. J. Exp. Zool. B Mol. Dev. Evol. 310, 650–667.

Jamieson, A.J., 2015. The Hadal Zone: Life in the Deepest Oceans. Cambridge University Press.

Jatkar, A.A., Brown, B.E., Bythell, J.C., Guppy, R., Morris, N.J., Pearson, J.P., 2010. Coral mucus: the properties of its constituent mucins. Biomacromolecules 11 (4), 883–888.

Johnson, B.R., 2018. Taxonomically restricted genes are fundamental to biology and evolution. Front. Genet. 9, 407.

Jones, N.P., Figueiredo, J., Gilliam, D.S., 2020. Thermal stress-related spatiotemporal variations in high-latitude coral reef benthic communities. Coral Reefs. https://doi.org/10.1007/s00338-020-01994-8.

Kahng, S.E., Akkaynak, D., Shlesinger, T., Hochberg, E.J., Wiedenmann, J., Tamir, R., Tchernov, D., 2019. Light, temperature, photosynthesis, heterotrophy, and the lower depth limits of mesophotic coral ecosystems. In: Mesophotic Coral Ecosystems. Springer, p. 801.

Kahng, S.E., Garcia-Sais, J.R., Spalding, H.L., Brokovich, E., Wagner, D., Weil, E., et al., 2010. Community ecology of mesophotic coral reef ecosystems. Coral Reefs 29, 255–275.

Khalturin, K., Hemmrich, G., Fraune, S., Augustin, R., Bosch, T.C., 2009. More than just orphans: are taxonomically-restricted genes important in evolution? Trends Genet. 25 (9), 404–413.

Koonin, E.V., 2009. Darwinian evolution in the light of genomics. Nucleic Acids Res. 37 (4), 1011–1034.

Kozhemyako, V.B., Veremeichik, G.N., Shkryl, Y.N., Kovalchuk, S.N., Krasokhin, V.B., Rasskazov, V.A., et al., 2010. Silicatein genes in spicule-forming and nonspicule-forming Pacific demosponges. Mar. Biotechnol. 12, 403–409.

Kozmik, Z., 2008. The role of pax genes in eye evolution. Brain Res. Bull. 75 (2–4), 335–339.

Krell, A., Beszteri, B., Dieckmann, G., Glöckner, G., Valentin, K., Mock, T., 2008. A new class of ice-binding proteins discovered in a salt-stress-induced cDNA library of the psychrophilic diatom *Fragilariopsis cylindrus* (Bacillariophyceae). Eur. J. Phycol. 43 (4), 423–433.

Krieger, K.J., Wing, B.L., 2002. Megafauna associations with Deepwater corals (Primnoa spp.) in the Gulf of Alaska. Hydrobiologia 471 (1–3), 83–90.

Lan, Y., Sun, J., Tian, R., Bartlett, D., Li, R., Wong, Y.H., et al., 2017. Molecular adaptation in the world's deepest-living animal: insights from transcriptome sequencing of the hadal amphipod *Hirondellea gigas*. Mol. Ecol. 26, 3732–3743.

Lee, B.P., Messersmith, P.B., Israelachvili, J.N., Waite, J.H., 2011. Mussel-inspired adhesives and coatings. Annu. Rev. Mater. Res. 41, 99–132.

Lejzerowicz, F., Gooday, A.J., Barrenechea Angeles, I., Cordier, T., Morard, R., Apothéloz-Perret-Gentil, L., et al., 2021. Eukaryotic biodiversity and spatial patterns in the clarion-Clipperton zone and other abyssal regions: insights from sediment DNA and RNA metabarcoding. Front. Mar. Sci. 8, 671033.

Lessios, H.A., 1988. Mass mortality of Diadema antillarum in the Caribbean: what have we learned? Annu. Rev. Ecol. Syst. 19 (1), 371–393.

Levin, L.A., Baco, A.R., Bowden, D.A., Colaco, A., Cordes, E.E., Cunha, M.R., et al., 2016. Hydrothermal vents and methane seeps: rethinking the sphere of influence. Front. Mar. Sci. 3, 72.

Levine, M.T., Jones, C.D., Kern, A.D., Lindfors, H.A., Begun, D.J., 2006. Novel genes derived from noncoding DNA in *Drosophila melanogaster* are frequently X-linked and exhibit testis-biased expression. Proc. Natl. Acad. Sci. 103 (26), 9935–9939.

Lewin, H.A., Richards, S., Lieberman Aiden, E., Allende, M.L., et al., 2022. The Earth BioGenome Project 2020: starting the clock. Proc. Natl. Acad. Sci. U. S. A. 119 (4), e2115635118. https://doi.org/10.1073/pnas.2115635118.

Licuanan, W.Y., Cabreira, R.W., Alino, P.M., 2019a. The Philippines. In: Sheppard (Ed.), World Seas: An Environmental Evaluation, Indian Ocean to the Pacific. Vol. II. Academic Press, London, pp. 515–537.

Licuanan, A.M., Reyez, M.Z., Luzon, K.S., Chan, M.A.A., Licuanan, W.Y., 2017. Initial findings of the Nationwide assessment of Philippine coral reefs. Philippine J. Sci. 146 (2), 177–185.

Licuanan, W.Y., Robles, R., Reyes, M., 2019b. Status and recent trends in coral reefs of the Philippines. Mar. Pollut. Bull. 142, 544–550.

Lighty, R.G., 1977. Relict shelf-edge Holocene coral reef: southeast coast of Florida. In: Proceedings: 3rd International Coral Reef Symposium. 2, pp. 215–221.

Lighty, R.G., Macintyre, I.G., Stuckenrath, R., 1978. Submerged early Holocene barrier reef south-east Florida shelf. Nature 275, 59–60.

Linse, K., Griffiths, H.J., Barnes, D.K.A., Clarke, A., 2006. Biodiversity and biogeography of Antarctic and Sub-Antarctic Mollusca. Deep-Sea Res. Pt. II 53, 985–1008. https://doi.org/10.1016/j.dsr2.2006.05.003.

Littler, M.M., Littler, D.S., Brooks, B.L., 2006. Harmful algae on tropical coral reefs: bottom-up eutrophication and top-down herbivory. Harmful Algae 5 (5), 565–585.

Macintyre, I.G., 1988. Modern coral reefs of western Atlantic: new geologic perspectives. Am. Assoc. Petrol Geol. Bull. 72, 1360–1369.

Macintyre, I.G., Milliman, J.D., 1970. Physiographic features on the outer shelf and upper slope, Atlantic continental margin, southeastern United States. Geol. Soc. Am. Bull. 81, 2577–2598.

Mallo, M., Alonso, C.R., 2013. The regulation of hox gene expression during animal development. Development 140 (19), 3951–3963.

Mao, F., Mu, H., Wong, N.K., Liu, K., Song, J., Qiu, J., et al., 2020. Hemocyte phagosomal proteome is dynamically shaped by cytoskeleton remodeling and interorganellar communication with endoplasmic reticulum during phagocytosis in a marine invertebrate, *Crassostrea gigas*. Sci. Rep. 10 (1), 1–16.

Marini, S., Bonofiglio, F., Corgnati, L.P., Bordone, A., Schiaparelli, S., Peirano, A., 2022. Long-term automated visual monitoring of Antarctic benthic fauna. Methods Ecol. Evol. 13 (8), 1746–1764.

Mayr, E., 1944. Systematics and the Origin of Species. Columbia University Press, New York.

McFadden, C.S., Brown, A.S., Brayton, C., Hunt, C.B., Van Ofwegen, L.P., 2014. Application of DNA barcoding in biodiversity studies of shallow-water octocorals: molecular proxies agree with morphological estimates of species richness in Palau. Coral Reefs 33, 275–286.

McGinnis, W., Krumlauf, R., 1992. Homeobox genes and axial patterning. Cell 68 (2), 283–302.

McKenna, V., Archibald, J.M., Beinart, R., Dawson, M.N., Hentschel, U., Keeling, P.J., Lopez, J.V., Martín-Durán, J.M., Petersen, J.M., Sigwart, J.D., Simakov, O., Sutherland, K.R., Sweet, M., Talbot, N., Thompson, A.W., Bender, S., Harrison, P.W., Rajan, J., Cochrane, G., Berriman, M., Lawniczak, M., Blaxter, M., 2021. The Aquatic Symbiosis Genomics Project: probing the evolution of symbiosis across the tree of life. [version 1; peer review: awaiting peer review]. Wellcome Open Res. 6, 254. https://wellcomeopenresearch.org/articles/6-254/v1.

McMahon, B.J., Teeling, E.C., Höglund, J., 2014. How and why should we implement genomics into conservation? Evol. Appl. 7, 999–1007. https://doi.org/10.1111/eva.12193.

McManus, J.W., Menez, L.B., 1998. Proceedings of the 8th International Coral Reef Symposium. Vol. 2 The Proposed International Spratly Island.

Meng, J., Zhu, Q., Zhang, L., Li, C., Li, L., She, Z., et al., 2013. Genome and transcriptome analyses provide insight into the euryhaline adaptation mechanism of *Crassostrea gigas*. PLoS One 8 (3), e58563.

Morin, J.G., Hastings, J.W., 1971. Energy transfer in a bioluminescent system. J. Cell. Physiol. 77 (3), 313–318.

Morishita, H., Yagi, T., 2007. Protocadherin family: diversity, structure, and function. Curr. Opin. Cell Biol. 19 (5), 584–592.

Moyer, R.P., Riegl, B., Banks, K., Dodge, R.E., 2003. Spatial patterns and ecology of benthic communities on a high-latitude South Florida (Broward County, USA) reef system. Coral Reefs 22 (4), 447–464.

Mu, Y., Bian, C., Liu, R., Wang, Y., Shao, G., Li, J., et al., 2021. Whole genome sequencing of a snailfish from the Yap Trench (~7,000 m) clarifies the molecular mechanisms underlying adaptation to the deep sea. PLoS Genet. 17 (5), e1009530.

Myers, N., 1990. The biodiversity challenge: expanded hot-spots analysis. Environmentalist 10 (4), 243–256.

Nei, M., Kumar, S., 2000. Molecular Evolution and Phylogenetics. Oxford University Press, United States.

Neuhaus, B., 2004. Description of Campyloderes cf. vanhoeffeni (Kinorhyncha, Cyclorhagida) from the central American East Pacific deep sea with a review of the genus. Meiofauna Mar 13, 3–20.

Nie, H., Jiang, L., Chen, P., Huo, Z., Yang, F., Yan, X., 2017. High throughput sequencing of RNA transcriptomes in *Ruditapes philippinarum* identifies genes involved in osmotic stress response. Sci. Rep. 7 (1), 4953.

Oakley, T.H., 2017. Furcation and fusion: the phylogenetics of evolutionary novelty. Dev. Biol. 431 (1), 69–76.

Obura, D.O., Aeby, G., Amornthammarong, N., Appeltans, W., Bax, N., Bishop, J., et al., 2019. Coral reef monitoring, reef assessment technologies, and ecosystem-based management. Front. Mar. Sci. 6, 580.

Ohno, S., 1970. Evolution by Gene Duplication. Vol. xv Allen & Unwin; Springer-Verlag, London, New York. 160 p.

Ortiz, M., Levins, R., Campos, L., Berrios, F., Campos, F., Jordán, F., et al., 2013. Identifying keystone trophic groups in benthic ecosystems: implications for fisheries management. Ecol. Indic. 25, 133–140.

Paoli, L., Ruscheweyh,, H.J., Forneris, C.C., Hubrich, F., Kautsar, S., et al., 2022. Biosynthetic potential of the global ocean microbiome. Nature 607 (7917), 111–118.

Pascal, N., Allenbach, M., Brathwaite, A., Burke, L., Le Port, G., Clua, E., 2016. Economic valuation of coral reef ecosystem service of coastal protection: a pragmatic approach. Ecosyst. Serv. 21, 72–80.

Pearson, J.C., Lemons, D., McGinnis, W., 2005. Modulating Hox gene functions during animal body patterning. Nat. Rev. Genet. 6 (12), 893–904.

Perry, C.T., Murphy, G.N., Kench, P.S., Smithers, S.G., Edinger, E.N., Steneck, R.S., Mumby, P.J., 2013. Caribbean-wide decline in carbonate production threatens coral reef growth. Nat. Commun. 4 (1), 1402.

Perry, C.T., Morgan, K.M., Yarlett, R.T., 2017. Reef habitat type and spatial extent as interacting controls on platform-scale carbonate budgets. Front. Mar. Sci. 4, 185.

Petersen, H.C., Knott, K.E., Banta, G.T., Hansen, B.W., 2022. Ultra-conserved elements provide insights to the biogeographic patterns of three benthic macroinvertebrate species in the Baltic Sea. Estuar. Coast. Shelf Sci. 271, 107863.

Phochanukul, N., Russell, S., 2010. No backbone but lots of Sox: invertebrate Sox genes. Int. J. Biochem. Cell Biol. 42 (3), 453–464.

Poore, G.C., Avery, L., Błażewicz-Paszkowycz, M., Browne, J., Bruce, N.L., Gerken, S., et al., 2015. Invertebrate diversity of the unexplored marine western margin of Australia: taxonomy and implications for global biodiversity. Mar. Biodivers. 45, 271–286.

Punnett, R.C., 1922. Mendelism. Macmillan, London (1st Pub. 1905) https://archive.org/details/mendelism00punn/page/n7/mode/2up?view=theater.

Putnam, N.H., Butts, T., Ferrier, D.E., Furlong, R.F., Hellsten, U., Kawashima, T., et al., 2008. The amphioxus genome and the evolution of the chordate karyotype. Nature 453, 1064–1071.

Pyle, R.L., Copus, J.M., 2019. Mesophotic coral ecosystems: introduction and overview. In: Loya, Y., et al. (Eds.), Mesophotic Coral Ecosystems, Coral Reefs of the World. Vol. 12, pp. 3–27, https://doi.org/10.1007/978-3-319-92735-0_1.

Ransome, E., Geller, J.B., Timmers, M., Leray, M., 2017. The importance of standardization for biodiversity comparisons: a case study using autonomous reef monitoring structures (ARMS) and metabarcoding to measure cryptic diversity on Mo'orea coral reefs, French Polynesia. PLoS One 12 (4), e0175066.

Reaka, M.L., Sealover, N., Semmler, R.F., Villalobos, S.G., 2016. A quantitative study of the biodiversity of coral and other groups of organisms in the mesophotic zone compared to shallower depths in the Gulf of Mexico,

USA. In: Baker, E.K., Puglise, K.A., Harris, P.T. (Eds.), Mesophotic Coral Ecosystems – A Lifeboat for Coral Reefs? The United Nations Environment Programme and GRID-Arendal, Nairobi and Arendal, pp. 61–62. http://www.grida.no/publications/mesophotic-coral-ecosystems/.

Reid, W.V., 1998. Biodiversity hotspots. Trends Ecol. Evol. 13 (7), 275–280.

Reimer, J.D., Albelda, R.L., Biondi, P., Hardianto, E., Huang, S., Masucci, G.D., et al., 2022. Literature review of coral reef restoration in and around the coral triangle from the viewpoint of marine biodiversity. Arq. Ciên. Mar, Fortaleza 55, 413–431.

Rex, M.A., McClain, C.R., Johnson, N.A., Etter, R.J., et al., 2005. A source-sink hypothesis for abyssal biodiversity. Am. Nat. 165 (2), 163–178.

Rex, M.A., 1973. Deep-sea species diversity: decreased gastropod diversity at abyssal depths. Science 181 (4104), 1051–1053.

Riegl, B., 2003. Climate change and coral reefs: different effects in two high-latitude areas (Arabian Gulf, South Africa). Coral Reefs 22, 433–446.

Rocha, L.A., Pinheiro, H.T., Shepherd, B., Papastamatiou, Y.P., Luiz, O.J., Pyle, R.L., Bongaerts, R., 2018. Mesophotic coral ecosystems are threatened and ecologically distinct from shallow water reefs. Science 361 (6399), 281–284.

Roda, A., 2010. Discovery and development of the green fluorescent protein, GFP: the 2008 Nobel Prize. Anal. Bioanal. Chem. 396 (5), 1619–1622.

Rooney, J., Donham, E., Montgomery, A., Spalding, H., Parrish, F., Boland, R., et al., 2010. Mesophotic coral ecosystems in the Hawaiian Archipelago. Coral Reefs 29, 361–367.

Roskov, Y., Kunze, T., Paglinawan, L., 2013. Species 2000 & ITIS Catalogue of Life, 2013 Annual Checklist. https://centaur.reading.ac.uk/34322/1/Catalogue_of_Life_2013.pdf.

Rothschild, L.J., Mancinelli, R.L., 2001. Life in extreme environments. Nature 409 (6823), 1092–1101.

Ruby, E.G., Morin, J.G., 1979. Luminous enteric bacteria of marine fishes: a study of their distribution, densities, and dispersion. Appl. Environ. Microbiol. 38 (3), 406–411.

Saeedi, H., Brandt, A., 2020. Biogeographic Atlas of the Deep NW Pacific Fauna. Pensoft Publishers, Sofia.

Saeedi, H., Simoes, M., Brandt, A., 2019. Endemicity and community composition of marine species along the NW Pacific and the adjacent Arctic Ocean. Prog. Oceanogr. 178, 102199.

Sánchez, N., González-Casarrubios, A., Cepeda, D., Khodami, S., Pardos, F., Vink, A., Arbizu, P.M., 2022. Diversity and distribution of Kinorhyncha in abyssal polymetallic nodule areas of the Clarion-Clipperton Fracture Zone and the Peru Basin, East Pacific Ocean, with the description of three new species and notes on their intraspecific variation. Mar. Biodivers. 52 (5), 52.

Sarrazin, J., Juniper, S.K., Massoth, G., Legendre, P., 1999. Physical and chemical factors influencing species distributions on hydrothermal sulfide edifices of the Juan de Fuca Ridge, Northeast Pacific. Mar. Ecol. Prog. Ser. 190, 89–112.

Schleyer, M.H., Floros, C., Laing, S.C., Macdonald, A.H., Montoya-Maya, P.H., Morris, T., et al., 2018. What can South African reefs tell us about the future of high-latitude coral systems? Mar. Pollut. Bull. 136, 491–507.

Serrano, X., Baums, I.B., O'reilly, K., Smith, T.B., Jones, R.J., Shearer, T.L., et al., 2014. Geographic differences in vertical connectivity in the Caribbean coral M ontastraea cavernosa despite high levels of horizontal connectivity at shallow depths. Mol. Ecol. 23 (17), 4226–4240.

Shendure, J., Ji, H., 2008. Next-generation DNA sequencing. Nat. Biotechnol. 26 (10), 1135–1145.

Sheppard, C. (Ed.), 2018. World Seas: An Environmental Evaluation: Volume I: Europe, The Americas and West Africa.

Shimomura, O., Johnson, F.H., Saiga, Y., 1962. Extraction, purification and properties of aequorin, a bioluminescent protein from the luminous hydromedusan, Aequorea. J. Cell. Comp. Physiol. 59 (3), 223–239.

Shinzato, C., Iguchi, A., Hayward, D.C., Technau, U., Ball, E.E., Miller, D.J., 2008. Sox genes in the coral *Acropora millepora*: divergent expression patterns reflect differences in developmental mechanisms within the Anthozoa. BMC Evol. Biol. 8 (1), 1–16.

166 Chapter 3 Diversity hotspots on the benthos

Shkryl, Y.N., Veremeichik, G.N., Kamenev, D.G., Gorpenchenko, T.Y., Yugay, Y.A., Mashtalyar, D.V., et al., 2018. Green synthesis of silver nanoparticles using transgenic *Nicotiana tabacum* callus culture expressing silicatein gene from marine sponge *Latrunculia oparinae*. Artif. Cells Nanomed. Biotechnol. 46 (8), 1646–1658.

Simpson, G.G., 1944. Tempo and Mode in Evolution. Columbia. University Press, New York.

Slattery, M., Lesser, M.P., 2021. Gorgonians are foundation species on sponge-dominated Mesophotic Coral Reefs in the Caribbean. Front. Mar. Sci. 8, 654268.

So, C.R., Fears, K.P., Leary, D.H., Scancella, J.M., Wang, Z., Liu, J.L., et al., 2016. Sequence basis of barnacle cement nanostructure is defined by proteins with silk homology. Sci. Rep. 6 (1), 36219.

Song, Y., Song, L., Soomro, M.A., Dong, X., Hu, G., 2021. Transcriptome analysis reveals immune-related differentially expressed genes in the hepatopancreas and gills of *Penaeus vannamei* after *Vibrio anguillarum* infection. Aquac. Res. 52 (9), 4303–4316.

Stępień, A., Pabis, K., Błażewicz, M., 2019. Small-scale species richness of the Great Barrier Reef tanaidaceans—results of the CReefs compared with worldwide diversity of coral reef tanaidaceans. Mar. Biodivers. 49, 1169–1185.

Stone, R.P., Shotwell, S.K., 2007. State of deep coral ecosystems in the Alaska region: Gulf of Alaska, Bering Sea and the Aleutian Islands. State Deep Coral Ecosyst. United States 365, 65–108.

Stracke, K., Hejnol, A., 2023. Marine animal evolutionary developmental biology—advances through technology development. Evol. Appl. 16 (2), 580–588.

Stroeve, J., Notz, D., 2018. Changing state of Arctic Sea ice across all seasons. Environ. Res. Lett. 13 (10), 103001.

Styfhals, R., Seuntjens, E., Simakov, O., Sanges, R., Fiorito, G., 2019. In silico identification and expression of protocadherin gene family in *Octopus vulgaris*. Front. Physiol. 9, 1905.

Subramaniam, S., Fletcher, C., 2018. Trimethylamine N-oxide: breathe new life. Br. J. Pharmacol. 175 (8), 1344–1353.

Sun, Z., Zhao, Z.X., Li, J.B., Zhou, D., Wang, Z.W., 2011. Tectonic analysis of the breakup and collision unconformities in the Nansha. Chin. J. Geophys. 54 (12), 3196–3209.

Tautz, D., Domazet-Lošo, T., 2011. The evolutionary origin of orphan genes. Nat. Rev. Genet. 12 (10), 692–702.

Teh, L.S., Witter, A., Cheung, W.W., Sumaila, U.R., Yin, X., 2017. What is at stake? Status and threats to South China Sea marine fisheries. Ambio 46, 57–72.

Teschke, K., Brey, T., 2019a. Presence and absence records of sea star species (class: Asteroidea) from trawl, grab and trap samples in the Weddell Sea and western Antarctic Peninsula region during POLARSTERN cruises ANT I/2, ANT II/4, ANT V/3, ANT VI/3, ANT XV/3 and ANT XVII/3. Alfred Wegener Institute, Helmholtz Centre for Polar and Marine Research, Bremerhaven, PANGAEA, https://doi.org/10.1594/PANGAEA.898629.

Teschke, K., Brey, T., 2019b. Abundance records of five most abundant brittle star species (Ophiuroidea) from trawl, grab and trap samples in the Weddell Sea and neighbouring seas during POLARSTERN cruises ANT I/2, ANT II/4, ANT V/3 and 4, ANT VI/3, ANT IX/3 and ANT X/3 between 1983 and 1992. Alfred Wegener Institute, Helmholtz Centre for Polar and Marine Research, Bremerhaven, PANGAEA, https://doi.org/10.1594/PANGAEA.898773.

Teschke, K., Brey, T., 2020. Semi-quantitative records of benthic taxa from trawl samples taken in the wider Weddell Sea (Antarctica) during POLARSTERN cruises ANT VII/4, ANT IX/3, ANT XIII/3, ANT XV/3 and ANT XXI/2 between 1989 and 2004. Alfred Wegener Institute, Helmholtz Centre for Polar and Marine Research, Bremerhaven, PANGAEA, https://doi.org/10.1594/PANGAEA.911801.

Teschke, K., Pehlke, H., Brey, T., 2019. Spatial distribution of zoobenthos (sponges, echinoderms) in the wider Weddell Sea (Antarctica) with links to ArcGIS map packages. PANGAEA, https://doi.org/10.1594/PANGAEA.899645.

Teschke, K., Pehlke, H., Siegel, V., Bornemann, H., Knust, R., Brey, T., 2020. An integrated compilation of data sources for the development of a marine protected area in the Weddell Sea. Earth Syst. Sci. Data 12, 1003–1023. https://doi.org/10.5194/essd-12-1003-2020.

Theede, H., Schneppenheim, R., Beress, L., 1976. Antifreeze glycoproteins in Mytilus-edulis. Mar. Biol. 36 (2), 183–189.

Theissinger, K., Fernandes, C., Formenti, G., Bista, I., Berg, P.R., et al., 2023. How genomics can help biodiversity conservation. Trends Genet. 39 (7845), 545. https://doi.org/10.1016/j.tig.2023.01.005.

Thrush, S., Dayton, P., Cattaneo-Vietti, R., Chiantore, M., Cummings, V., Andrew, N., et al., 2006. Broad-scale factors influencing the biodiversity of coastal benthic communities of the Ross Sea. Deep-Sea Res. II Top. Stud. Oceanogr. 53 (8–10), 959–971.

Ulanowicz, R.E., 2012. Growth and Development: Ecosystems Phenomenology. Springer Science & Business Media.

UNESCO, 2009. Global Open Oceans and Deep Seabed (GOODS) – Biogeographic Classification. UNESCO-IOC, Paris (IOC Technical Series 84).

Valentine, J.W., Jablonski, D., 2015. A twofold role for global energy gradients in marine biodiversity trends. J. Biogeogr. 42 (6), 997–1005.

Van Dover, C.L., Humphris, S.E., Fornari, D., Cavanaugh, C.M., Collier, R., Goffredi, S.K., et al., 2001. Biogeography and ecological setting of Indian Ocean hydrothermal vents. Science 294 (5543), 818–823.

Van Oss, S.B., Carvunis, A.R., 2019. De novo gene birth. PLoS Genet. 15 (5), e1008160.

Van Woesik, R., 2000. Modelling processes that generate and maintain coral community diversity. Biodivers. Conserv. 9 (9), 1219–1233.

Vannier, T., Hingamp, P., Turrel, F., Tanet, L., Lescot, M., Timsit, Y., 2020. Diversity and evolution of bacterial bioluminescence genes in the global ocean. NAR Genomics Bioinform. 2 (2), lqaa018.

Veron, J.E., Devantier, L.M., Turak, E., Green, A.L., Kininmonth, S., Stafford-Smith, M., Peterson, N., 2009. Delineating the coral triangle. Galaxea J. Coral Reef Stud. 11 (2), 91–100.

Veron, J.C.E., DeVantier, L.M., Turak, E., Green, A.L., Kininmonth, S., Stafford-Smith, M., Peterson, N., 2011. The coral triangle. In: Coral Reefs: An Ecosystem in Transition. Springer, Dordrecht, pp. 47–55.

Von Damm, K.L., 1990. Seafloor hydrothermal activity: black smoker chemistry and chimneys. Annu. Rev. Earth Planet. Sci. 18 (1), 173–204.

Walker, B.K., 2012. Spatial analyses of benthic habitats to define coral reef ecosystem regions and potential biogeographic boundaries along a latitudinal gradient. PLoS One 7, e30466. https://doi.org/10.1371/journal.pone.0030466.

Walker, B.K., Gilliam, D.S., 2013. Determining the extent and characterizing coral reef habitats of the northern latitudes of the Florida Reef Tract (Martin County). PLoS One 8 (11), e80439. https://doi.org/10.1371/journal.pone.0080439.

Walker, B.K., Jordan, L.K.B., Spieler, R.E., 2009. Relationship of reef fish assemblages and topographic complexity on Southeastern Florida coral reef habitats. J. Coast. Res. 53 (Suppl. 1), 39–48. Retrieved from http://www.dx.doi.org/10.2112%2FSI53-005.1.

Walker, B.K., Riegl, B., Dodge, R.E., 2008. Mapping coral reef habitats in Southeast Florida using a combined technique approach. J. Coast. Res. 24 (5), 1138–1150.

Warwick, R.M., Clarke, K.R., 2001. Practical measures of marine biodiversity based on relatedness of species. Oceanogr. Mar. Biol. Annu. Rev. 39, 207–231.

Washburn, T.W., Menot, L., Bonifácio, P., Pape, E., Błażewicz, M., et al., 2021. Patterns of macrofaunal biodiversity across the Clarion-Clipperton Zone: an area targeted for seabed mining. Front. Mar. Sci. 8, 250.

Watling, L., Guinotte, J., Clark, M.R., Smith, C.R., 2013. A proposed biogeography of the deep ocean floor. Prog. Oceanogr. 111, 91–112.

Wilkinson, C., Salvat, B., Eakin, C.M., Brathwaite, A., Francini-Filho, R., Webster, N., et al., 2016. Tropical and Sub-Tropical Coral Reefs. United Nations, pp. 1–42.

Wilson, G.A., Bertrand, N., Patel, Y., Hughes, J.B., Feil, E.J., Field, D., 2005. Orphans as taxonomically restricted and ecologically important genes. Microbiology 15 (8), 2499–2501.

168 **Chapter 3** Diversity hotspots on the benthos

Witman, J.D., Etter, R.J., Smith, F., 2004. The relationship between regional and local species diversity in marine benthic communities: a global perspective. Proc. Natl. Acad. Sci. 101 (44), 15664–15669.

Wolf, Y.I., Rogozin, I.B., Grishin, N.V., Koonin, E.V., 2002. Genome trees and the tree of life. Trends Genet. 18 (9), 472–479.

Wright, S., 1986. *Evolution*: Selected Papers. University of Chicago Press, Chicago.

Wright, R.M., Strader, M.E., Genuise, H.M., Matz, M., 2019. Effects of thermal stress on amount, composition, and antibacterial properties of coral mucus. PeerJ 7, e6849.

Yancey, P.H., 2020. Cellular responses in marine animals to hydrostatic pressure. J. Exp. Zool. Part A: Ecol. Integr. Physiol. 333 (6), 398–420.

Yoshioka, Y., Yamashita, H., Suzuki, G., Zayasu, Y., Tada, I., Kanda, M., et al., 2021. Whole-genome transcriptome analyses of native symbionts reveal host coral genomic novelties for establishing coral–algae symbioses. Genome Biol. Evol. 13 (1), evaa240.

Yuan, J., Zhang, X., Gao, Y., Zhang, X., 2020. Adaptation and molecular evidence for convergence in decapod crustaceans from deep-sea hydrothermal vent environments. Mol. Ecol. 29 (20), 3954–3969.

Zheng, P., Wang, M., Li, C., Sun, X., Wang, X., Sun, Y., Sun, S., 2017. Insights into deep-sea adaptations and host–symbiont interactions: a comparative transcriptome study on Bathymodiolus mussels and their coastal relatives. Mol. Ecol. 26 (19), 5133–5148.

Zhuang, X., Yang, C., Murphy, K.R., Cheng, C.H.C., 2019. Molecular mechanism and history of non-sense to sense evolution of antifreeze glycoprotein gene in northern gadids. Proc. Natl. Acad. Sci. 116 (10), 4400–4405.

CHAPTER

Threats to benthic biodiversity

4

Death is one thing. An end to birth is another.
Michael Soulé

The provocative essay "Tragedy of the Commons," written by Garrett Hardin, synthesizes ecology, economics, sociology, and demographics, describing how untethered open access to natural resources can lead to eventual environmental collapse (Hardin, 1968). The premise was that common resources, such as a shared pasture or a fishery, become depleted when individuals act in their own self-interest and overuse them. He explains that each person has the incentive to extract as much benefit as possible from the resource before someone else does, leading to its eventual degradation. Hardin claims that the tragedy of the commons occurs because individuals lack a sense of responsibility for the resource and do not consider the effects of their actions on future generations. He asserts that private ownership of a resource helps prevent its overuse, but only if the owner has a long-term interest in preserving it. He also elaborates on the pros and cons of free will, or different types of public regulation or "coercion," which may take the form of taxes, fees, and fines. Hardin concludes by calling for greater social and political intervention to manage shared resources and prevent their degradation, such as regulation and the establishment of property rights. The composite messages of the article were both compelling and controversial, and thus some have experienced modulations, including those made by Hardin over the years (Hardin, 1991; Diamond, 2011). In a type of epiphany regarding possible actions to rectify or avoid the tragedies, Hardin states near the end of the article:

> …we can never do nothing. That which we have done for thousands of years is also action. it also produces evils. Once we are aware that the status quo is action, we can then compare its discoverable advantages and disadvantages with the predicted advantages and disadvantages of the proposed reform, discounting as best we can for our lack of experience. On the basis of such a comparison we can make a rational decision which will not involve the unworkable assumption that only perfect systems are tolerable.

Interestingly, this view unintentionally resembles some offshoots of Hinduism, whereupon one's passive interest, even without a specific activity, but rather the angle of approach to a problem, is thought to produce a type of "force" (e.g., shakti).

The commons also referred mostly to farms and pastures and did not directly refer to common marine areas, although the criterion of "low population" sizes can apply to benthic habitats such as oceanographic basins rarely visited by people. As in the earlier example, we can think about a "commons" as an abstraction, but which can be easily realized when we stroll over a sidewalk or public square and

Assessments and Conservation of Biological Diversity From Coral Reefs to the Deep Sea. https://doi.org/10.1016/B978-0-12-824112-7.00001-7
Copyright © 2024 Elsevier Inc. All rights reserved.

169

170   **Chapter 4** Threats to benthic biodiversity

notice too many used chewing gums carelessly dropped on the pathway. This is quantifiable and not precisely related to Hardin's use of the term, but a sidewalk or street is public and an openly accessible resource, nonetheless.

I begin this chapter with this primary concept because the ocean and seafloor are common resources and should belong to many individuals and society—e.g., the earth's inhabitants and not just those who may live on coastlines. Whether shallow or deep, reefs, seamounts, and other benthic habitats are permeated with a cerulean continuum of shading from bright to dark. In this context, we also continually need reminders about the relevance of our stewardship and responsibilities. Deliberate actions are required to conserve what remains of the natural benthos. However, this chapter firstly focuses on the several threats that result in tangible tragedies for too many benthic habitats. The specific threats and potential tragedies stemming from drilling, pollution, climate change, and oil and mineral exploration are described briefly here, whereas the possible solutions will appear in the last chapter.

What kind of "information age" do we live in?

Before going to some of the obvious physical threats, an abstract threat that may or may not affect benthic conservation efforts directly is the application/misapplication of *information*. In earlier chapters, I have alluded to the promise and potential benefits of the information-rich data sciences, especially in the context of genomics enterprises and the bases of biodiversity. We can view this book as an extension of the campaign to educate ourselves about assessing and reviewing the current status of the world's benthos. The case has been made plainly that although many studies have been conducted to reveal the secrets of the deep, multiple questions remain or can be generated. Thus, much more research and efforts can be expended to address the lack of sufficient information on many benthic ecosystems and organismal functions within them. We need more comprehensive data, monitoring, and coordination of these efforts to yield the most effective management plans. Moreover, because the scientific community generates data and interprets this into information, the community concomitantly has the obligation to accurately convey and educate the public with the results.

Unfortunately, we also live in an era of widespread and active "disinformation" campaigns against scientific evidence and basic facts (Farrell et al., 2019; DiResta, 2020; Dahlstrom, 2021). We could expect this in commercial media and forums where companies vie for customers or in politics where candidates may be compelled to fluff their nest with unverified credentials. We have mentioned the high stakes and likely profits that stem from benthic resources such as rare earth metals and nodules, so the profit incentives appear present to affect benthic habitats and well-being. Moreover, we also cannot ignore the occasional, and luckily few, cases when fraud occurs within scientific circles.

However, even more broadly, we find ourselves in a time when the disambiguation of information, unfiltering its sources, or how and when to apply evidence appears vital. We could even take Hardin's metaphor to the digital commons since the idea of "open access" is consistent with a commons. Sadly, the rise in disinformation accompanied the COVID-19 pandemic and public health institutions' responses to tamp down the spread of the SARS-2 coronavirus. Wang et al. (2019) carried out a systematic review and found that instantaneous communication and powerful amplification via the Internet and social media have brought about a quantum change in the speed and volume of misinformation. They recommend more studies to determine who and why certain parts of the demographic may be more easily swayed by disinformation and away from publicly confirmed knowledge. A salient rhetorical question that should concern all societies is "Why and how has public health or objective scientific information become so politicized?" Various empirical studies have tried to pursue answers (Bolsen and Druckman, 2015; Sosa et al., 2019). Because of its universal relevance, I now apply the following quote by former University of Chicago physicist Leo Szilard as it is appropriate to include here:

> "Political issues were often complex, but they were rarely anywhere as deep as the scientific problems which had been solved… with amazing rapidity because they had been constantly exposed to discussion among scientists, and thus it appeared reasonable to expect that the solution of political problems could be greatly sped up also if they were subjected to the same kind of discussion. The discussions of political problems by politicians were much less productive because they differed in one important respect from the discussions of scientific problems: when a scientist says something, his colleagues must ask themselves only whether it is true. But when a politician says something, his colleagues must first ask, 'Why does he say it?'"

Scientists, of course, are people who could be affected by the (multiple) maelstroms because they can be so pervasive and emotional. However, scientists' inclinations, e.g., their choice of career and then rigorous and sometimes extensive training in critical thinking, can provide strong bulwarks against misinformation and outright fraud. It may also be a good idea to include more formal ethics training for scientists. Skepticism and some enhanced objectivity already accompany many scientists as a favorite traveling companion.

The oversight and skepticism helped uncover perhaps one of the most serious cases of intentional fraud perpetrated by a scientist (Rao and Andrade, 2011). This involved the infamous case of Andrew Wakefield and colleagues who cherry-picked data and falsified facts in their *Lancet* publication, which promoted the idea that the measles, mumps, and rubella (MMR) vaccine could promote autism developmental disorders in children (Wakefield et al., 1998). Wakefield was eventually found to have clear conflicts of interest due to payments from lawyers who carried out lawsuits against vaccine-producing companies. In the end, the original paper was retracted, and Wakefield was removed from the UK medical register, which prevented him from practicing medicine in the UK due to his deliberate falsification of research and ethical violations. Unfortunately, since the original debacle, Wakefield remains free to spread unsubstantiated anti-vaccine propaganda, and even directed the 2016 film *Vaxxed: From Cover-Up to Catastrophe*. Together with a new cadre of anti-vaccine activists and misinformation campaigns sparked by the coronavirus pandemic, vaccine skepticism among the lay public may still not have reached its zenith, tracking its origins to the Wakefield case. One recent study found that countries with a higher trust in science and scientists showed lower levels of susceptibility to vaccine misinformation (Roozenbeek et al., 2020).

Other cases of fraud and unethical practices have occurred since. The second case is one of lack of disclosure when Chinese scientist He Jiankui who applied novel CRISPR technologies to manipulating the genomes within human embryos, making them heritable (Normile, 2018). He was sentenced to 3 years in prison in China for this offense. We can likely find other ethical violations not yet discovered among scientists, postdocs, or students who are under pressure to provide results to their supervisors and shareholders. Vigilance among ethical scientists and regulators is required, and indeed advocated (Brokowski and Adli, 2019). From this case and others, we see that few major penalties exist to prevent scientists from carrying out similar acts (McIntyre, 2019). Though it would seem that the possibility of

172 **Chapter 4** Threats to benthic biodiversity

a catastrophic career finale would be a sufficient reason to avoid falsification of data, apparently, this is not always the case. Benthic scientists should stay vigilant, as CRISPR/CAS applications in the sea and elsewhere promise to grow (Cleves, 2022).

Ecosystem stability and disturbance

For many years, stability was the common perception of the deeper ocean floor. Frequently, organisms need a stable framework. Of course, multiple activities occur in both the water column and seabed. For thousands of years, or at least before humanity could exploit the oceans, one can argue that most parts of the deep seabed have enjoyed fairly high stability. The converse of stability in ecological terms is disruption or *disturbance*, which can be defined as "any discrete event in time that disrupts ecosystem, community or population structure and changes resources, substrate availability or the physical environment" (Meurant, 2012). We are now seeing more examples of disturbance as population and human activities increase (Thrush and Dayton, 2002; Stark et al., 2014; Sciberras et al., 2018; Chatzinikolaou et al., 2018).

A difficulty with characterizing the extent of disturbances on benthic habitats, which has been well considered, is the lack of agreed-upon "baselines" upon which to base comparisons and use as reference points. This stems from the fact that the "pristine" state (prior to human disturbance) of many benthic habitats has rarely been documented or measured (e.g., for species diversity) (Knowlton and Jackson, 2008). This then leads to the conundrum whereupon true baseline characteristics may never be fully or realistically determined, but rather "shift" or change over time based upon the varying amounts of degradation or improvement (Pauly, 1995). At this point, I will not delve deeply into the mechanics and problems of *shifting baselines* and its derivative syndromes, but acknowledge its importance and relevance to the ecology and biodiversity assessments for many of the habitats we have been discussing (Duprey et al., 2016; Braverman, 2020). One idea I will reiterate is that much of the effort and literature referenced here focuses on shallow coral reef ecosystems, while many bathyal and abyssal habitats remain relatively pristine, perhaps stable, but also lacking in data. Therefore, the potential problems of shifting baseline syndrome may not apply to current deep sea research and assessments. With this assumption, an opportunity for ecologists and conservation biologists focused in biodiversity can be seized to collect as much baseline data as possible, prior to the likely eventual disruptions at the seafloor.

Disturbances may have greater impacts when they occur without warnings. Primary disturbances to the benthos could encompass relatively rare natural processes such as storms, earthquakes, spreading tectonic centers and sliding plates, and marine volcanoes. In the last few centuries, mankind has discovered the sea floor and this has led to more frequent disturbances to it. Disruptions to benthic ecosystems can elicit multiple types of responses, depending on the type and degree of disruption, habitat types, and community compositions. Species abundances may decrease (Schratzberger and Warwick, 1999; Huxham et al., 2000; Whomersley et al., 2010) or the disruptions and effects may be multiple, sequential, or cascading (Butler IV et al., 1995). For example, invasive species may gain footholds in exotic locales when resident species are weakened by disruptions and stress. The bearded fireworm *Hermodice carunculata* (Annelida: Amphinomidae) appears poised to become a continual nuisance species on certain reefs due to its omnivorous diet and its ability to withstand variable temperatures, salinity, low oxygen, and ocean acidification (OA) conditions (Schulze et al., 2017).

Extreme disturbances have been predicted and manifested by the long-term effects of climate change, such as high-energy cyclones, raging wildfires, and extended droughts. Bauman et al. (2013) indicated how extreme environmental swings can affect reef growth in two regions of the Persian Gulf: the southern Gulf (Dubai and Abu Dhabi) and eastern Gulf (western Musandam). Significant and consistent differences appeared in the mean colony sizes and size distributions between the two regions in the most abundant corals (*Acropora downingi*, *Favia pallida*, *Platygyra daedalea*, and massive *Porites* spp.). Another interesting study harkens back to secondary metabolism, because a high number of biosynthetic gene clusters (BGCs) have been found in necrotic octocoral samples with ≥ 70 clusters per metagenome compared to healthy tissues from *Eunicella gazella* (Keller-Costa et al., 2021). The BGCs encompassed compound classes such as NRPSs, bacteriocins, homoserine lactone, aryl-polyene, and terpenes, which also underscores the activity and shifting of the microbiome in disrupted, dysbiotic states (Sweet and Bulling, 2017).

These examples bring to mind a question "Is stasis or disruption a driving factor for biodiversity"? Estes and Vermeij (2022) provide an interesting view of disruption of several ecosystems by suggesting a greater integration of long-term historical perspectives to the ecological consequences. One of their examples is the state of turbidity and eutrophication of N. American east estuaries attributed to land-based sources of pollution and nutrients. However, a historical analysis pointed to the depletion of oyster beds that contributed high filtration rates and facilitated benthic primary production in Chesapeake Bay and Pamlico Sound in 1800s and early 1900s, as a major cause of turbidity in modern times. We discussed this situation in Chapter 1, but in the context of benthic-pelagic coupling. That relationship still apparently holds in this case as well.

Climate change continues to heat up

We must eventually confront perhaps the greatest physical threat to ourselves and the world's ecosystems, which is climate change. This feature was once called "global warming," formally refers to the long-term increase in Earth's average surface temperature caused by human activities, particularly the burning of fossil fuels such as coal, oil, and gas (Heede, 2014). This phenomenon began in the early 1900s, when the concentration of greenhouse gases in the atmosphere started rising rapidly due to human activities. Several gases such as methane, NO, and carbon dioxide, along with water vapor, trap heat from the sun in the atmosphere, leading to warming effects.

The idea of climate change was proposed in the late 19th century by scientists such as Svante Arrhenius, who first suggested that burning fossil fuels could lead to global warming (Arrhenius, 1896). Since then, a large body of evidence for global warming has accumulated, including direct measurements of temperature, gradual atmospheric greenhouse gas levels, and multiple data points for sea level rise (Bathiany et al., 2018; Heckendorn et al., 2009; Heede, 2014; IPCC, 2014; Sand et al., 2015; Rubin et al., 2015; Magnan et al., 2022; Dangendorf et al., 2023).

The Intergovernmental Panel on Climate Change (IPCC), established in 1988 by the United Nations (UN), has become a primary reference for scientific consensus on climate change, summarizing and synthesizing research from thousands of scientists around the world. Another important reference is the Paris Agreement, an international treaty adopted in 2015 by the United Nations Framework Convention on Climate Change (UNFCCC). The Paris Agreement aims to limit global warming to well below 2°C above preindustrial levels and to pursue efforts to limit it to 1.5°C. It has been ratified by 191 countries

174 Chapter 4 Threats to benthic biodiversity

as of 2021, indicating global recognition of the urgency of addressing climate change. Multiple studies verify the increasing amounts of carbon dioxide and methane. The data and predictions are reiterated by the IPCC in their final synthesis report in March 2023 (IPCC, 2023). Unfortunately, efforts to hide and confuse the issue of climate change also began around the same time period of the late 1980s (Rich, 2018). Wider awareness of various environmental problems spread to the wider masses of people and coincided with cultural upheavals in the 1960s and 1970s, sometimes dubbed as the "raising of consciousness."

Our primary focus here will be the direct and indirect effects of climate change on benthic systems. For example, warming of our climate can cause melting of polar ice, long-term snowpack, and glaciers. Clear qualitative evidence stems from simple photography and satellite images. The melting ice eventually reaches the sea and will gradually cause sea level rise. Current estimates show sea level rising around 5–8 mm/year with a possible total increase of 5–17 cm of sea level rise by 2100 (Huss and Hock, 2015). Another possible side effect of melting ice and permafrost raised in recent years is that pathogens locked away for millennia could soon wreak havoc on contemporary and extant populations of animals and plants (Yarzábal et al., 2021). Although this is possible, the headlines may be overly alarmist and lean too much toward hyperbole. We should remind ourselves and understand that we are already surrounded now by millions of species of microbes, and most of them (>90%) do not cause disease and/or may be positively beneficial to organismal and ecosystem health (Blaser, 2011; Hevia et al., 2015). Exceptional pathogens do occur and are discussed later in this chapter.

For another climate change-related threat, we have mentioned several times how increased sea surface temperatures (SST) cause thermal stress at shallow coral reef habitats. This fundamentally disrupts the stability of symbiosis via oxidative damage to Symbiodiniaceae and/or host cells, which expels the algae to cause coral bleaching (Oliver et al., 2018). One way to crystallize the various effects of climate change would be to again focus on a few biodiversity hotspots or events. For example, Heron et al. (2016) quantitated increasing SST on reefs at a scale of 4 km pixels and showed 97% of reefs with positive SST trends, while 60% warmed significantly during the period of 1998–2015. Fortunately, SST has less severe effects on deeper reefs. Bleaching incidences worldwide are continually monitored and have recently been summarized, supporting, for example, positive correlations between the standard deviation of SST and bleaching in both global assessments and in the Pacific Ocean (Shlesinger and van Woesik, 2023) (Fig. 4.1).

The 1997–98 El Niño–La Niña cycle precipitated one of the worst mass coral bleaching events ever recorded and it had variable impacts on reefs around the world (Harii et al., 2014; Heron et al., 2016). Reefs that had mesophotic components were able to recover faster. At the time of writing, reefs in the Florida Keys are suffering the combined effects of a predicted 2023 El Niño and extreme marine heat wave (Dennis et al., 2023). The extreme summer heat wave of 2023 resulted in some of the hottest days on Earth over the last 125,000 years: the global average temperature climbed to ~62.92°F (17.18°C) around July 4–6, 2023 (https://climatereanalyzer.org/clim/t2_daily/). The unfortunate convergence with an impending El Niño-Southern Oscillation (ENSO) in the Pacific Ocean sparked a marine heat wave with sustained sea surface temperatures over 31°C in the Florida Keys Marine National Sanctuary. The concomitant coral bleaching and high mortality in the sanctuary and nearby nurseries should make adaptive management proposals described in a National Academy of Sciences, Engineering and Medicine Ocean Studies Board report, such as moving threatened corals to cooler waters (e.g., Bermuda) or artificial upwelling of cooler waters, appear less radical and instead more urgent (NASEM, 2019).

Climate change continues to heat up   **175**

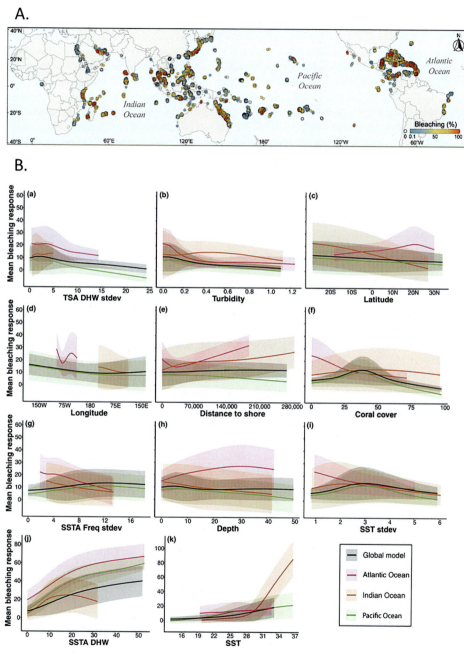

FIG. 4.1 See figure legend on next page.

(Continued)

176 Chapter 4 Threats to benthic biodiversity

FIG. 4.1, CONT'D

(A) Prevalence of coral bleaching. Coral bleaching is presented as a percentage of bleached corals at the time of the survey ($n = 23,288$), measured at 11,058 sites in 88 countries, and three oceans, from 1980 to 2020, based on data compiled by van Woesik and Kratochwill (2022). *Colored circles* indicate >0% bleaching *(blue)* through 100% bleaching *(red)*, and empty circles indicate no bleaching. (B) Partial dependency plots illustrate the effect of each environmental variable on the bleaching responses of corals. Relationships between the bleaching responses of corals and the 11 environmental variables were estimated using a deep-learning model—a neural-network analysis with data from 23,288 coral-reef surveys at 11,058 sites in 88 countries, and three oceans, from 1980 to 2020. The *different colors* depict the effects, and their standard errors, for the Atlantic, Indian, and Pacific Oceans together with the global model in black. *TSA*, thermal-stress anomaly; *DHW*, degree heating weeks; *stdev*, standard deviation; *SST*, sea surface temperature (°C); *SSTA*, sea surface temperature anomaly; *SSTA freq stdev*, the standard deviation of the frequency of SSTA.

Reproduced with permission from Shlesinger, T., van Woesik, R. 2023. Oceanic differences in coral-bleaching responses to marine heatwaves. Sci. Total Environ. 871, 162113.

For illustrative purposes, we revisit Philippine reefs that are part of the Coral Triangle introduced in the last chapter. This region is both iconic and unique for several environmental disturbances to reefs. The Philippines has experienced warming waters at a rate of 0.2°C per decade from 1985 to 2006, although Philippine waters have experienced a greater rate of increase (Peñaflor et al., 2009; Aliño, 2019). The 1998 mass bleaching caused an 8.07% decline in coral cover on Philippine reefs (Magdaong et al., 2014). Global warming may also raise sea level by a projected 0.19–1.04 m by 2080 (Villanoy et al., 2012). In addition, the Philippines experienced some of the largest cyclones in recorded history, such as Category 5 Typhoon Haiyan (Marler, 2014). Besides climate change, Philippine reefs feel pressure from large human population increases (pollution and urban runoff), as the Philippines is the world's 12th most populous country, with approximately 101 million people in 2015 (Aliño, 2019). Moreover, cyanide and dynamite fishing practices add to the threats that Philippine reefs must encounter (Hampton-Smith et al., 2021; Tahiluddin and Sarri, 2022).

Going back to climate change effects, surface ocean pH has had a stable pH between 8.0 and 8.3 for the last 25 million years. However, increased carbon dioxide levels in the atmosphere drive the following reaction:

$$CO_2 + H_2O \leftrightarrow H_2CO_3 \leftrightarrow HCO_3^- + H^+ \leftrightarrow CO_3^{2-} + 2H^+$$

This reaction produces excess H^+ ions, which can overwhelm the ocean buffering system and eventually lead to more H^+ and higher acidity (Widdicombe and Spicer, 2008). Thus, increased CO_2 causes OA with predictions of an average pH of 7.8 by the year 2100. This increased acidity has been termed hypercapnia (elevated levels of dissolved CO_2). In turn, hypercapnia has been shown to reduce calcification rates of key marine taxa such as echinoderms, corals, and coralline algae affecting sediments and reef structures (Anthony et al., 2008; Kroeker et al., 2013; Kornder et al., 2018).

Cornwall et al. (2021) have provided a recent comprehensive synopsis of coral reefs with regard to climate change and OA. Unfortunately, and as expected, not all of the conclusions are hopeful: "Net carbonate production and accretion of most the world's coral reefs will be fundamentally reduced by ongoing climate change." Overall, the meta-analysis of 183 reefs worldwide (49% in the Atlantic Ocean, 39% in the Indian Ocean, and 11% in the Pacific Ocean) indicated that the rate of coral accretion could not keep up with ongoing bioerosion in the face of the dual threats of climate change with OA. Responses of coral reef taxa calcification and bioerosion rates to predicted changes in coral cover driven by climate change to estimate the net carbonate production rates of 183 reefs worldwide by 2050 and 2100. Bioerosion itself is a complex phenomenon that entails both biotic degradation processes (by endolithic algae or sponges) and coral growth rates that can be exacerbated by OA, sedimentation, or climate change (Glynn and Manzello, 2015).

Due to the inherent physiology, metabolism, and calcium carbonate exoskeletal compositions, OA will likely affect some benthic taxa more than others. For a number of years, echinoderms, calcifying algae, and coccolithophores were viewed as the most vulnerable to OA, followed by mollusks, crustaceans, and annelids (Widdicombe and Spicer, 2008). This disparity would be caused by differing capacities to regulate extracellular maintenance of acid-base balances via ion regulation based on metabolism or mobility. For example, recent experiments with the calcareous sponge *Paraleucilla magna* have shown that their primmorphs can synthesize their skeletons under low pH conditions (Ribeiro et al., 2023). With these effects, OA has the potential to change the composition of multiple benthic habitats. Some of these dire forecasts may be countered by additional data and initiatives.

Various studies, such as by Baumann (2019), have reviewed the difficulties of OA assessments due to multiple stressor interactions brought about by climate change and the need for new experimental approaches. However, new assessments show the possibility that OA could also select for more tolerant species (Pascal et al., 2010) and that life history stage should be considered since larval stage calcifiers appear more sensitive than adults (Leung et al., 2022). For coccolithophores, recent studies indicate that OA had negative effects on the calcification of two abundant coccolithophore species: *Emiliania huxleyi* and *Gephyrocapsa oceanica*, whereas more heavily calcified species *Coccolithus braarudii* did not show a distinct response when exposed to elevated $p\text{CO}_2$/reduced pH (Meyer and Riebesell, 2015). The mixed results suggest more research on specific taxa is required. Coralline algae appear more robust to the impacts of thermal stress than corals.

Habitat loss and destructive extraction methods

Loss of benthic habitats can arise from various human-based routes. This includes sedimentation that lowers the radiance but enhances burying communities on the seabed (Miller et al., 2016), sea level rise affecting intertidal communities (Kaplanis et al., 2020), and loss to introduced invasive species (Bax et al., 2003). Trawling and dredging represent another direct threat to benthic habitats (Thrush and Dayton, 2002; Hiddink et al., 2017). Various scenarios of deep sea mining (DSM) are shown in Fig. 4.2 and also discussed earlier with the DISCOL experiment. In 2012, the International Seabed Authority approved an Environmental Management Plan (EMP) for the Clarion-Clipperton Fracture Zone (CCZ) in the Eastern Central Pacific (Lodge et al., 2014). DSM is typically focused on three main deposit types: polymetallic nodules, seafloor massive sulfides (SMS), and cobalt-rich crusts (Childs, 2019; Niner et al., 2018). DSM cannot be considered as a constructive biotechnology since it blatantly extracts, disrupts, or irreversibly destroys habitats and does not generate new biomass (as in terrestrial farming) (Miller et al., 2018). Mining the deep sea with large modern trawlers for the nodules will produce at least two types of plumes that will have effects far from the actual extraction sites. Several studies have been modeling plume movements, sometimes in large experimental setups

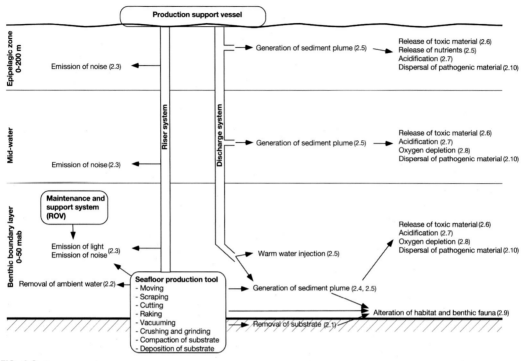

FIG. 4.2

Graphical representation of the main primary and secondary mining-related processes associated with deep sea mining and which will potentially affect with pelagic and benthopelagic biota. Numbers in brackets designate chapters in the original publication with detailed descriptions (Christiansen et al., 2020).

Reproduced with permission from Christiansen, B., Denda, A., Christiansen, S. 2020. Potential effects of deep seabed mining on pelagic and benthopelagic biota. Mar. Policy 114, 103442.

(Spearman et al., 2020; Drazen et al., 2020). However, although some tools of the marine mining industry can be scientific in nature, this does not guarantee the full safety of the occupation (Halfar and Fujita, 2007).

Pollution and eutrophication

Eutrophication represents nutrient overloading (of total nitrogen, phosphorous, and carbon) typically from agricultural or urban runoff, which in turn can deplete oxygen levels in the water column, harming fish and other aquatic life. This phenomenon also has the unpleasant effects of causing higher turbidity, which blocks sunlight. The anoxia can result in foul odors and tastes, making it unsuitable for recreational activities or as a source of drinking water. Excessive nutrient inputs are known to have adverse effects, especially in natural habitats such as oligotrophic coral reefs (Lapointe, 1997; Lapointe et al., 1997; De'ath and Fabricius, 2010; Vega Thurber et al., 2014). Nutrients can cause phase shifts from corals to algal-dominated reefs or may increase the presence of algal symbionts within coral tissue, making those hosts more susceptible to bleaching stresses (Shantz et al., 2016). The extent of eutrophication effects on coral reefs has been debated (Hughes et al., 1999), but overall, the negative effects on benthic biodiversity have been documented. In the highly biodiverse Indian River Lagoon on the east coast of Florida, aging municipal treatment facilities and $>300,000$ septic tanks have contributed to higher ammonium and nitrate concentrations which fuel algal blooms (e.g., *Aureoumbra lagunensis*) and diminish sea grass cover (Lapointe et al., 2020). Thus, improving water quality has been a major focus of environmental managers across many habitats (Duprey et al., 2016). Conversely, habitats with higher biodiversity exhibit more resilience and capabilities to withstand deteriorating water quality (Cardinale, 2011).

Not all of the organic materials that reach the seafloor are nutrients. One of the more recent yet significant threats to marine organisms is human-made "microplastics and nanoplastics" (Shen et al., 2020; Wilcox et al., 2020). These artificial entities encompass plastic particles and textile fibers, or films less than 5 mm (0.20 in) in length. They arise from either (1) degradation of larger plastic items or (2) preexisting components, such as from synthetic clothes. Notwithstanding the famous line in the film *The Graduate* (1967), where unsolicited advice was doled out that a stable financial future lay in "plastics," we truly live and are surrounded by plastic-based items, which include textiles, shoes, cosmetics, exfoliating soaps, containers, housing tires, and city dust (Andrady, 2015; Nielsen et al., 2020). Based on just this partial list, one can easily ascertain that plastic products have wide utility and distributions. In 2019, plastic world production was >350 million tons, with 31% of the production by China alone. On the other end, global plastic waste is expected to triple to 270 million tons from 2015 to 2060.

These common items are composed of polyethylene, polypropylene, polyethylene terephthalate (PET), polystyrene, and nylon. Most of the compounds derive from petroleum sources. For example, Phillips Petroleum chemists first synthesized polypropylene, the second most common plastic commodity. Regarding economics, plastics may be inseparable from our daily lives for many years. In 2019, the global market for polypropylene was worth $126 billion.

Microplastics are insidious because they withstand natural biodegradation (Wang et al., 2018). When the particles work their way into organismal bodies via respiration, they may stay permanently in the gas exchange organs. After entering the body via the gastrointestinal tract, plastics can remain for periods of days to weeks before excretion. If this occurs in smaller fauna, then bioaccumulation up the food chain can occur. However, the full effects of microplastic presence in organisms remain equivocal. In humans, body burdens of microplastics through table salt, drinking water, and inhalation were estimated to be $(0–7.3) \times 104$, $(0–4.7) \times 103$, and $(0–3.0) \times 107$ items per person per year, respectively, although the actual effects on human physiology remain to be characterized (Zhang et al., 2020). Interestingly, the microplastic occurrence in table salt varied by location. The highest abundance was reported in Croatia $(1.4 \times 104–2.0 \times 104 \text{ items·kg}^{-1})$, followed by Indonesia $(1.4 \times 104 \text{ items·kg}^{-1})$, Italy $(1.6 \times 103–8.2 \times 103 \text{ items·kg}^{-1})$, the United States $(0.5 \times 102–8.0 \times 102 \text{ items·kg}^{-1})$, and China $(5.5 \times 102–6.8 \times 102 \text{ items·kg}^{-1})$.

Evidence is accumulating of microplastic incorporation and its impacts across the phylogenetic spectrum of organisms. For the benthos, studies with the marine ciliate *Uronema marinum*, Zhang et al. (2021) have recently shown that this protozoan ingests polystyrene beads. This leads to a decline in their grazing and biomass, which negatively affects the microbial loop at an important juncture. Among different annelids, microplastics have been found embedded in the gastrointestinal tracts of

180 **Chapter 4** Threats to benthic biodiversity

deposit-feeding lugworms (*Arenicola marina*) and shore crab (*Carcinus maenas*) (Watts et al., 2014). Foley et al. (2018) carried out an extensive meta-analysis of microplastics on marine organisms. Despite finding the presence of plastic in a wide spectrum of taxa, mostly neutral to weakly negative effects were posited.

Seagrass habitats

Algae and seagrasses are the foundations of shallow coastal communities. Macroalgae can help stabilize marine sediments. Overfishing of herbivorous fish and increased eutrophication both increase the abundance of macroalgae (Sandin et al., 2008). Alternatively, the increase of turf algae (or "algal turfs") signals problems for coral reef health. Turf algae can become dense and are comprised of multispecies assemblages of filamentous benthic algae and cyanobacteria that are typically less than 1 cm in height (Steneck and Dethier, 1988).

Perhaps some of the relatively neglected benthic habitats are the seagrasses. They are distributed worldwide in shallow areas requiring much sunlight similar to reefs but only cover 0.1%–0.2% of the global ocean. Marine seagrasses encompass about 60 seagrass species worldwide, which belong to four families (Posidoniaceae, Zosteraceae, Hydrocharitaceae, and Cymodoceaceae) (Duarte, 2002). Seagrasses are more ephemeral than calcium carbonate structures, but they can also sustain a wide biodiversity. The importance of seagrass habitats can be appreciated when we notice their decline (Morris et al., 2022). Besides being a source of primary production for the benthos, seagrass meadows can act as bioindicators for water quality and turbidity (Lapointe et al., 2020). Burkholder et al. (2007) have described the primary stressors on seagrass such as light reduction from nutrient overloading as well as anoxia stemming from macroalgal growth with excess hydrogen sulfide production. Florida manatees (*Trichechus manatus latirostris*) are dependent upon seagrasses for their nutrition, but the mammal species has recently relapsed to a severe threatened status as the seagrass meadows on Florida coastlines experience steady contractions (Allen et al., 2022). Recent assessments of seagrass meadow distributions point to a global area of 160,387 km^2 across 103 countries/territories with moderate to high confidence. However, these abundance estimates may also be off by an order of magnitude (McKenzie et al., 2020). Interestingly, recent studies by Chin et al. (2021) show the connection of seagrass habitat health to the occurrence of lucinid bivalves, which can act as "ecological engineers" and lower potentially toxic levels of sulfides in sediment porewater.

Marine diseases that afflict the benthos

Because we do not closely monitor most benthic habitats on a regular basis, many are probably blissfully unaware of most pathogens affecting and arising from these habitats. Moreover, humans' long-range effects and activities on the welfare of benthic habitats and organisms remain mostly unknown. For example, coral diseases were first recorded in the 1970s and, over time, have increased in occurrence, prevalence, and severity (Porter et al., 2001; Harvell et al., 2007). Generally, the increase in marine diseases has reduced reef health, resilience, diversity, and fecundity, while also endangering the services and resources that reef ecosystems supply to local communities. A recent synthesis of various coral diseases underscores our collective ignorance about, counted 22 diseases across 165 coral species (though *Acropora* sp. was the most studied), and the deficiency of experimental approaches

(19.9% of total studies) and specific regions such as the Red Sea or Persian Gulf (Morais et al., 2022) (Fig. 4.3). What is more, the etiologies of these diseases are complex and often marked by "imbalances of the natural microbial consortia" instead of the presence of a single pathogen (Angermeier et al., 2011). From our current understanding of more accessible habitats and human examples, several diseases often arise from this imbalance and disruption of an organism's microbial equilibrium. In

FIG. 4.3

Photographs showing examples of the six most frequent coral diseases in the literature: (A) Black band, (B) White plague, (C) White Syndromes, (D) Dark spot, (E) Skeletal eroding, and (F) Yellow band.

Photograph credit: Sara Williams (D, E), Aiara P.L.R Cardoso (C, F), Juliano Morais (B), and Christina Kellogg (A). Reproduced with permission from Morais, J., Cardoso, A.P., Santos, B.A. 2022. A global synthesis of the current knowledge on the taxonomic and geographic distribution of major coral diseases. Environ. Adv. 100231.

182 **Chapter 4** Threats to benthic biodiversity

humans, the term "dysbiosis" was applied to gut microbiome imbalances or a significant shift from the balanced normal microbiome. This led Zaneveld and Vega Thurber to propose an "Anna Karenina" hypothesis for coral microbiomes, named for the Tolstoy adage "all happy families are alike but unhappy families [microbial communities] are unhappy in their own distinct way" (Zaneveld et al., 2017). Previously, commensal or outright pathogenic organisms may exploit these imbalances, the consequent downturn in a host's (or holobiont's) immunological defenses and thus eventually overwhelm a host. Pollution, in the form of synthetic and toxic industrial chemicals and compounds released into the environment, can affect host immune system defenses and tolerance or cause mutations in the genetic material. It is likely that some diseases are driven by the combinations of all of these factors, e.g., the presence of a given pathogen and contextual environmental and host conditions. Identification of the specific synergisms will be critical in understanding how disease epizootics contribute to organismal and reef decline.

With an increasing number of pathogens afflicting people topside, we are gaining better knowledge of disease dynamics. In this thinly veiled reference to the COVID-19 pandemic again, we now have more support that diverse epidemics probably occur more frequently, roughly within 20–30 years, or almost every generation (Piret and Boivin, 2021). Pathogens could possibly run amok in other metazoan populations and below the surface waves as well, while we remain unaware.

Several examples of marine pandemics have been documented and appear to be increasing in recent years. Harvell and Lamb (2020) have recently written a chapter summarizing four major marine diseases and epidemics. The case histories included seagrasses and corals, abalone, and keystone predators such as sea stars. I will only touch upon some of these now and reach back to some of the earlier known marine diseases.

At one time, the long spiny sea urchin *Diadema antillarum* ranged across the tropical Western Atlantic Ocean, including the Caribbean, Gulf of Mexico, and the coasts of South America (as far south as Brazil). However, beginning around 1983, a marine pathogen wiped out 97% of *D. antillarum* populations in these areas but was never identified (Lessios, 1988). Because of the important role this species had by grazing macroalgae on reefs, the sudden demise of *Diadema* could have possibly led or accelerated Caribbean coral reef deterioration in the 1990s (Edmunds and Carpenter, 2001).

Sea star wasting disease (SSWD) has occurred more recently and refers to the rapid deterioration and death of starfish populations along the West Coast of North America. The affliction first appeared around 2013 and then spread along the Pacific coast from Alaska to Mexico. Scientists observed the following symptoms: sea stars lose turgor, arms twist and break away from the body, and within days the entire sea star turns into a white, gooey mass, appearing to "melt." Subsequent studies have identified a sea star-associated densovirus (SSaDV) (*Parvoviridae*) (Hewson et al., 2014) and that warm temperature anomalies (unusual 2–3°C) were coincidental with the summer 2014 ochre starfish *Pisaster ochraceus* mortalities along South Puget Sound and Washington's outer coasts (Eisenlord et al., 2016).

We can revisit reef organisms and Porifera again since sponges have been fished from Mediterranean seabeds since ancient Greek antiquity (Pronzato, 1999). The overfishing did more to cull the populations than disease, but pathogens do take their toll. For example, the bath sponge industry grew in the Caribbean, especially at locales such as Red Bays on northern Andros Island due to the proximity to shallow flat banks (Lopez et al., 2000). Besides being used for bathing, natural sponges could also serve to bolster car upholstery and lower the costs of recycling since it was a natural product. However, a severe epidemic caused the disappearance of between 70% and 95% of sponge specimens in 1938–39 in the Caribbean (Galstoff, 1942).

In parallel with other marine diseases, sponge disease frequency and severity appear linked to higher seawater temperatures and various stressors (Webster, 2007; Slaby et al., 2019). The author's laboratory encountered sponge disease off the coast of Broward County in 2012 and carried out a cursory study of sponge diseases after an outbreak of sponge orange band disease (SOB). *Xestospongia muta* from the Caribbean had been known to experience episodic disease outbreaks from the island of Curacao and the Belize Barrier Reef (Nagelkerken et al., 2000). This affliction in particular has been named for the characteristic orange coloration that borders a growing area of bleaching that stems from localized lesions. Gross symptoms of the disease are bleaching, softening and necrosis of the affected area, and rapid death of most or the whole sponge individual. Bleached *X. muta* tissue shows necrosis and then disintegrates, leaving behind an exposed skeleton (Angermeier et al., 2011). Bleaching progression can occur over days or weeks. SOB within the Caribbean most often occurs in early spring and late summer (Cowart et al., 2006) and is often associated with coral bleaching (McMurray et al., 2008). For example, pathology and histological analysis characterized the photosynthetic symbiont microbial community in healthy *X. muta*, which was dominated by *Synechococcus* and *Prochlorococcus*. Interestingly, diseased *X. muta* exhibited a shift of its microbial community to a more heterogeneous mixture of filament-forming cyanobacterial genera: *Limnothrix*, *Plectonema*, and *Leptolyngbya* (Angermeier et al., 2011). Reaction to SOB included an increase in phagocytosis of cyanobacteria in the pinacoderm. Culturing of resident bacteria from the disease transition zones was also attempted, resulting in the isolation of several *Gammaproteobacteria* that did not fulfill Koch's postulates. In another study, SOB was found to increase the presence of a *Crenarchaeota*, thought to be responsible for nitrification and ammonia oxidation in *X. muta* and also found in surrounding sediment and sand (Lopez-Legentil et al., 2008). *X. muta* also showed an increased expression of the ammonia oxidation gene expression, which has been linked to ammonia release due to tissue death. This study further illustrated the SOB's ability to disrupt *X. muta* homeostasis. Again, however, this study did not confirm the presence of a suspected pathogen as these *Crenarchaeota* are commonly found in this species of sponge.

On May 11, 2012, the author's Nova Southeastern University laboratory collected *X. muta* specimens ($n = 40$) from Gulf Stream reef (Boynton Beach, FL) found to be highly afflicted by or associated with SOB syndrome. Clear examples of this condition can be viewed by videos posted by Palm Beach's Reef Rescue (https://www.youtube.com/watch?v=rz6Q7wt7Cx8). The collected specimens included the following subsamples/health states from the same sponges: (1) diseased (D: bleached, completely dead), (2) healthy (H: still retaining normal brown pigmentation), and (3) intermediate (I) regions between live and diseased regions. Samples also were taken from external neighboring healthy (HH) sponges that did not show any outward signs of SOB, as well as ambient seawater and reef sediment samples. Although continued reef monitoring at the time, provided another 10 samples from the region as part of nondisease-monitoring surveys, we did not know transmission details; e.g., whether the same contagion was passed on between different S. Florida reefs. For example, throughout May 2012, several reports have been posted on the National Oceanic and Atmospheric Administration (NOAA) coral list of similar SOB observations in northern Broward County and in the northern end of Florida Keys.

Of the original samples taken from the height of the 2012 SOB disease outbreak, 14 diseased individuals and 6 nondiseased healthy control sponges were subjected to 16S rDNA amplicon sequencing analyses via pyrosequencing technology. In unpublished results, we found that the dominant Operational Taxonomic Unit (OTU) in the intermediate stage of the disease (healthy portions of the diseased animals) was a *Verrucomicrobiaceae* (light orange) in the genus *Rubritalea*. Although the relative proportion of

184 **Chapter 4** Threats to benthic biodiversity

this taxa was not significantly different from the other sample types due to its presence in only 2 of the 4 intermediate samples, this taxon interestingly produces carotenoids of orange and red (Yoon et al., 2011), like the discoloration patterns seen in this disease. Furthermore, this *Verrucomicrobia* taxon occurred at very low relative abundances in both the healthy and diseased tissues, suggesting that this could be an alternative culprit in the onset of the disease or an opportunist pathogen/commensal that emerges during the infection. It is interesting that Angermeier et al. (2011) also reported detecting *Verrucomicrobia* in diseased tissues. Other candidates revealed by the sequences from this SOB outbreak were a Flavobacteriaceae, Rhodobacteraceae, and unidentified γ-proteobacteria.

The aforementioned sponge outbreak directly preceded one of the most devastating coral disease outbreaks on the Florida Reef Tract in recent times. Since 2014, a white syndrome called stony coral tissue loss disease (SCTLD) has proven lethal for more than 20 species of reef-building scleractinian corals. Highly susceptible species included *Colpophyllia natans*, *Dendrogyra cylindrus*, *Dichocoenia stokesii*, *Diploria labyrinthiformis*, *Eusmilia fastigiata*, *Meandrina meandrites*, *Pseudodiploria strigose*, and *Pseudodiploria clivosa*. Up to 30% loss of regional coral cover had occurred in 3 years (Walton et al., 2018; Aeby et al., 2019; Precht et al., 2016). SCTLD has exhibited very high rates of transmission and rapid transmission, and has some similarities to white band disease (Cróquer et al., 2021). However, at the time of writing, a SCTLD pathogen has not been unequivocally identified. Accumulating evidence is pointing toward a possible viral infection of *Symbiodinium* (Work et al., 2021). The morphology appeared as filamentous-positive single-stranded RNA viruses similar to those of plants termed anisometric viral-like particles (AVLP). The emergence of SCTLD coincided with a deep dredging event in the Port of Miami, which included rock-chopping via a cutterhead, and a regional sea-surface temperature thermal anomaly (Precht et al., 2019). This generated a large amount of turbidity and sedimentation that could have been carried out to adjacent reefs (Miller et al., 2016; Cunning et al., 2019). In 2014, SCTLD was limited mostly to Miami-Dade County reefs and found nowhere else on the Florida Reef Tract (FRT). However, local epidemiological or microbiome data from before the outbreak were not taken for systematic comparisons, and so only circumstantial evidence currently links the events. This also brings up the problem of baseline data deficiencies mentioned earlier. Moreover, the current SCTLD outbreak has already radiated out of Florida to other Atlantic and Caribbean reef regions. These examples further underscore the stresses on this ecosystem.

This chapter could continue to list the many, unfortunately, growing number of threats to benthic spaces and organisms. However, I mention and treat the variety of maladies and threats to the benthos in a rudimentary fashion and trust the reader to pursue them in more depth when desired. This will allow us to move on to more hopeful topics of finding possible solutions and success stories related to benthic assessment and conservation in the next chapter. Overall, biologists commit to studying life and thus strive to prolong or permit the possibility for the "birth" or radiation of new species, continuing the evolutionary process, which would avoid Soule's opening warning.

References

Aeby, G.S., Ushijima, B., Campbell, J., Jones, S., Williams, G.J., Meyer, J., Hase, C., Paul, V., 2019. Pathogenesis of a tissue loss disease affecting multiple species of corals along the Florida reef tract. Front. Mar. Sci. 6, 678. https://doi.org/10.3389/fmars.2019.00678.

Aliño, P.M., 2019. The Philippines. In: Licuanan, W.Y., Cabreira, R.W., Sheppard (Eds.), World Seas: An Environmental Evaluation, Indian Ocean to the Pacific. vol. II. Academic Press, London, pp. 515–537.

References

Allen, A.C., Beck, C.A., Sattelberger, D.C., Kiszka, J.J., 2022. Evidence of a dietary shift by the Florida manatee (*Trichechus manatus latirostris*) in the Indian River Lagoon inferred from stomach content analyses. Estuar. Coast. Shelf Sci. 268, 107788.

Andrady, A.L., 2015. Plastics and Environmental Sustainability. John Wiley & Sons.

Angermeier, H., Kamke, J., Abdelmohsen, U.R., Krohne, G., Pawlik, J.R., et al., 2011. The pathology of sponge orange band disease affecting the Caribbean barrel sponge *Xestospongia muta*. FEMS Microbiol. Ecol. 75, 218–230.

Anthony, K.R., Kline, D.I., Diaz-Pulido, G., Dove, S., Hoegh-Guldberg, O., 2008. Ocean acidification causes bleaching and productivity loss in coral reef builders. Proc. Natl. Acad. Sci. 105 (45), 17442–17446.

Arrhenius, S., 1896. XXXI. On the influence of carbonic acid in the air upon the temperature of the ground. Lond. Edinb. Dublin Philos. Mag. J. Sci. 41 (251), 237–276.

Bathiany, S., Dakos, V., Scheffer, M., Lenton, T.M., 2018. Climate models predict increasing temperature variability in poor countries. Sci. Adv. 4 (5), eaar5809. https://doi.org/10.1126/sciadv.aar5809.

Bauman, A.G., Pratchett, M.S., Baird, A.H., Riegl, B., Heron, S.F., Feary, D.A., 2013. Variation in the size structure of corals is related to environmental extremes in the Persian Gulf. Mar. Environ. Res. 84, 43–50.

Baumann, H., 2019. Experimental assessments of marine species sensitivities to ocean acidification and co-stressors: how far have we come? Can. J. Zool. 97 (5), 399–408.

Bax, N., Williamson, A., Aguero, M., Gonzalez, E., Geeves, W., 2003. Marine invasive alien species: a threat to global biodiversity. Mar. Policy 27 (4), 313–323.

Blaser, M., 2011. Stop the killing of beneficial bacteria. Nature 476 (7361), 393–394.

Bolsen, T., Druckman, J.N., 2015. Counteracting the politicization of science. J. Commun. 65 (5), 745–769.

Braverman, I., 2020. Shifting baselines in coral conservation. Environ. Plan. E Nat. Space 3 (1), 20–39.

Brokowski, C., Adli, M., 2019. CRISPR ethics: moral considerations for applications of a powerful tool. J. Mol. Biol. 431 (1), 88–101.

Burkholder, J.M., Tomasko, D.A., Touchette, B.W., 2007. Seagrasses and eutrophication. J. Exp. Mar. Biol. Ecol. 350 (1–2), 46–72.

Butler IV, M.J., Hunt, J.H., Herrnkind, W.F., Childress, M.J., Bertelsen, R., Sharp, W., et al., 1995. Cascading disturbances in Florida Bay, USA: cyanobacteria blooms, sponge mortality, and implications for juvenile spiny lobsters *Panulirus argus*. Mar. Ecol. Prog. Ser. 129, 119–125.

Cardinale, B.J., 2011. Biodiversity improves water quality through niche partitioning. Nature 472 (7341), 86–89.

Chatzinikolaou, E., Mandalakis, M., Damianidis, P., Dailianis, T., Gambineri, S., Rossano, C., et al., 2018. Spatio-temporal benthic biodiversity patterns and pollution pressure in three Mediterranean touristic ports. Sci. Total Environ. 624, 648–660.

Childs, J., 2019. Greening the blue? Corporate strategies for legitimizing deep sea mining. Polit. Geogr. 74, 102060.

Chin, D.W., de Fouw, J., van der Heide, T., Cahill, B.V., Katcher, K., Paul, V.J., et al., 2021. Facilitation of a tropical seagrass by a chemosymbiotic bivalve increases with environmental stress. J. Ecol. 109 (1), 204–217.

Christiansen, B., Denda, A., Christiansen, S., 2020. Potential effects of deep seabed mining on pelagic and bentho-pelagic biota. Mar. Policy 114, 103442.

Cleves, P.A., 2022. A need for reverse genetics to study coral biology and inform conservation efforts. In: Coral Reef Conservation and Restoration in the Omics Age. Springer International Publishing, Cham, pp. 167–178.

Cornwall, C.E., Comeau, S., Kornder, N.A., Perry, C.T., van Hooidonk, R., et al., 2021. Global declines in coral reef calcium carbonate production under ocean acidification and warming. Proc. Natl. Acad. Sci. 118 (21), e2015265118.

Cowart, J.D., Henkel, T.P., McMurray, S.E., Pawlik, J.R., 2006. Sponge orange band (SOB): a pathogenic-like condition of the giant barrel sponge, *Xestospongia muta*. Coral Reefs 25, 513.

Cróquer, A., Weil, E., Rogers, C.S., 2021. Similarities and differences between two deadly Caribbean coral diseases: white plague and stony coral tissue loss disease. Front. Mar. Sci. 8, 1331.

186 Chapter 4 Threats to benthic biodiversity

Cunning, R., Silverstein, R.N., Barnes, B.B., Baker, A.C., 2019. Extensive coral mortality and critical habitat loss following dredging and their association with remotely-sensed sediment plumes. Mar. Pollut. Bull. 145 (18084), 185–199. https://doi.org/10.1016/j.marpolbul.2019.05.027.

Dahlstrom, M.F., 2021. The narrative truth about scientific misinformation. Proc. Natl. Acad. Sci. 118 (15), e1914085117.

Dangendorf, S., Hendricks, N., Sun, Q., Klinck, J., Ezer, T., Frederikse, T., et al., 2023. Acceleration of US Southeast and Gulf coast sea-level rise amplified by internal climate variability. Nat. Commun. 14 (1), 1935.

De'ath, G., Fabricius, K., 2010. Water quality as a regional driver of coral biodiversity and macroalgae on the Great Barrier Reef. Ecol. Appl. 20 (3), 840–850.

Dennis, B., Ajasa, A., Mooney, C., 2023. As Florida Ocean Temperatures Soar, a Race to Salvage Imperiled Corals. https://www.washingtonpost.com/climate-environment/2023/07/26/florida-coral-reef-ocean-temperatures-heat/?wpisrc=nl_most. (Accessed 26 July 2023).

Diamond, J., 2011. Collapse: How Societies Choose to Fail or Succeed, revised ed. Penguin.

DiResta, R., 2020. The Supply of Disinformation Will Soon Be Infinite. The Atlanic. https://www.theatlantic.com/ideas/archive/2020/09/future-propaganda-will-be-computer-generated/616400/. (Accessed 5 September 2023).

Drazen, J.C., Smith, C.R., Gjerde, K.M., Haddock, S.H., Carter, G.S., Choy, C.A., et al., 2020. Midwater ecosystems must be considered when evaluating environmental risks of deep-sea mining. Proc. Natl. Acad. Sci. 117 (30), 17455–17460.

Duarte, C.M., 2002. The future of seagrass meadows. Environ. Conserv. 29 (2), 192–206.

Duprey, N.N., Yasuhara, M., Baker, D.M., 2016. Reefs of tomorrow: eutrophication reduces coral biodiversity in an urbanized seascape. Glob. Chang. Biol. 22 (11), 3550–3565.

Edmunds, P.J., Carpenter, R.C., 2001. Recovery of *Diadema antillarum* reduces macroalgal cover and increases abundance of juvenile corals on a Caribbean reef. Proc. Natl. Acad. Sci. 98 (9), 5067–5071.

Eisenlord, M.E., Groner, M.L., Yoshioka, R.M., Elliott, J., Maynard, J., Fradkin, S., et al., 2016. Ochre star mortality during the 2014 wasting disease epizootic: role of population size structure and temperature. Philos. Trans. R. Soc. B Biol. Sci. 371 (1689), 20150212.

Estes, J.A., Vermeij, G.J., 2022. History's legacy: why future progress in ecology demands a view of the past. Ecology 103 (11), e3788.

Farrell, J., McConnell, K., Brulle, R., 2019. Evidence-based strategies to combat scientific misinformation. Nat. Clim. Chang. 9 (3), 191–195.

Foley, C.J., Feiner, Z.S., Malinich, T.D., Höök, T.O., 2018. A meta-analysis of the effects of exposure to microplastics on fish and aquatic invertebrates. Sci. Total Environ. 631, 550–559.

Galstoff, P.S., 1942. Wasting disease causing mortality of sponges in the West Indies and Gulf of Mexico. In: Proceedings of the VIII American Science Congress. 3, pp. 411–421.

Glynn, P.W., Manzello, D.P., 2015. Bioerosion and coral reef growth: a dynamic balance. In: Coral Reefs in the Anthropocene. Springer, pp. 67–97.

Halfar, J., Fujita, R.M., 2007. Danger of deep-sea mining. Science 316 (5827), 987.

Hampton-Smith, M., Bower, D.S., Mika, S., 2021. A review of the current global status of blast fishing: causes, implications and solutions. Biol. Conserv. 262, 109307.

Hardin, G., 1968. The tragedy of the commons: the population problem has no technical solution; it requires a fundamental extension in morality. Science 162 (3859), 1243–1248.

Hardin, G., 1991. Paramount positions in ecological economics. In: Ecological Economics: The Science and Management of Sustainability. Columbia University Press, pp. 47–57.

Harvell, C.D., Lamb, J.B., 2020. Disease outbreaks can threaten marine biodiversity. In: Marine Disease Ecology. Oxford University Press, pp. 141–158.

Harii, S., Hongo, C., Ishihara, M., Ide, Y., Kayanne, H., 2014. Impacts of multiple disturbances on coral communities at Ishigaki Island, Okinawa, Japan, during a 15 year survey. Mar. Ecol. Prog. Ser. 509, 171–180.

Harvell, D., Jordan-Dahlgren, E., Merkel, S., Rosenberg, E., Raymundo, L., et al., 2007. Coral disease, environmental drivers, and the balance between coral microbial associates. Oceanography 20, 172–195.

Heckendorn, P., et al., 2009. The impact of geoengineering aerosols on stratospheric temperature and ozone. Environ. Res. Lett. 4 (4), 045108. https://doi.org/10.1088/1748-9326/4/4/045108.

Heede, R., 2014. Tracing anthropogenic carbon dioxide and methane emissions to fossil fuel and cement producers, 1854-2010. Clim. Chang. 122 (1), 229–241. https://doi.org/10.1007/s10584-013-0986-y.

Heron, S.F., Maynard, J.A., Van Hooidonk, R., Eakin, C.M., 2016. Warming trends and bleaching stress of the world's coral reefs 1985–2012. Sci. Rep. 6 (1), 1–14.

Hevia, A., Delgado, S., Sánchez, B., Margolles, A., 2015. Molecular players involved in the interaction between beneficial bacteria and the immune system. Front. Microbiol. 6, 1285.

Hewson, I., Button, J.B., Gudenkauf, B.M., Miner, B., Newton, A.L., Gaydos, J.K., et al., 2014. Densovirus associated with sea-star wasting disease and mass mortality. Proc. Natl. Acad. Sci. 111 (48), 17278–17283.

Hiddink, J.G., Jennings, S., Sciberras, M., et al., 2017. Global analysis of depletion and recovery of seabed biota after bottom trawling disturbance. Proc. Natl. Acad. Sci. 114 (31), 8301–8306.

Hughes, T., Szmant, A.M., Steneck, R., Carpenter, R., Miller, S., 1999. Algal blooms on coral reefs: what are the causes? Limnol. Oceanogr. 44, 1583–1586.

Huss, M., Hock, R., 2015. A new model for global glacier change and sea-level rise. Front. Earth Sci. 3, 54.

Huxham, M., Roberts, I., Bremner, J., 2000. A field test of the intermediate disturbance hypothesis in the soft-bottom intertidal. Int. Rev. Hydrobiol. 85 (4), 379–394.

IPCC, 2014. In: Core Writing Team, Pachauri, R.K., Meyer, L.A. (Eds.), Climate Change 2014: Synthesis Report. Contribution of Working Groups I, II and III to the Fifth Assessment Report of the Intergovernmental Panel on Climate Change. IPCC. 169 pp.

IPCC, 2023. Synthesis Report of the Ipcc Sixth Assessment Report (Ar6). https://report.ipcc.ch/ar6syr/pdf/IPCC_AR6_SYR_SPM.pdf.

Kaplanis, N.J., Edwards, C.B., Eynaud, Y., Smith, J.E., 2020. Future sea-level rise drives rocky intertidal habitat loss and benthic community change. PeerJ 8, e9186.

Keller-Costa, T., Lago-Lestón, A., Saraiva, J.P., Toscan, R., Silva, S.G., Gonçalves, J., et al., 2021. Metagenomic insights into the taxonomy, function, and dysbiosis of prokaryotic communities in octocorals. Microbiome 9 (1), 1–21.

Knowlton, N., Jackson, J.B.C., 2008. Shifting baselines, local impacts, and global change on coral reefs. PLoS Biol. 6 (2), e54.

Kornder, N.A., Riegl, B.M., Figueiredo, J., 2018. Thresholds and drivers of coral calcification responses to climate change. Glob. Chang. Biol. 24 (11), 5084–5095.

Kroeker, K.J., Kordas, R.L., Crim, R., Hendriks, I.E., Ramajo, L., Singh, G.S., et al., 2013. Impacts of ocean acidification on marine organisms: quantifying sensitivities and interaction with warming. Glob. Chang. Biol. 19 (6), 1884–1896.

Lapointe, B.E., 1997. Nutrient thresholds for bottom-up control of macroalgal blooms on coral reefs in Jamaica and Southeast Florida. Limnol. Oceanogr. 42 (5part2), 1119–1131.

Lapointe, B.E., Littler, M.M., Littler, D.S., 1997. Macroalgal overgrowth of fringing coral reefs at Discovery Bay, Jamaica: bottom-up versus top-down control. Proc. 8th Int. Coral Reef Sym. 1, 927–932.

Lapointe, B.E., Herren, L.W., Brewton, R.A., Alderman, P.K., 2020. Nutrient over-enrichment and light limitation of seagrass communities in the Indian River Lagoon, an urbanized subtropical estuary. Sci. Total Environ. 699, 134068.

Lessios, H.A., 1988. Mass mortality of *Diadema antillarum* in the Caribbean: what have we learned? Annu. Rev. Ecol. Syst. 19 (1), 371–393.

Leung, J.Y., Zhang, S., Connell, S.D., 2022. Is ocean acidification really a threat to marine calcifiers? A systematic review and meta-analysis of 980+ studies spanning two decades. Small 18 (35), 2107407.

188 Chapter 4 Threats to benthic biodiversity

Lodge, M., Johnson, D., Le Gurun, G., Wengler, M., Weaver, P., Gunn, V., 2014. Seabed mining: international seabed authority environmental management plan for the Clarion–Clipperton zone. A partnership approach. Mar. Policy 49, 66–72.

Lopez, J.V., Peterson, C.L., Morales, F., Brown, L., 2000. Andros Island flora and fauna in the new millennium. Bahamas J. Sci. 8 (1), 32.

Lopez-Legentil, S., Song, B., McMurry, S.E., Pawlik, J.R., 2008. Bleaching and stress in coral reef ecosystems: hsp70 expression by the giant barrel sponge *Xestospongia muta*. Mol. Ecol. 17, 1840–1849.

Magdaong, E.T., Fujii, M., Yamano, H., Licuanan, W.Y., Maypa, A., Campos, W.L., et al., 2014. Long-term change in coral cover and the effectiveness of marine protected areas in the Philippines: a meta-analysis. Hydrobiologia 733, 5–17.

Magnan, A.K., Oppenheimer, M., Garschagen, M., Buchanan, M.K., Duvat, V.K., Forbes, D.L., et al., 2022. Sea level rise risks and societal adaptation benefits in low-lying coastal areas. Sci. Rep. 12 (1), 10677.

Marler, T.E., 2014. Pacific island tropical cyclones are more frequent and globally relevant, yet less studied. Front. Environ. Sci. 2, 42.

McIntyre, L., 2019. The Scientific Attitude: Defending Science from Denial, Fraud, and Pseudoscience. Mit Press.

McKenzie, L.J., Nordlund, L.M., Jones, B.L., Cullen-Unsworth, L.C., Roelfsema, C., Unsworth, R.K., 2020. The global distribution of seagrass meadows. Environ. Res. Lett. 15 (7), 074041.

McMurray, S.E., Blum, J.E., Pawlik, J.R., 2008. Redwood of the reef: growth and age of the giant barrel sponge *Xestospongia muta* in the Florida Keys. Mar. Biol. 155, 159–171.

Meurant, G., 2012. The Ecology of Natural Disturbance and Patch Dynamics. Academic Press.

Meyer, J., Riebesell, U., 2015. Reviews and syntheses: responses of coccolithophores to ocean acidification: a meta-analysis. Biogeosciences 12 (6), 1671–1682.

Miller, M.W., Karazsia, J., Groves, C.E., Griffin, S., Moore, T., Wilber, P., Gregg, K., 2016. Detecting sedimentation impacts to coral reefs resulting from dredging the PEI of Miami, Florida USA. PeerJ 4 (12), e2711. https://doi.org/10.7717/peerj.2711.

Miller, K.A., Thompson, K.F., Johnston, P., Santillo, D., 2018. An overview of seabed mining including the current state of development, environmental impacts, and knowledge gaps. Front. Mar. Sci. 4, 418.

Morais, J., Cardoso, A.P., Santos, B.A., 2022. A global synthesis of the current knowledge on the taxonomic and geographic distribution of major coral diseases. Environ. Adv. 8, 100231.

Morris, L.J., Hall, L.M., Jacoby, C.A., Chamberlain, R.H., Hanisak, M.D., Miller, J.D., Virnstein, R.W., 2022. Seagrass in a changing estuary, the Indian River Lagoon, Florida, United States. Front. Mar. Sci. 8, 2121.

Nagelkerken, I., Aerts, L., Pors, L., 2000. Barrel sponge bows out. Reef Encount. 28, 14–15.

NASEM, 2019. A Decision Framework for Interventions to Increase the Persistence and Resilience of Coral Reefs. National Academy Press, Washington DC.

Nielsen, T.D., Hasselbalch, J., Holmberg, K., Stripple, J., 2020. Politics and the plastic crisis: a review throughout the plastic life cycle. Wiley Interdiscip. Rev. Energy Environ. 9 (1), e360.

Niner, H.J., Ardron, J.A., Escobar, E.G., Gianni, M., Jaekel, A., Jones, D.O., et al., 2018. Deep-sea mining with no net loss of biodiversity—an impossible aim. Front. Mar. Sci. 5, 53.

Normile, D., 2018. Shock greets claim of CRISPR-edited babies. Science 362, 978–979.

Oliver, J.K., Berkelmans, R., Eakin, C.M., 2018. Coral bleaching in space and time. Coral bleaching: Patterns, Processes, Causes and Consequences. Springer, pp. 27–49.

Pascal, P.Y., Fleeger, J.W., Galvez, F., Carman, K.R., 2010. The toxicological interaction between ocean acidity and metals in coastal meiobenthic copepods. Mar. Pollut. Bull. 60 (12), 2201–2208.

Pauly, D., 1995. Anecdotes and the shifting base-line syndrome of fisheries. Trends Ecol. Evol. 10, 430.

Peñaflor, E.L., Skirving, W.J., Strong, A.E., Heron, S.F., David, L.T., 2009. Sea-surface temperature and thermal stress in the coral triangle over the past two decades. Coral Reefs 28, 841–850.

Piret, J., Boivin, G., 2021. Pandemics throughout history. Front. Microbiol. 11, 631736.

Precht, W.F., Gintert, B.E., Robbart, M.L., Fura, R., Van Woesik, R., 2016. Unprecedented disease-related coral mortality in Southeastern Florida. Sci. Rep. 6 (1), 1–11.

Porter, J.W., Dustan, P., Jaap, W.C., Patterson, K.L., Kosmynin, V., Meier, O.W., et al., 2001. Patterns of spread of coral disease in the Florida Keys. In: The Ecology and Etiology of Newly Emerging Marine Diseases. Springer, pp. 1–24.

Precht, W.F., Gintert, B.E., Fura, R., Robbart, M.L., Rogers, K., Dial, R.S., 2019. Miami Harbor deep dredge project: a reappraisal reveals same results. In: Dredging Summit & Expo '19 Proceedings.

Pronzato, R., 1999. Sponge-fishing, disease and farming in the Mediterranean Sea. Aquat. Conserv. Mar. Freshwat. Ecosyst. 9 (5), 485–493.

Rao, T.S., Andrade, C., 2011. The MMR vaccine and autism: sensation, refutation, retraction, and fraud. Indian J. Psychiatry 53 (2), 95.

Ribeiro, B., Lima, C., Pereira, S.E., et al., 2023. Calcareous sponges can synthesize their skeleton under short-term ocean acidification. Sci. Rep. 13, 6776. https://doi.org/10.1038/s41598-023-33611-3.

Rich, N., 2018. Losing Earth: The Decade we Almost Stopped Climate Change. New York Times Magazine. August 1 https://www.nytimes.com/interactive/2018/08/01/magazine/climate-change-losing-earth.html.

Roozenbeek, J., Schneider, C.R., Dryhurst, S., Kerr, J., Freeman, A.L., Recchia, G., et al., 2020. Susceptibility to misinformation about COVID-19 around the world. R. Soc. Open Sci. 7 (10), 201199.

Rubin, E.S., Davison, J.E., Herzog, H.J., 2015. The cost of CO_2 capture and storage. Int. J. Greenhouse Gas Control 40, 378–400. https://doi.org/10.1016/j.ijggc.2015.05.018.

Sand, M., et al., 2015. Response of Arctic temperature to changes in emissions of short-lived climate forcers. Nat. Clim. Chang. 6 (3), 286–289. https://doi.org/10.1038/nclimate2880.

Sandin, S.A., Smith, J.E., DeMartini, E.E., Dinsdale, E.A., Donner, S.D., Friedlander, A.M., et al., 2008. Baselines and degradation of coral reefs in the Northern Line Islands. PLoS One 3 (2), e1548.

Schratzberger, M., Warwick, R.M., 1999. Differential effects of various types of disturbances on the structure of nematode assemblages: an experimental approach. Mar. Ecol. Prog. Ser. 181, 227–236.

Schulze, A., Grimes, C.J., Rudek, T.E., 2017. Tough, armed and omnivorous: *Hermodice carunculata* (Annelida: Amphinomidae) is prepared for ecological challenges. J. Mar. Biol. Assoc. U. K. 97 (5), 1075–1080.

Sciberras, M., Hiddink, J.G., Jennings, S., Szostek, C.L., Hughes, K.M., Kneafsey, B., et al., 2018. Response of benthic fauna to experimental bottom fishing: a global meta-analysis. Fish Fish. 19 (4), 698–715.

Shantz, A.A., Lemoine, N.P., Burkepile, D.E., 2016. Nutrient loading alters the performance of key nutrient exchange mutualisms. Ecol. Lett. 19 (1), 20–28.

Shen, M., Ye, S., Zeng, G., Zhang, Y., Xing, L., Tang, W., et al., 2020. Can microplastics pose a threat to ocean carbon sequestration? Mar. Pollut. Bull. 150, 110712.

Shlesinger, T., van Woesik, R., 2023. Oceanic differences in coral-bleaching responses to marine heatwaves. Sci. Total Environ. 871, 162113.

Slaby, B.M., Franke, A., Rix, L., Pita, L., Bayer, K., Jahn, M.T., Hentschel, U., 2019. Marine sponge holobionts in health and disease. In: Symbiotic Microbiomes of Coral Reefs Sponges and Corals. Springer, pp. 81–104.

Sosa, B., Fontans-Álvarez, E., Romero, D., da Fonseca, A., Achkar, M., 2019. Analysis of scientific production on glyphosate: An example of politicization of science. Sci. Total Environ. 681, 541–550.

Spearman, J., Taylor, J., Crossouard, N., Cooper, A., Turnbull, M., Manning, A., et al., 2020. Measurement and modelling of deep sea sediment plumes and implications for deep sea mining. Sci. Rep. 10 (1), 5075.

Stark, J.S., Kim, S.L., Oliver, J.S., 2014. Anthropogenic disturbance and biodiversity of marine benthic communities in Antarctica: a regional comparison. PLoS One 9 (6), e98802.

Steneck, R.S., Dethier, M.N., 1988. A functional group approach to the structure of algal-dominated communities. Oikos 69, 476–498.

Sweet, M.J., Bulling, M.T., 2017. On the importance of the microbiome and pathobiome in coral health and disease. Front. Mar. Sci. 4 (9), 00009.

Tahiluddin, A.B., Sarri, J.H., 2022. An overview of destructive fishing in the Philippines. Acta Nat. Sci. 3 (2), 116–125.

Thrush, S.F., Dayton, P.K., 2002. Disturbance to marine benthic habitats by trawling and dredging: implications for marine biodiversity. Annu. Rev. Ecol. Syst., 449–473.

van Woesik, R., Kratochwill, C., 2022. A global coral-bleaching database, 1980–2020. Sci. Data 9 (1), 20.

Vega Thurber, R.L., Burkepile, D.E., Fuchs, C., Shantz, A.A., McMinds, R., Zaneveld, J.R., 2014. Chronic nutrient enrichment increases prevalence and severity of coral disease and bleaching. Glob. Chang. Biol. 20 (2), 544–554.

Villanoy, C., David, L., Cabrera, O., Atrigenio, M., Siringan, F., Aliño, P., Villaluz, M., 2012. Coral reef ecosystems protect shore from high-energy waves under climate change scenarios. Clim. Chang. 112, 493–505.

Wakefield, A.J., Murch, S.H., Anthony, A., Linnell, J., Casson, D.M., Malik, M., et al., 1998. RETRACTED: ileal-lymphoid-nodular hyperplasia, non-specific colitis, and pervasive developmental disorder in children. Lancet 351 (9103), 637–641.

Walton, C.J., Hayes, N.K., Gilliam, D.S., 2018. Impacts of a regional, multi-year, multi-species coral disease outbreak in Southeast Florida. Front. Mar. Sci. 5, 323. https://doi.org/10.3389/fmars.2018.00323.

Wang, J., Wang, M., Ru, S., Liu, X., 2018. High levels of microplastic pollution in the sediments and benthic organisms of the South Yellow Sea, China. Sci. Total Environ. 651 (2019), 1661–1669. https://doi.org/10.1016/j.scitotenv.2018.10.007.

Wang, Y., McKee, M., Torbica, A., Stuckler, D., 2019. Systematic literature review on the spread of health-related misinformation on social media. Soc. Sci. Med. 240, 112552.

Watts, A.J.R., Lewis, C., Goodhead, R.M., Beckett, S.J., Moger, J., Tyler, C.R., Galloway, T.S., 2014. Uptake and retention of microplastics by the shore crab Carcinus maenas. Environ. Sci. Technol. 48 (15), 8823–8830.

Webster, N.S., 2007. Sponge disease a global threat? Environ. Microbiol. 9 (6), 1363–1375.

Whomersley, P., Huxham, M., Bolam, S., Schratzberger, M., Augley, J., Ridland, D., 2010. Response of intertidal macrofauna to multiple disturbance types and intensities – an experimental approach. Mar. Environ. Res. 69 (5), 297–308.

Widdicombe, S., Spicer, J.I., 2008. Predicting the impact of ocean acidification on benthic biodiversity: what can animal physiology tell us? J. Exp. Mar. Biol. Ecol. 366 (1–2), 187–197.

Wilcox, C., Hardesty, B.D., Law, K.L., 2020. Abundance of floating plastic particles is 442 increasing in the western North Atlantic Ocean. Environ. Sci. Technol. 54, 790–796.

Work, T.M., Weatherby, T.M., Landsberg, J.H., Kiryu, Y., Cook, S.M., Peters, E.C., 2021. Viral-like particles are associated with endosymbiont pathology in Florida corals affected by stony coral tissue loss disease. Front. Mar. Sci. 8, 750658.

Yarzábal, L.A., Salazar, L.M.B., Batista-García, R.A., 2021. Climate change, melting cryosphere and frozen pathogens: should we worry…? Environ. Sustain. 4 (3), 489–501.

Yoon, J., Matsuda, S., Adachi, K., Kasai, H., Yokota, A., 2011. Rubritalea halochordaticola sp. nov., a carotenoid-producing verrucomicrobial species isolated from a marine chordate. Int. J. Syst. Evol. Microbiol. 61 (Pt 7), 1515–1520. https://doi.org/10.1099/ijs.0.025031-0.

Zhang, Q., Xu, E.G., Li, J., Chen, Q., Ma, L., Zeng, E.Y., Shi, H., 2020. A review of microplastics in table salt, drinking water, and air: direct human exposure. Environ. Sci. Technol. 54 (7), 3740–3751.

Zaneveld, J.R., McMinds, R., Vega Thurber, R., 2017. Stress and stability. applying the Anna Karenina principle to animal microbiomes. Nat. Microbiol. 2 (9), 1–8.

Zhang, Y., Wang, J., Geng, X., Jiang, Y., 2021. Does microplastic ingestion dramatically decrease the biomass of protozoa grazers? A case study on the marine ciliate Uronema marinum. Chemosphere 267, 129308.

CHAPTER

Possible solutions for the conservation of benthic habitats and organisms

5

The great aim of education is not knowledge, but action.
Herbert Spencer

Exploration, education, engagement, and optimism

At the time this book was being written, humanity was slowly emerging from the collective, protective two-year shell of isolation from a pandemic scourge, caused by coronavirus (SARS-CoV-2), also known as COVID-19, named for the year that the pandemic started. The easily transmissible respiratory virus (delta variant) had infected over 38 million and killed over 1.5 million people worldwide within 10 months of the first infections. Even when tested and effective vaccines were rolled out in the spring of 2021, many people took tentative steps to normalcy. Indeed, provocative suggestions to take this opportunity to reset much of society, establishing a new normal and several predictions, were made (Politico, 2020). One set of predictions claimed that there will be "(1) *revived trust in institutions*, (2) *a new civic federalism*, (3) [rejuvenation of] *government service cachet*, and (4) *greater trust in science….*" Some of the rosier, yet not altogether unrealistic, predictions have not yet materialized, and perhaps trends and expectations for scientific progress shamefully went in the opposite direction (Hotez et al., 2021; Balog-Way and McComas, 2022). This is unfortunate.

The COVID-19 pandemic converged with simmering or sudden cultural upheavals to exacerbate turmoil during this time. These have all possibly scarred the collective psyche deeper than expected, and we are still assessing the damage. Emotions and passions understandably ran high due to the public health emergency, forced quarantines, and disruption of work routines, in combination with racial, economic, and political tensions.

However, when trying to protect habitats and promote constructive environmental policies and awareness, it is also important to avoid fatalistic and defeatist attitudes, even in the face of a rising anti-science or informational tide. Science and other progress did not completely stand still during the pandemic, as productivity continued in several sectors that were not directly biomedically related. The oldest DNA at 2 million years was discovered, zero hydrogen-based power has been implemented, and the Webb Telescope began gazing across the universe (Kjær et al., 2022; Witze, 2022; https://www.smithsonianmag.com/smart-news/hydrogen-powered-passenger-trains-are-now-running-in-germany-180980706). Many scientists stayed active during mandatory work lockouts, slowdowns, and remote work, and I would venture to say that several discoveries may yet be hatched after marinating in relative isolation or patient focus on the problem.

Assessments and Conservation of Biological Diversity From Coral Reefs to the Deep Sea. https://doi.org/10.1016/B978-0-12-824112-7.00005-4
Copyright © 2024 Elsevier Inc. All rights reserved.

192 **Chapter 5** Conservation of benthic habitats and organisms

Herbert Spencer's opening quotation fits the restlessness of the post-COVID mentality and instructs on where people can possibly channel pent-up ideas and kinetic energy. New motivational mottos may be required to lure the wary back out into the race, and so we propose one that would fit this chapter— "Exploration, education, and engagement" for benthic conservation. An underlying question is how we can best apply ourselves. Happily, we are seeing more action each day, and this chapter aims to capture just a few examples that should uplift the cause of benthic conservation.

In the previous chapter, I presented some of the most acute, large problems facing benthic habitats and their conservation. This chapter remains expansive but also takes a broader, slightly philological approach to applying the necessary concrete solutions. Many may agree that climate change is a huge global issue which no single country is responsible for or can fix alone. The latter will take massive coordination, which is aimed at helping and hopefully saving threatened species and ecosystems for future human generations. The hyperbole is all too real. But in times of crisis, large, multinational, or multiethnic groups have rallied. For example, the specter of chlorofluorocarbons (CFCs) was recognized to jeopardize large swaths of the Southern Hemisphere by widening a huge hole in the atmosphere (Manzer, 1990). If allowed to grow unabated, this hole could have significantly heightened human exposure to harmful ultraviolet (UV) rays from the sun. However, after 35 years of concerted efforts, including awareness-raising campaigns and effective regulation of industry production and emissions, a United Nations Environmental Programme (UNEP) report has indicated (World Meteorological Organization (WMO), 2022) a decline of nearly 99% of banned ozone-depleting substances (ODSs), which has restored much of the damage in the ozone layer. This major progress was summarized in the "Montreal Protocol":

- Actions taken under the Montreal Protocol continued to decrease atmospheric abundances of controlled ODSs and advance the recovery of the stratospheric ozone layer. The atmospheric abundances of both total tropospheric chlorine and total tropospheric bromine from long-lived ODSs have continued to decline since the 2018 Assessment. New studies support previous assessments in that the decline in ODS emissions due to compliance with the Montreal Protocol avoids global warming of approximately 0.5–1°C by mid-century compared to an extreme scenario with an uncontrolled increase in ODSs of 3%–3.5% per year. (A cautionary footnote to this success story is that the fix could be easily undone, similar to a bandage being ripped off too soon or harshly.)

The International Space Station (ISS) provides another prime example of a major engineering and scientific feat through collaboration and curiosity (https://www.nasa.gov/mission_pages/station/main/index.html). The ISS grew out of American Space Stations *Skylab* and *Freedom* and Soviet *Mir* station projects began in the 1980s. Soon after, the European Space Agency and a Japanese Experiment module were added to the growing ISS project. With construction beginning in 1998, the ISS was intended to be a multinational "laboratory, observatory, and factory while providing transportation, maintenance, and a low Earth orbit staging base for possible future missions" at an altitude of approximately 400 km. The ISS has been continuously occupied by rotating crews of astronauts and cosmonauts since November 2000. The primary goals of the ISS are to conduct scientific research in a microgravity environment, to develop and test new technologies for space exploration, and to serve as a platform for international cooperation and diplomacy. The ISS has hosted thousands of scientific experiments in various fields, including biology, physics, astronomy, medicine, and earth science. Some notable scientific achievements of the ISS include the study of the human body's adaptation to long-term spaceflight and the detection of cosmic rays. The ISS is also a vital component of NASA's long-term plans for human space exploration, serving as a stepping stone for future missions to the Moon and Mars. In addition, the ISS provides opportunities for international collaboration and cooperation, with crew members from different countries working together on various research projects.

Another large-scale, global endeavor called the "One Health Initiative (OHI)" (https://onehealthinitiative.com/) is equally as ambitious as the previous examples. The OHI grew out of the severe acute respiratory disease (SARS) in early 2003–04, then was followed up with official collaborations between the Wildlife Conservation Society and, in 2008, the American Veterinary Medical Association in collaboration with the American Medical Association (Atlas, 2012; Mackenzie and Jeggo, 2019). The ideas of interconnectedness stem farther back to researchers such as Louis Pasteur and Robert Koch, who crossed the boundaries between animal and human health. OHI goals also aim to link the health of the environment, humans, and wildlife together, so that all could be addressed holistically. The utility and potential of OHI can be poignantly realized in the aftermath of zoonotic outbreaks like the recent COVID-19 pandemic (Bonilla-Aldana et al., 2020). The initiative also intends to remind researchers and the public just how dependent all components are. Who can predict where the next pathogen may emerge, and can marine samples or organisms be fully ruled out? Thinking about the breadth and pervasiveness of the oceans, its biology and habitats should thus be implicitly included in One Health strategies and approaches.

These previous examples provide strong evidence that large, global-scale efforts can lead to success. Each major environmental crisis can be approached in multiple ways but often requires coordination between diverse academic sectors (natural sciences, psychology, sociology, etc.) (Tàbara et al., 2018). Shrivastava et al. (2020) elaborate a transdisciplinary, integrative model (for sustainability) needed for global-wide change and which can be applied to biodiversity conservation goals, which recognize the difficulties for large-scale changes (Fig. 5.1). Promoting biodiversity fits within these approaches, and its conservation runs counter to monoculture establishment and their higher levels of homozygosity, whose dangers should be amplified. For example, some factions of contemporary society, across multiple nations, appear to opine for conformity and a homogenization of cultures, which are antithetical to diversification. Solutions to counter a status quo often require more creative thinking or unusual boldness that may stem from the margins, sometimes pejoratively referred to as the fringe. However, environmental activists such as Rachel Carson, Chico Mendes, Barry Commoner, Greta Thunberg, Paulo Paulino Guajajara, Nazildo dos Santos Brito, and others took significant personal risks and thus represented more than old clichés. Their bravery to speak out resulted in several tangible benefits to protect the environment. Perhaps, a rallying cry is needed to focus on the issues that resound across generations and peoples again. Facile terminologies and language are needed to describe the tangible plight of biodiversity loss, for the destruction of a major habitat or ecosystem before most people have the chance to know about it, see it, or even visit it. "Ignorant destruction" or blind loss demarcate this topic, but cannot convey the gravity of the actual loss. A merger of ideas could yield new slogans like *bioloss or bloss* to designate significant "biodiversity loss" at a site. The popular media seems to recognize the urgency, but translating the awareness and information to action remains slow and problematic (Gilbert, 2022; https://www.worldatlas.com/articles/10-people-who-died-to-save-the-amazon.html).

The great seal of the United States of America reads *E pluribus unum* that translates to "Out of many, one." The well-known Latin phrase has 13 letters, which represent the original 13 American colonies. This concept dates back to the origin of the country, being suggested in 1776 by Pierre Eugene

194 **Chapter 5** Conservation of benthic habitats and organisms

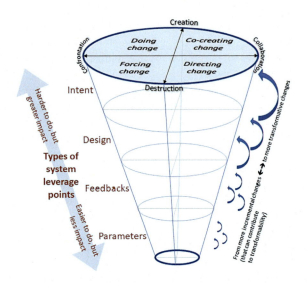

FIG. 5.1

Pathways for transformative change for sustainability. The cone of increasingly transformative change: systems interventions with increasing leverage (from bottom to top on left side—see also Meadows, 1999; Abson et al., 2017) as the cone opens out have greater potential to cause transformational change and require a larger application of the four strategies for systems change *(top plane)*; easier but directed incremental changes can contribute to the transformability of the system *(right side)* (Folke et al., 2010).

Reproduced with permission from Shrivastava, P., Smith, M.S., O'Brien, K., Zsolnai, L. 2020. Transforming sustainability science to generate positive social and environmental change globally. One Earth 2 (4), 329–340.

du Simitiere, but the cooperative mentality will be required to cope with our current large-scale environmental crises.

Marine efforts have initiated various environmental causes and promotions such as the Year of the Ocean, Year of the Reef, and, of course, Earth Day and, recently, a United Nations (UN)-sponsored Oceans Day (https://www.un.org/en/observances/oceans-day) (Rome, 2013). Events will likely be set up specifically for the seafloor and benthos, though they are naturally encompassed in the earlier two. One special Year of Biodiversity has occurred (in 2010), and this was then followed by the "Decade of Biodiversity." Unfortunately, this ended inconclusively since none of the Aichi Biodiversity Targets established in 2010 were met during the UN Decade on Biodiversity (Rees et al., 2018; United Nations, 2020). These initiatives all help the causes of conservation and awareness. However, a whole month per year may be more appropriate for something as important as the ocean, which produces at least 40% of the world's oxygen while absorbing 30% of its carbon dioxide. In the context of benthic conservation, a phrase like "bottom-up approaches" can take on new meaning and galvanize new movements.

As previously mentioned, "being out of sight" can certainly be a factor for benthic habitats being "out of mind," especially for the lay public. How does anyone appreciate or even know of something that is mostly unseen? This question underscores the need for engagement and also underlies how the ocean can be so paradoxical. Many of the people can "see" the ocean a great deal, almost constantly, when they live and work on the coasts. According to several studies, more than 600 million people (around 10% of the world's population) live in coastal areas that are less than 10 m above sea level, and at least 2.5 billion people (about 40% of the world's population) live within 100 km (60 miles) of the coast (Small and Nicholls, 2003; Neumann et al., 2015; United Nations, 2017). While living on the coasts, people enjoy recreational activities such as boating and fishing, or participate in marine-focused labors—commercial fisheries, fossil fuel extractions, cruise ships, ocean cargo, etc. I had already mentioned cruise ship escapades earlier. Overall, these activities represent many eyes and hands across the water, but did their hearts and minds also follow? Moreover, a much smaller percentage of this population plunge into the oceanic water per se, dive and view benthic habitats by snorkeling or scuba. We can safely presume that much of the ocean benthos remains a mystery to most of the world's citizens. But should this always be?

If the inventors and engineers could make undersea travel easier, safer, and less expensive, would the lay public flock to the opportunity to visit the benthos? It would be an adventure, but underwater dives could create the same awe that sparks many young dreamers to fly above the earth in a jet or Blue Horizon rocket. These are hypotheticals now, but plausible. The *Avatar* movie director James Cameron built his own deep-sea submersible, but few of us have this capacity. However, creative inventors and engineers have successfully overcome the intense pressures of the Challenger Deep and hadal trenches and could possibly develop routine transit to the aphotic zone for curious tourists. To protect heterogeneous and large habitats such as on the seafloor, humanity must be inventive again, and in a way that does little or no harm. Caution would be required since we are asking for this capacity, but also know that more traffic often brings about greater disturbances to a habitat. Future ferrymen would also have to assure safe and secure passage to tourists and the public on submersibles which are fully certified, to avoid the unnecessary tragic 2023 accident of OceanGate's *Titan* submersible. A current model for this type of deep sea enterprise would be the commercial *Idabel* submersible run by Stanley Karl in Roatan (https://stanleysubmarines.com/). This vessel has an excellent safety record and taken more than 2500 dives to a maximum depth of 915 m. While transiting and enjoying the seabed, we could try to avoid making Jean-Jacques Rousseau's warning come true ("Everything is good as it comes from the hands of the Maker of the world, but degenerates once it gets into the hands of man"). Yet, these futuristic scenarios could benefit humanity by raising higher awareness of the benthos' value and beauty, providing imperative education, outreach and research. The convergence of these factors in the last few decades to advance the prospects and knowledge of benthic habitats gives encouraging signs for the future. This chapter highlights how several dedicated scientists take the lead to protect these benthic zones that very few people know about.

Ocean optimism arrived as one strategy to educate and galvanize popular support for marine protections in the early 2010s. The message was to help propel a marine conservation movement that focuses on "solutions rather than problems" with the aim of creating a new narrative of hope for our oceans. Coral reef biologist Nancy Knowlton helped mature the concept of ocean optimism and has been its champion for nearly 20 years (http://www.oceanoptimism.org/reasons-to-be-cheerful/ (Knowlton, 2021). Knowlton helped spearhead this approach with in-person workshops, seminars, and a popular public service campaign. A popular book, *Citizens of the Sea* (Knowlton, 2010), was written based partially on data from the Census of Marine Life. Published by the National Geographic, stunning photographs of marine organisms including hermit crabs (*Ciolopagurus strigatus*), gooseneck barnacles (*Pollicipes polymerus*), tube worms (*Riftia pachyptila*) and many other sea creatures abound through 200+ pages. The book capitalizes on the photogenicity of its subjects and helps propel popular engagement.

196 Chapter 5 Conservation of benthic habitats and organisms

It was a pleasure having Nancy Knowlton as a postdoctoral advisor since she allowed me to plunge headfirst into the benthos from a totally terrestrial academic start. However, she provided a good example of leading by inspiration for many. As a National Academy of Science member, Knowlton had long been known as one of the major experts on marine biodiversity (Knowlton, 2010). Therefore, she would often be called to give seminars on the topics of biodiversity, coral reefs, etc. When doing so, in some of these series, she would sometimes travel and present with her spouse, another National Academy member, Jeremy Jackson, and introduce themselves as "Drs. Gloom and Doom," referring to the subject of the deteriorating state of coral reefs around the world, which they would discuss.

And yet they would balance that self-deprecation and the voluminous data about coral reefs' general decline with more hopeful messages. These stories and the positive spin have led to becoming a champion of "ocean and earth optimism" (Knowlton, 2017, 2021). The movement has continued to bubble up, with indicators like "#oceanoptimism" reaching over 76 million Twitter users in just 3 years (McAfee et al., 2019). With her ocean optimism work and mindset, Knowlton and colleagues had tapped into a vein of energy that I will also try to highlight throughout this chapter. In a recent collection of *PLOS* papers, major problems of ocean ecosystem degradation were constructively addressed (Knowlton and Di Lorenzo, 2022). A major part of the optimism strategy is to show evidence of successes for restoring and protecting marine habitats in the face of increasing pessimism. I list a few points that can provide more hope as follows:

- Ocean biodiversity losses appear less than on land, and more resilient (McCauley et al., 2015).
- Portions of the Great Barrier Reef, which completely bleached in 2016–17, are showing signs of returning.
- Awareness of society's dependence on the ocean is growing, along with concerted efforts to protect it (see below).
- UN's Green Climate Fund has mobilized US$10.3 billion annually to assist developing countries to adapt to climate change, with a goal of US$100 billion per year in 2020 (https://www.greenclimate.fund/how-we-work/resource-mobilization).
- To reverse and halt some of the deterioration confronting the Florida Reef Tract (FRT), innovative programs such as the 2020 "Iconic Reefs" were implemented (NOAA Fisheries, 2019).

Duarte et al. (2020) continue these assertions in their review: "We argue that the focus should be on increasing the abundance of key habitats and keystone species and restoring the three-dimensional complexity of benthic ecosystems." The optimistic outlook is that marine life can be "rebuilt" and some of its biodiversity preserved, if certain major environmental pressures (overfishing, climate change, and habitat loss) could be halted and reversed (Jackson et al., 2001).

Several prominent coral reef scientists, including Knowlton, Stephen Palumbi, and others, explore possible "interventionist" methods that could save deteriorating coral species for the long term. The 2019 National Academy of Sciences, Engineering and Medicine consensus report (NASEM, 2019) described the benefits, risks and consequences of adaptive management which spanned multiple types of interventions: (a) *genetic and reproductive* (e.g., managed selection and breeding, gamete and larval seeding, coral cryopreservation, and genetic manipulation); (b) *physiological* (preexposure, symbiont (zooxanthellae and bacteria) manipulation, phage therapy, antibiotics, antioxidants); (c) *population and community-wide* (managed relocation, coral predators and competitors) and *environmental* (coral reef shading, managing water temperatures, and acidity). Since the report is fairly recent, most of these methods have not yet been fully tested or implemented at large scales. These potential new programs provide hope and should help continue to galvanize marine students and the lay public, many of whom are already involved in various *citizen science* coastal enhancement projects. These projects are often focused on local habitats and depend on volunteer novices and experts. A few examples appear below:

- Crowdsourced bathymetry is the collection of depth measurements from vessels, using standard navigation instruments, while engaged in routine maritime operations. This was applied recently in the Great Barrier Reef (GBR) (https://www.deepreef.org/2022/02/18/crowdsourced-hydrospatial/).
- Chris Roelfsema, from the University of Queensland, leads volunteer divers to survey reefs. He says that "Citizen science plays a crucial role filling information gaps and offering timely reporting on the environmental health of sites, when government agencies don't have time or funding to do so."
- Rachel Meyer implements the CALeDNA program that organizes group sampling efforts and applies environmental DNA (eDNA) to assess the biodiversity of microbes, fungi, plants, and animals found in California (https://ucedna.com/).
- Reef Check applies globally recognized and standardized monitoring methods (Hodgson et al., 1998; Habibi et al., 2007; Done et al., 2017).
- Artificial intelligence has been added as a final stage of citizen science projects to assist data collection and analyses, facilitating management decisions (McClure et al., 2020).
- Cigliano et al. (2015) implemented citizen science projects to monitor queen conch populations in Belize, and also described the development of toolkits that citizens can use in conjunction with scientists or on their own. These cover strategies to affect policy or for education. Effective participation sometimes requires a "sense of place," which has been emphasized in this volume (Haywood, 2014).

Another dimension that could be considered as a form of citizen science are museums and aquaria since they are visited by people from all over the world. These venerated institutions typically have long-standing relationships with their communities and thus have built up a cache of trust and respect. Most aquaria emphasize environmental problems of the natural habitats of their living residents. Besides the public displays, some well-funded aquariums, such as the L'Aquarium de Barcelona in Spain, Osaka Aquarium Kaiyukan in Japan, Shedds Aquarium in Chicago, and the Georgia Aquarium in Atlanta, have extensive research programs (Che-Castaldo et al., 2018; Kaufman et al., 2019; Van Bonn et al., 2022). Many public aquaria have energetic, creative, and eye-fetching public outreach programs, staff, and social media, and thus would be ripe for conveying research results from in house or innovative collaborations (Fauville et al., 2015). Lastly, institutions like the Smithsonian promote ocean awareness at the national level (https://ocean.si.edu/conservation/gulf-oil-spill/bacteriums-super-powers).

Biodiversity policies from governmental and nongovernmental organizations

To reemphasize that when the scope of many conservation problems is literally as large as the oceans, coordinating efforts between multiple groups makes good sense. For example, the International Seabed Authority (ISA) has formulated the Sustainable Seabed Knowledge Initiative (SSKI), which will focus on the international deep seabed. This follows the ambitious goals of the UN Decade of Ocean Science for Sustainable Development (DOSSD) plans adopted by all 168 members in 2020. As one of multiple DOSSD initiatives, the Challenger 150 project aims "to advance understanding of the diversity, distribution, function and services provided by deep-ocean biota" and includes a Deep-Ocean Genomes Program, led by the Woods Hole Oceanographic Institution (https://oceandecade.org/actions/challenger-150-a-decade-to-study-deep-sea-life/). Moreover, and much earlier, the Convention on International Trade in Endangered Species of Wild Flora and Fauna (CITES) has effectively diminished the trade of rare and endangered organisms, which preserves biodiversity, since 1975. All member states of the United Nations comply with CITES. Similarly, multinational coordinating bodies like the United Nations remain pivotal in implementing solutions to global and benthic biodiversity crises (https://www.un.org/sustainabledevelopment/wp-content/uploads/2017/05/Ocean-fact-sheet-package.pdf; https://press.un.org/en/2020/ga12274.doc.htm). They have been a motivating entity for biodiversity since 1992, when the United Nations Conference on Environment and Development (UNCED), also known as the Rio Conference or the Earth Summit, was held (Gupta and Singh, 2023). Besides discussing policies on climate change and fossil fuels, emphasis was made on protecting the world's oceans and the multilateral treaty, known as the Convention on Biological Diversity (CBD), was introduced for debate and signatures (https://www.cbd.int/convention/) (Secretariat of the Convention on Biological Diversity, United Nations, 2011).

Participants of the first and subsequent summits have put forth ambitious plans to protect earth's natural resources through policy and practice, and to their credit have met many of their goals. These efforts have been well documented (Miller et al., 2013). Besides the adoption of the Cartagena biosafety protocol, and the Global Strategy for Plant Conservation, the Nagoya Protocol (NP) on Access to Genetic Resources and the Fair and Equitable Sharing of Benefits Arising from their Utilization to the Convention on Biological Diversity, also known as the Nagoya Protocol on Access and Benefit Sharing (ABS) a 2010 supplementary agreement to the CBD, was passed. The NP's aim is the implementation of one of the three objectives of the CBD: the fair and equitable sharing of benefits arising out of the utilization of genetic resources, thereby contributing to the conservation and sustainable use of biodiversity. The NP sets out obligations for its contracting parties to take measures in relation to access to genetic resources, benefit-sharing and compliance. To this day, the policies and practice of the NP have permeated through many aspects of biological research (Ambler et al., 2021; Lawniczak et al., 2022). This also includes the "contention and resistance based on specific disciplines" (Overmann and Scholz, 2017) or national policies based on countries who have not signed onto the protocol, such as the United States, based on the grounds that the NP would have adverse economic impacts. Viewed as a regulatory nuisance by some, the spirit of the NP has been accepted and provides an essential buffer against wanton exploitation.

Signatories to the Convention on Biodiversity set 2030 to meet the UN's sustainable development goals. For example, the UN-proposed Sustainable Development Goal 14 (UN SDG 14 or "life below water") aims to "conserve and sustainably use the oceans, seas and marine resources for sustainable development" (https://sustainabledevelopment.un.org/sdg14) (Duarte et al., 2020).

The COP 15 Biodiversity Conference met December 7–19, 2022, in Montreal, Canada, and in spite of the usual posturing and hype, it ended with a historic concrete statement of prominent 2030 goals for biodiversity shown below (https://prod.drupal.www.infra.cbd.int/sites/default/files/2022-12/221219-CBD-PressRelease-COP15-Final_0.pdf):

Biodiversity policies from governmental and nongovernmental organizations **199**

UN biodiversity Goals A to D

GOAL A

- The integrity, connectivity, and resilience of all ecosystems are maintained, enhanced, or restored, substantially increasing the area of natural ecosystems by 2050;
- Human-induced extinction of known threatened species is halted, and, by 2050, the extinction rate and risk of all species are reduced tenfold, and the abundance of native wild species is increased to healthy and resilient levels;
- The genetic diversity within populations of wild and domesticated species is maintained, safeguarding their adaptive potential.

GOAL B

- *Biodiversity* is sustainably used and managed, and nature's contributions to people, including ecosystem functions and services, are valued, maintained, and enhanced, with those currently in decline being restored, supporting the achievement of sustainable development for the benefit of present and future generations by 2050.

GOAL C

- The monetary and nonmonetary benefits from the utilization of genetic resources, digital sequence information on genetic resources, and traditional knowledge associated with genetic resources, as applicable, are shared fairly and equitably, including, as appropriate with indigenous peoples and local communities, and substantially increased by 2050, while ensuring traditional knowledge associated with genetic resources is appropriately protected, thereby contributing to the conservation and sustainable use of biodiversity, in accordance with internationally agreed access and benefit-sharing instruments.

GOAL D

- Adequate means of implementation, including financial resources, capacity-building, technical and scientific cooperation, and access to and transfer of technology to fully implement the Kunming-Montreal global biodiversity framework, are secured and equitably accessible to all Parties, especially developing countries, in particular, the least-developed countries and small island developing States, as well as countries with economies in transition, progressively closing the biodiversity finance gap of $700 billion per year, and aligning financial flows with the Kunming-Montreal Global Biodiversity Framework and the 2050 Vision for Biodiversity.

These targets appear realistic and contemporary, such as the reference to preserving important genetic resources, the basis of all biodiversity. Goal A clearly incorporates major genetics principles espoused by many and discussed in this book. Goal C advocates equitable sharing. Governmental buy-in and commitment to the policy recommendations will be essential for ultimate success. There is little point in restoring benthic habitats, outplanting millions of coral juveniles, marshalling local stakeholders etc., if these labor intensive efforts are undercut from more massive pressures coming from above—e.g., greenhouse gases, uncontrolled pollution inputs, or relaxed governmental protections.

The environmental movement currently has large, established entities leading the way, such as Greenpeace, World Wildlife Fund, Environmental Defense Fund, and The Nature Conservancy, which

200 Chapter 5 Conservation of benthic habitats and organisms

carry out important local and regional environmental protection projects (Kritzer et al., 2016; Anbleyth-Evans, 2018; Collins et al., 2022). For example, the International Union for the Conservation of Nature (IUCN), a global consortium of 1400 member organizations and 15,000 experts since 1948, compiles the Red List, a comprehensive list of threatened species worldwide (Mace et al., 2008). However, the sheer number of these and smaller groups reflects a type of activist balkanization, which itself could create some inertia and hindrances or at least a dilution of energy. A more unified front could be more impactful and effective for benthic conservation, but harder to direct and coordinate. The bigger the ship, the harder it is to steer, and yet many of these nongovernmental organizations have similar and overlapping goals and interests. For example, in Chapter 3, we found that conservation studies and resources for the Coral Triangle appear woefully deficient in proportion to the abundance of biodiversity of the area and compared to other benthic hotspots. Efforts have begun to delineate priority areas, but this is only the beginning (Asaad et al., 2019). Moreover, more collaborations between independent groups provide a remedy to the excessive splintering. When various organizations can congregate, join forces, and merge into a larger unified body (like single amoeboid cells of a *Dictyostelium* slime mold converging to form the fruiting body), then a single agreed-upon mission and message may be more effectively broadcasted. One example for unifying efforts in the United States is the Cooperative Ecosystem Studies Units (CESU) Network, which "is a national consortium of federal agencies, tribes, academic institutions, state and local governments, nongovernmental conservation organizations, and other partners working together to support informed public trust resource stewardship" (Lapointe et al., 2016). Table 5.1 shows a cross section of active NGO's carrying out work that benefit benthic habitats.

Table 5.1 Nonprofit organizations and projects involved in benthic habitat conservation or assessment.

NGO name	Focus	Group website
Allen Coral Atlas	Coral reef maps	https://www.allencoralatlas.org/
Biodiversity Collaborative	Biodiversity and human interface	https://www.biodiversitycollaborative.org/
The Climate Group	Climate change	https://www.theclimategroup.org/
Cooperative Ecosystem Studies Units (CESU) Network	Ecological and conservation networking	http://www.cesu.psu.edu/
Coral Reef Alliance (CORAL)	Coral reef protection, climate change	https://coral.org/en/
Deep-Ocean Stewardship Initiative (DOSI)	Deep-ocean sustainable, deep-ocean governance, and management of resources	https://www.dosi-project.org/
Environmental Defense Fund	Climate change, environmental protection	https://www.edf.org/
Global Invertebrate Genomics Alliance	Research and training in invertebrate genomics	http://www.gigacos.org/
Greenpeace	Environmental protection	https://www.greenpeace.org/usa/
International Union for the Conservation of Nature	Conservation, endangered species	https://www.iucn.org/
The Nature Conservancy	Marine spatial planning	https://marineplanning.org/
International Seabed authority	UN mandates to characterize new species	https://www.isa.org.jm/SSKI/
Water.org	Water and sanitation	https://water.org/

Art within science, and science with an art

Novel approaches and projects to more effectively meld the natural sciences with the social sciences, humanities, and arts are needed to increase awareness and amplify conservation and science messages (Hopwood, 2022). The cleverness and intelligence of individual scientists and researchers, regulatory measures of large administrations, or even the genuine passion of minions of environmentalists may not be sufficient to advance the necessary conservation practices to protect benthic habitats, when carried out separately. However, a chasm still exists between disciplines: We surveyed peer-reviewed papers published between 2018 and 2023 and searched for intersections between "social sciences or humanities" and "benthic/marine biodiversity" keywords. This yielded only 10 out of 3709 articles within multiple literature databases (e.g., Web of Science, Academic Search Premier, and ProQuest Central). Thus new approaches such as STEAM (Science, Technology, Engineering, Arts, and Math), which sprout from formal STEM programs (Sousa and Pilecki, 2013), can provide the missing interdisciplinary bridges. Multiple examples of how art buoyed humanity's sinking spirits during the COVID-19 pandemic paralleled the research efforts in the laboratories: outdoors, neighbors serenaded and played songs from balconies of the isolated and quarantined, television shows provided distractions from the daily news, and multimedia ensembles formed and performed via Zoom. A worthwhile adage was professed by author Mo Willems, who reflected on the complementarity: "Science is going to get us out of this. Art is going to get us through this" (De Lisle, 2021).

Art is large, and it is old. Art (such as in the Paleolithic caves of Altamira, and many subsequent artists thereafter) preceded the development of sciences. Yet, partnerships between the arts and sciences are still in early development but could be expanded. Conversely, many scientists are expending more effort and resources to produce the clearest visualizations of complex data (O'Donoghue et al., 2018; Nusrat et al., 2019). Some argue that modernizing visualization methods will be essential for biology to advance in the era of Big Data (Nayak and Iwasa, 2019). If communication is tantamount to success, then artists can challenge scientists to hone their messages and have specific target audiences rather than just the "general public" (Khoury et al., 2019). We can thus continue with this theme of unexpected juxtapositions. Earlier, I briefly introduced science fiction as one mode of communication that can stimulate imagination, foreshadow, and sometimes predict societal trends and sea change developments. There are many good examples of this, such as Verne's submarine, Robert Heinlein's foreshadowing of iPad-like electronic paper devices, or space travel and knowledge-based preservation by Isaac Asimov (Asimov, 1966), and I will not have the capacity to touch upon them all. Before writing his most popular *Foundation* series, Asimov was a biochemist. We can also reach back farther in time and credit HG Wells as the likely father of science fiction. Yet he had a solid footing in history and facts, a characteristic modern people should never abandon. Without history, we may never be able "step back to leap" (otherwise known as "reculer pour mieux sauter"). For example, his still popular book *The Outline of History*, wherein he described the history of the Earth to his time (ca. World War I) (Wells, 1921), can be contrasted to contemporary cynicism about history in general. Overall, whenever possible, scientists should learn to tell good stories to promote useful knowledge to the public (Leslie et al., 2013). When some of the fictional ideas are based within a realistic framework, they can be very compelling. Science and its practitioners should not be afraid to step into the unknown, and perhaps writing a grant proposal can be viewed as some sort of fiction until the hypotheses and experiments are actually tested, proven, or nullified. Borrowing allegory or metaphor may not be fully scientific, but it can still spark the imagination of readers, educate and serve a greater purpose.

202   Chapter 5 Conservation of benthic habitats and organisms

The performing arts can also support the natural sciences. For example, what would be the impact of a composer who creates an opera or ballet based on ecology or DNA, which has the dancers acting out the activities of different community or molecular components, such as individual nucleotides or the genetic code, and its replication into double strands? Science would just have to find the right composer and then help choreograph the biological processes. This portrayal could be instructive. The pandemic lockdowns did create the opportunity for unusual collaborations, as in this World Ocean Day 2021 song called, "The Song of the Ocean," sung by a virtual choir and partially sponsored by the Deep-Ocean Stewardship Initiative (DOSI) (https://www.youtube.com/watch?v=5JHzWPgtmOM). Although not a professional singer, the late Harvard professor and evolutionary biologist Stephen J. Gould added his baritone to the Boston Cecilia, along with voicing his scientific messages. In the latter, Gould exemplified effective and erudite communication across diverse demographics and genres. One of his research foci centered on paleontology. This work resulted in the provocative theory of "punctuated equilibrium," which asserted that once a species appears in the fossil record, its population becomes more stable (entering into stasis) and showing little evolutionary change throughout most of its geological history (Gould and Eldredge, 1993). Gould was also an eloquent speaker and essay writer, contributing up to 300 regular articles for *Natural History* magazine alone. For example, he avidly blended anecdotes of baseball and architecture to convey messages and lessons on evolution and the history of science (Gould and Lewontin, 1979; Gould, 1983). In a similar context, another scientist named "Stephen" comes to mind in the form of Stephen J. O'Brien. As the author's doctoral mentor, he is another accomplished evolutionary biologist, though more focused on genetics compared to Gould. However, similar to the former, O'Brien had an uncanny talent to convey sometimes complex topics on genetics and evolutionary phenomena across a wide audience with engaging language. O'Brien also championed conservation genomics for multiple endangered terrestrial species, especially in the family Felidae. This was also accomplished in lectures, articles, and books. The relevant articles and references that come to mind are *Tears of the Cheetah* and related articles in *Scientific American* (O'Brien et al., 1986; O'Brien, 1987, 2003).

The success of "nature" documentaries and films over the past few decades shows that a large appetite for poignant nature and beauty exists among the public. Part of this began in the 1960s with "Wild Kingdom," and ocean lovers had "The Undersea World of Jacques Cousteau" (1968–1976), as a weekly television series. These were soon followed by nature shows from the Public Broadcasting Service (PBS) and British Broadcasting Company (BBC), whose production and cinematography improved year after year. The extended documentaries, *Our Planet* (2019) and *Blue Planet II* (2017) by Netflix or Saving Atlantis (2018) other streaming companies, have reportedly mesmerized millions of viewers, and surely have raised the awareness of environmental plights. The visceral visuals of natural biodiversity and its wonders can be trumpeted and leveraged for education and conservation purposes. However, assessments of these popular shows indicate complex reactions, and generally unquantifiable translations to correlated actions or changes in behavior following the presentations (Jones et al., 2019). The documentaries are also critiqued for not fully recognizing the threats caused by humans, who are not often portrayed but tacitly known to be the cause of environmental impacts. Of course, the key to any efficacy is consistent follow-through and action. The *Our Planet* producers, for example, provided a condensed video of their documentary to continually engage their audience (https://www.ourplanet.com/en/video/how-to-save-our-planet/).

The documentation of natural phenomena, imagery, and film has also progressed on to individual practitioners and amateur productions that can post onto social media platforms such as YouTube, Instagram, and TikTok. These purportedly represent the next waves to democratize of technology, envisioned by some of the early proponents of the "world wide web" (a.k.a., the Web and Internet). The beautiful digital images of nature conveyed there and along other electronic media can create positive emotions and benefits, though we should now perhaps be wary about digital forgeries and embellishments.

Also, sociologists have long identified factors that could blunt or negatively affect intended education or retention of conservation lessons and complex messages. It is well known that socioeconomic factors affect education, and a full treatment is beyond the scope of this book (but see Tonkiss et al., 1993; Strupp and Levitsky, 1995). Nonetheless, we can mention how videos can affect different parts of the viewing audience and keep in mind the existence of likely "knowledge gaps": "The heterogeneous distribution of knowledge results from the heterogeneous socioeconomic status of the recipients" (Kohler and Dietrich, 2021). Furthermore, "… as the infusion of mass media information into a social system increases, segments of the population with higher socioeconomic status tend to acquire this information at a faster rate than the lower status segments, so that the gap in knowledge between these segments tends to increase rather than decrease" (Tichenor et al., 1970, 159f). Educational videos, such as those available online, can be highly informative and artistically rendered. They have the potential to reach large audiences, as the accessibility to smartphones and streaming media increases. Although hemispheric asymmetries in the brain may not be as pronounced as first proposed, the artistic approaches may appeal to the right hemisphere of the brain, which has been associated as a center for visualization, imagination, and conceptualization, and specialized for metaphoric thinking, playfulness, solution finding, and synthesizing (Rubenzer, 1979; Demarin et al., 2016; Corballis, 2019). However, as mentioned earlier, follow-up is needed, and effectiveness appears to depend on the levels of a priori existing knowledge (Kohler and Dietrich, 2021). Unintended cognitive overload may confound the main messages of a production. Overall, imagery, when used proportionally and in proper context, carries wide power to educate about objects and concepts unknown and unseen (Zheng and Waller, 2017; Goodsell and Jenkinson, 2018; Patterson et al., 2022) (Fig. 5.2). These resources have been used as didactic tools by the educators for benthic conservation and other sciences. Whether illustrations and videos can be taken too far without proper context or oversight, or perceived as part of an inevitable collective march to a future "post-literate" society as envisioned by some (McLuhan, 1962) also remains to be proven and assessed. Nonetheless, this stage of development could be hastened with the increasing rise of shared experiences on social media.

In the same vein, less technical, publicly accessible books such as those in Table 5.2 could be widely disseminated to reveal the wonders of the sea and benthos. These are rare jewels and not trivially produced. For the public, the media can flaunt the most prominent examples of benthic biodiversity. But why not allow some of the experts to gush and wax poetic about the habitats and their own unique experiences with them? These are the stories to be told.

Local energy engages and educates

With respect to strategies for energizing environmental and conservation movements, the regional approach carries substantial weight. Stakeholders get to see the harvest of fruitful efforts immediately. The above optimism helps people address some of the current and looming problems of biodiversity conservation and protection. Because they involve large goals, strategizing and coordinating can also seem overwhelming, but when done locally the burden may appear lighter. For this reason, dissecting

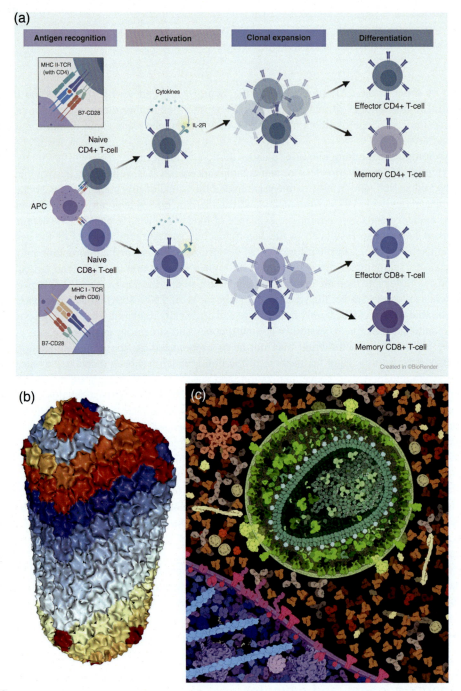

FIG. 5.2

Examples of elaborate graphics and illustrations that can provide informative examples of complex biological phenomena. (A) BioRender (https://biorender.io) is a growing creative tool for scientists and used to create this schematic representation to show molecular interactions in immune recognition. (B) NGL (http://proteinformatics.charite.de/ngl) uses a coarse molecular surface to display very large structures such as HIV capsid (PDB entry 3j3q). (C) CellPAINT (http://cellpaint.scripps.edu) uses smooth surfaces in an illustration of HIV, blood plasma, and the surface of a T cell.

(A) Illustration by Eli Lee, used with permission. (B and C) Reproduced with permission from Goodsell, D.S., Jenkinson, J. 2018. Molecular illustration in research and education: past, present, and future. J. Mol. Biol. 430 (21), 3969–3981.

Table 5.2 Marine biodiversity books for the lay pubic.

Title, publication year	Author(s)	Subject
Citizens of the Sea (2010)	Nancy Knowlton	Biological diversity
Corals in a Microbial Sea (2010)	Forest Rohwer	Coral reefs and marine microbiology
Below the Edge of Darkness: A Memoir of Exploring Light and Life in the Deep Sea (2021)	Edie Widder	Deep-sea diving, bioluminescence
The Extreme Life of the Sea (2021)	Palumbi and Palumbi (2021)	Extreme marine habitats
The Secret Life of Corals (2023)	Dave Vaughan	Coral biology
The Underworld: Journeys to the Depths of the Ocean (2023)	Susan Casey	Ocean exploration
Margaret and Christine Wertheim: Value and Transformation of Corals (2022)	Udo Kittelmann and Christine Wertheim	Art and science partnerships for coral reef conservation

the larger problems and forging more focused projects on specific localities can prove more effective in the long run. The author's home is in Florida, surrounded on three sides by water and the longest coastline of 2170 km (1350 miles) in the lower 48 United States. The state cannot ignore water or marine issues. Thus, South Florida PBS network in Miami, Florida, has been broadcasting the *Changing Seas* television series since 2008 (https://www.changingseas.tv/), with a focus not just on Florida but also globally. Produced by Alexa Elliot, the show covers various engaging marine topics such as living and diving from the Aquarius habitat, invasive species or marine mammal lifestyles. South Florida and the Florida Keys have multiple groups working on behalf of the reefs: Our Florida Reefs (https://ourfloridareefs.org/), Reef Relief (https://www.reefrelief.org/), Plant a Million Corals (https://plantamillioncorals.org/), Angari Foundation (https://angari.org/), and Miami Waterkeepers (https://www.miamiwaterkeeper.org/).

As another example, the South Florida Association of Environmental Professionals has actively engaged in a Community Coral Nursery Project (SFAEP CCNP) in the S.E. Florida Upper Reef Tract (https://www.sfaep.org/). The SFAEP Community Coral Nursery Program was set up as Reef Discovery Center (RDC) by Kirk Dotson, and presented many opportunities for conducting valuable rescue and research activities, as well as for its ability to restock and enhance the coral reefs of Broward County. The RDC is a 501(c)3 entity that holds permits for the collection of endangered species for the projects. The SFAEP derives from the larger National Association of Environmental Professionals, which aims to be a primary peer group to represent the "Environmental Professional." The SFAEP CCNP was created in 2015 to combine and pool the resources from several Broward County nonprofit organizations to advance a community-centered approach to educate the more than 5 million residents who live adjacent to the S.E. Florida reef tract, the largest continuous coral reef in the continental United States. SFAEP encompasses the region of Broward County, Miami-Dade County, and the Florida Keys, and also focuses on oyster reefs. The SFAEP CCNP and Community Oyster Gardening Program emulate successful programs existing elsewhere in Florida. SFAEP CCNP was initially conceived by board members who were serving as unpaid volunteers in the University of Florida Institute of Food and Agricultural Services' Florida Master Naturalist Program. There were no existing local coral nurseries or oyster gardening programs that could be incorporated into the curriculum.

206 **Chapter 5** Conservation of benthic habitats and organisms

The need for these educational resources also became apparent in many other professional development and public education activities, which are a core function of the SFAEP. Plans for CCNP will include the following:

- Environmental outreach: 10,000 local resident visitors attracted to the RDC in response to SFAEP CCNP public outreach activities; Geographic Information System (GIS) mapping of existing eastern oyster beds in Broward County 50% complete.
- Coastal resilience: 1000 corals rescued from the recovered tires and placed in our coral nursery; 1000 rescued corals that are successfully outplanted on reefs off of Broward County. The tires stem from a failed attempt of constructing an artificial reef in 1967 from unballasted waste tires (Morely et al., 2008). Some of the tires have attached corals, which RDC plans to save.
- Water quality impacts: Analyses for, and preparation of, environmental resource permit applications to replace eastern oyster bed habitats 100%.

Carbon sequestration could alleviate and offset increased atmospheric CO_2. For example, several studies have evaluated the possibility of viewing macroalgae as one possible sink for the greenhouse gas (Ortega et al., 2019). Perhaps, remedies to use this biomass need to be mobilized more quickly. Large masses of *Sargassum* seaweed have grown larger the past few years and threaten recreational shorelines and waterways. This is now dubbed the "Great Atlantic Sargassum Belt" and in June 2022, Hu has estimated belt sizes to reach 24.2 m tonnes, but this amount is still a minor fraction of the total phytoplankton carbon fixation in the Atlantic (Hu et al., 2021; https://www.theguardian.com/environment/2023/mar/07/great-atlantic-sargassum-belt-seaweed-visible-from-space). Sosa-Gutierrez indicate that tropical cycles can cause the sinking of the large patches (Sosa-Gutierrez et al., 2022). Boyd et al. (2019) have proposed particle-injection pumps (PIPs) to bring carbon to deeper ocean depths. The costs of carbon sequestration also need to be considered carefully (Rubin et al., 2015; National Academies of Sciences, Engineering, and Medicine (NASEM), 2021).

Coral nurseries ex situ and in situ

Optimism and the can-do spirit pervade another occupation taken up by many—growing corals in nurseries with the long-range intention to replant juveniles back on the reef. Some species of coral reef invertebrates grow very well in aquarium tanks, and this has spawned a thriving aquarium trade (Muka, 2022). Much has been learned from hobbyists. However, transferring this knowledge and success which enables the growth of the same species back out into their natural environments is easier said than done. This is because the natural (in situ) habitats can often be more degraded, polluted, and stressful than what the aquarist can provide.

Growing corals in nurseries (both in situ and ex situ) has been a valiant and evolving occupation for the past twenty years or more (Vaughan 2021; Vaughan and Nedimeyer, 2021). As a whole, coral researchers and "farmers" have tried to better understand how to efficiently grow an animal, which typically takes its time. Growth rates for coral species in shallow waters have been well studied globally with branching species (e.g., Acroporidae and Pocilloporidae) generally exhibiting higher rates of linear extension ($50–150 \, \text{mm yr}^{-1}$) than massive species ($3–15 \, \text{mm yr}^{-1}$) (reviewed in Dullo, 2005; Pratchett et al., 2015). Light is the likely primary environmental factor controlling the growth rate of zooxanthellate corals via light-enhanced calcification (reviewed in Allemand et al., 2011). Lower temperatures at higher latitudes and at depth have also been correlated with slower intraspecies growth rates (Lough et al., 2016; Kahng et al., 2019). Development of coral nurseries has proven very successful with several single coral species for the past decade and more, such as *Acropora cervicornis* and *Orbicella* (Nedimyer et al., 2011; Ware et al., 2020). This activity is gaining momentum, although this type of research initiative concentrates on isolated dynamics, inevitably treating the symptom (i.e., coral loss) rather than the fundamental cause. Recently, Wee et al. (2019) studied coral nurseries from a *community* perspective. Their results report that coral nurseries support various invertebrates and function as effective tools to conserve overall invertebrate diversity.

In parallel with the traditional and new coral nurseries is the application of beneficial microorganisms as a form of "microbiome stewardship" for the protection of individual species and their communities. Peixoto et al. (2017) promoted the efficacy of applying microbes as a potential prophylactic against some diseases. Subsequently, experiments with newly cultured bacteria consortia in Brazil, including *Cobetia marina*, showed that they could provide some benefits to the host coral *Pocillopora damicornis*, and even possible protection against inoculated pathogens such as *Vibrio coralliilyticus*. Conversely, other bacteria, such as *Halomonas* sp. provided no effects. This approach has now melded with a growing weltanschauung that proactively nurturing healthy microbiomes which will protect and bolster an organism's or species' overall health. To accomplish this ambitious stewardship would be complex and require multiple issues such as (1) carrying out detailed risk assessments and potential environmental impacts since the eventual application of the methods would be ecosystem wide, (2) selecting the proper beneficial microbes that fit the organism/community, (3) determining the proper dosage to apply for efficacy, (4) determining the proper delivery system for the probiotics, and (5) assessing the effects of probiotics on the target organism's different life stages, surrounding nontarget organisms, and collateral environments (Rosado et al., 2019; Peixoto et al., 2022). These tasks will involve commitment and investment to describe microbial communities across the globe, which has already begun in several quarters. The public has become increasingly informed about microbiome linkages to health (Górska et al., 2019), and so can support these approaches. For example, building upon the increasing capacity of high-throughput sequencing advanced in the early part of this century (Gilbert et al., 2010, 2011, 2014; Caporaso et al., 2012), the Earth Microbiome Project (EMP) (https://earthmicrobiome.org/), led by Jack Gilbert, Rob Knight, and others, was launched around the same time as the human microbiome project, though with more modest resources to apply high throughput sequencing (HTS). Nonetheless, this diverse, dedicated group of scientists welcomed others to join and build a consortium, which eventually generated a robust view of the planet's microbiomes across multiple diverse environments (Thompson et al., 2017; Thomas et al., 2016). Keys to EMP success included the development of innovative analyses and platforms as well as the establishment of experimental and analytical *standards* across the various sample types (Knight et al., 2012; Gonzalez et al., 2018; Chivian et al., 2022). These baseline microbial datasets represent a beginning, and will provide a platform to launch large-scale efforts such as microbiome manipulation and stewardship. Moreover, several signs of increasing interest and willingness from funding agencies and stakeholders have already appeared.

Marine-protected areas—"Wildness is the preservation of the world"

It would have been very interesting if 19th-century naturalist Henry David Thoreau (quoted in the header) had the opportunity to don scuba gear and describe dives to benthic or protected habitats. In 2008, CBD Parties further defined "scientific guidance for selecting areas to establish a

208 **Chapter 5** Conservation of benthic habitats and organisms

representative network of marine protected areas (MPAs), including in open ocean waters and deep sea habitats" (UNEP/CBD/COP/DEC/IX/20/Annex II) (Rees and Cranston, 2017). An MPA is a region that is designated to conserve marine ecosystems and biodiversity, as well as to protect human uses of the sea. MPAs can be established in both coastal and offshore areas, and they can include a variety of marine habitats, such as coral reefs, seagrass beds, and kelp forests. Maestro et al. (2019) summarize some of the most successful examples of MPAs. There are recommended priorities for Marine Protected Area Network Expansion (http://www.marine.auckland.ac.nz/CTMAPS). The specific goals of MPAs will vary depending on their location, organizing agency, and the resources they are designed to protect, but MPAs aspire to preserve and restore the primary functions of marine ecosystems. This includes maintaining trophic structure, habitat, and keystone species of the habitat.

MPAs protect threatened and endangered species and provide a means of managing human activities in the sea in a way that is sustainable and consistent with the conservation goals of the MPA. MPAs are often established by national or regional government decree, and they can be managed in a variety of ways, including through the development of management plans, the implementation of regulations and zoning, and the use of enforcement measures to ensure compliance with the rules of the MPA.

It is also important to recognize that the effectiveness of MPAs relies on their potential for genetic connections, as asserted in Convention on Biological Diversity (CBD), which aimed to develop representative networks of MPAs (CBD, 2010). This premise is further based on the concept of genetic and population connectivity, which has been applied widely for the optimal conservation of various populations. To ensure demographic connectivity is maintained among various marine animal populations, studies recommend that on average 20%–30% of the coastal seas be set aside as MPAs (Gaines et al., 2010; Magris et al., 2018; President Joe Biden made a historic commitment to escalate the pace of U.S. conservation and put the nation on a trajectory to conserve 30 percent of its lands, waters, and oceans by 2030 (https://www.whitehouse.gov/briefing-room/presidential-actions/2021/01/27/executive-order-on-tackling-the-climate-crisis-at-home-and-abroad/). Questions and debates about MPA placement typically occur, however, and can be based on both scientific parameters, and policy and practicality (Lubchenco and Grorud-Colvert, 2015). Lastly the type of benthic community, species content, diversity, etc., will vary between different MPAs. For example, as scleractinian corals continue to dwindle in coverage due to the problems discussed in Chapter 3, attention could appropriately turn to preserving and studying the successors, such as octocorals and gorgonians (Borgstein et al., 2020; Lasker et al., 2020).

Deepwater *Oculina* coral reefs

This would be an appropriate juncture to discuss the origin story of one particular MPA placement off the east coast of Florida. The deepwater *Oculina* coral reefs represent a quintessential example of how to establish a marine-protected zone, and highlights the convergence of timing, proximity, and at least one scientist's zeal and passion to protect a unique benthic resource. It can be quite amazing to observe how the determined will of one person can alter the course of a set of human affairs or the fate of a single ecosystem.

The deepwater *Oculina* coral reefs are unique, occurring solely off eastern Florida, but carry out parallel ecosystem services to their shallow counterparts (Reed et al., 2005). As discussed previously, deepwater coral reefs are slow-growing and fragile, but yet develop large three-dimensional structures without zooxanthellae and support unique biodiversity. Unfortunately throughout the world, deepwater

Deepwater *Oculina* coral reefs **209**

reefs have been severely impacted by destructive fishing practices, primarily from trawling. NOAA CRCP recently published their latest deep-sea coral report to Congress (https://www.fisheries.noaa.gov/s3/2023-02/deep-sea-coral-research-technology-program-report-congress-2022-Fisheries-OHC.pdf).

Thus I now add at this point, a short biography of a marine biologist who was instrumental in establishing the deep *Oculina* reef-protected area and whom I have had the pleasure to know personally. John Reed is a tall man and yet has a very modest demeanor. This is refreshing when we fully acknowledge all of his accomplishments. John was born in 1948 and raised in Middletown, Ohio. Because of frequent vacations to Florida, much time was spent in the water and John learned to snorkel. His scuba training began early in high school as a boy scout, and as expected the checkout dives were in very cold, low-visibility quarries. Nonetheless, with a scheduled trip to the Florida panhandle, John made his first ocean dives. At this time, Reed also recounted that while swimming back to the boat, people in the boat kept yelling to "Keep swimming, don't look back!" This was because once on the boat, he saw that a large shark had been following closely behind him. In addition, as a teenager, like many, John was influenced by the popular television show "The Wonderful World of Jacques Cousteau" every Sunday evening. Cousteau would travel around the globe on his ship, scuba diving, and promoting conservation, and showing the incredible diversity and beauty of coral reefs. These early thrills in and under the ocean most likely led John to where he is today—a respected marine biologist who traveled around the world with thousands of dives under his belt.

John Reed has recounted interesting anecdotes about his early science days getting started into benthic research, which would dominate most of his career. In the early 1970s, as a student at Florida Atlantic University (FAU), his master's thesis characterized the benthic fauna of Lake Worth lagoon (an estuary off West Palm Beach, Florida). At this time, much of the lagoon had been developed and bulkheaded, but the northern part was still pristine mangroves and seagrass beds. An investor wanted to develop the land while Reed and other students at FAU wrote a detailed report documenting the habitats, benthos, fish, and birds; and sent this report to the Florida Department of Environmental Protection. When the property owner caught wind of the project, he became irate, and asked that John and the university students be fired. However, this was not possible because they were students. Ironically, later in the same decade, this region was made a Florida State Park and was named after the land owner but incorrectly credited the biological characterizations to Florida State University, not FAU.

John's first job out of graduate school in 1976 took him to Harbor Branch Oceanographic Institution (HBOI) which was really gearing up to start extensive deepsea work with the *Johnson-Sea-Link* (JSL) submersibles. (HBOI was an interesting place in itself, and the author had the pleasure to work there from 1997 to 2007.)

HBOI was founded in 1971 by J. Seward Johnson Sr. and Edwin Link as a nonprofit research organization dedicated to exploring and preserving the ocean. Johnson, a millionaire philanthropist, was passionate about the ocean and recognized its importance to the planet's ecosystem. He established HBOI on a 640-acre campus in Fort Pierce, Florida, equipped with state-of-the-art laboratories, research vessels, and submersibles. Edwin Link was a marine engineer who designed and built at HBOI the *Johnson-Sea-Link* (JSL) submersibles which carried 4 people to depths of 3000 ft. In the early years, HBOI focused on developing deep-sea submersibles and conducting groundbreaking research on marine life, including the discovery of new species with biomedical properties and the study of bioluminescence of midwater organisms. Dr. Bob Avent, a deepsea oceanographer, started the submersible reconnaissance program at HBOI in 1975 and hired Reed a year later. Avent had the subs outfitted with video and 35 mm cameras, manipulator, and collection bins. Their first project was to explore and

210 Chapter 5 Conservation of benthic habitats and organisms

characterize the continental shelf from depths of 30–300 m off central eastern Florida. It was during these transects that the deepwater Oculina coral reefs were discovered. JSLsubmersibles logged nearly 9000 dives before they were retired by FAU in 2010.

Over the years, HBOI has expanded its reach and impact through collaborations with other organizations and partnerships with government agencies. In 2007 HBOI merged with FAU to become a research institute within the university. The merger has provided HBOI with greater resources and opportunities for collaboration with other research institutions. Today, HBOI boasts a team of world-class researchers, engineers, and scientists who are dedicated to understanding and protecting the ocean's ecosystems, carrying out marine aquaculture, and producing stimulating tools for marine education. The institution continues to make important contributions to marine science, with a focus on developing sustainable solutions for managing ocean resources and preserving marine biodiversity.

The *Oculina* reef tract was first discovered near HBOI from Fort Pierce to Cape Canaveral in 1977–78. This reef is comprised of the slow-growing ivory tree coral, *Oculina varicosa*, which thrives at 50–100 m depths along the continental shelf edge about 42–80 km (26–50 miles) off of Florida's central east coast. *Oculina* sp. forms conical mounds, up to 20 m tall, entirely covered with living corals (Reed, 1980, 2002a,b). The mounds are built up of coral rubble over a period of thousands of years, growing 1–2 cm per year, like other deep-water corals. Reed found that *Oculina* colonies the size of a soccer ball can host over 2000 other organisms—what lives inside of a coral, similar to Irwin's study of gassing the tropical rainforests. In spite of the uniqueness of the habitat, establishing an MPA had several hurdles. The shrimp fishery and its remoteness would make enforcement difficult. At the onset, the growing shrimp trawling industry began threatening *Oculina* habitat. The industry had targeted rock shrimp (*Sicyonia brevirostris*), which lived in coral rubble and required bottom trawling. In the Gulf of Mexico, sandy bottoms are trawled for pink and white shrimp. A typical tickler chain was used to hold the net down. The destructive practices became even worse, when rubber wheels, instead of a chain, were attached to their trawls, so they did not get hung up on the coral.

In 1981 Reed gave a preliminary report to the South Atlantic Fishery Management Council (SAFMC), which supervises fisheries from North Carolina to Dry Tortugas. A first step toward MPA status was to establish the *Oculina* tract as a "Habitat of Particular Concern (HAPC)." This breakthrough was accomplished when Reed and his allies appealed to the SAFMC, local public and Florida residents with a public relations campaign of videos from HBOI's JSL dives and marine education. Thus, the HAPC for *Oculina* was eventually set in 1984 through an amendment of the Magnusun Stevens Fisheries Act, and then expanded to 1029 km^2 (300 nmi^2) in 2000. This designation prohibited trawling, dredging, bottom longlines, and anchoring to protect the coral habitat (Reed, 2002a). After range expansions in 2000 and 2015, the OHAPC now extends over 130 nautical miles long, stretching from Fort Pierce northward to St. Augustine. Due to Reed's diligence and efforts, the *Oculina* reefs were the first deepwater coral reef in the world to become an MPA.

John ran marine operations and cruise planning as a senior scientist with the Division of Biomedical Marine Research at HBOI for nearly 35 years. He kept meticulous collections records for both scuba and JSL dives during this time. I would have to say that he may be one of the most organized people that the author has ever met. Part of the results of this work are evident in the recently published Division of Biomedical Marine Research (DBMR) collection database (Table 5.3). On the few Harbor Branch cruises on the R/V *Seward Johnson*, which I happened to sail on, John was always pivotal in the selection of collection sites. One long-lasting memory will be how the choice of the scuba dive sites

Table 5.3 Partial listing of biodiversity related databases relevant to benthic studies.

Database name	Scope	URL
Abyssal	South Atlantic Abyssal plains	http://abyssal.io.usp.br/mapper/
AGRRA	Reef conditions	https://www.agrra.org/resources/
Arctic traits	Arctic benthic ecology.	https://www.univie.ac.at/arctictraits
Barcode of Life (BOLD)	Taxonomy	https://v3.boldsystems.org/index.php/databases
COPO	Metadata	https://copo-project.org/
Darwin Tree of Life	Genome data of UK	https://portal.darwintreeoflife.org/
Digital Bathymetry	Global Seafloor mapping	https://www.ngdc.noaa.gov/iho/
Environmental Data initiative	LTER data sharing	https://portal.edirepository.org/nis/advancedSearch.jsp
GenBank	Genetic sequence database	https://www.ncbi.nlm.nih.gov/genbank/
Genomes on a Tree (GoaT)	Phylogenetic, genomics metadata	https://goat.genomehubs.org/
Genomes online	Publid genome projects	https://gold.jgi.doe.gov/index
GEOTRACE	Global Biogeochemical cyclin	https://www.geotraces.org/new-biogeotraces-data/
Global Biodiversity Information Facility (GBIF)	Biodiversity portal	http://gbif.org
GRIIDC	Gulf of Mexico	https://data.gulfresearchinitiative.org/
HBOI Marine Biotechnology Reference Collection	Benthic invertebrate sampling	https://hboi-marinebio-reference-collection.fau.edu/app/data-portal
Invert E Base	Invertebrate specimen data portal	https://invertebase.org/portal/index.php
NCBI Taxonomy Browser	Taxonomy of samples with sequences	https://www.ncbi.nlm.nih.gov/guide/taxonomy/
NCRMP	US Coral reef monitoring	https://www.coris.noaa.gov/monitoring/
NOAA	Database of expeditions	https://www.ncddc.noaa.gov/website/google_maps/OE/mapsOE.htm
Ocean Biodiversity Information System (OBIS)	Marine biodiversity	https://obis.org/
SCAR Antarctic Biodiversity Portal	Antarctic biodiversity	https://data.biodiversity.aq/
Sea Life Base	Marine Biodiversity	https://www.sealifebase.ca/
Species2000	Taxonomy	https://www.sp2000.org
WoRMS	Marine Taxonomy	https://obis.org/

happened. Although we could see the reef under us through crystal-clear waters in the Bahamas from our dive boat, John would still get into the water with a snorkel to ensure the site was worth a dive. So, we threw a line in, and slow towed John behind us. When he held up his arm, then we knew that would be the dive spot. John recently retired after 47 years at HBOI, living out the dream that began in those early years—to travel around the world, scuba diving and diving in submersibles. John recently completed two 10-year NOAA grants.

212   Chapter 5 Conservation of benthic habitats and organisms

In the same "Treasure Coast" area of Florida, Mary Rice had started directing a marine lab at the Smithsonian Institution's modest barge docked at HBOI premises in Ft Pierce in the early 1970s. Soon known as the Smithsonian Marine Station (SMS) (https://www.sms.si.edu/), the research facility focused on the biology, ecology, and geology of estuaries and near-shore environments, mostly in the Indian River Lagoon, a 120-mile waterway from Daytona Shores to West Palm Beach. SMS scientists conduct research on a variety of topics, including the impacts of human activities on coastal ecosystems, the ecology of coral reefs and seagrass beds, and the behavior and physiology of marine animals. The station also offers educational programs for students and the general public, including tours of its facilities and research vessels. Rice grew the SMS with a gentle hand, focused on and was a leading expert on Sipunculun worms (Rice, 1993).

Before leaving this topic, I will make one other nod to a longtime colleague at our home institution at the Nova Southeastern University Guy Harvey Oceanographic Center. Charles (Chuck) Messing spent much of his professional career studying the benthos and Crinoidea in particular. This class of echinoderms encompasses about 700 species and dates back to the Orodivician, 480 million years ago. Because crinoids can range from shallow to hadal depths of 9000 m, they could possibly be considered as a cosmopolitan or indicator species. Messing took his first D/V *Alvin* dive in 1975 and subsequently has led 14 deep-sea submersible expeditions funded by NSF and NOAA (https://oceanexplorer.noaa.gov/explorations/15biolum/logs/july21/july21.html). Along the way, he identified and named new echinoderm species (Messing, 2003, 2020). Besides his taxonomic and benthic work, Messing has also been a strong supporter of the arts. His illustrations of various invertebrate taxa appear in various texts, posters, and websites with examples shown at—https://www.charlesmessing.com/illustrations. The exquisite detail and high quality of Chuck's work rivals professional illustrators, including Haeckel (1904) in his *Art Forms in Nature*. Through his musical talents, he has provided much song and laughter to his colleagues at informal or scientific gatherings, such as the 2009 International Coral Reef Symposium (ICRS) and the 2010 World Sponge conference.

Biodiversity databases: Pooling of knowledge or a tangled web?

Our contemporary knowledge is only as good as our collective wisdom (e.g., databases, libraries, individuals), technology, and the brilliant minds who both generate and interpret the data. Knowledge tends to be dispersed across different research communities and laboratories. From an "ivory tower" perspective, knowledge on certain topics can sometimes stay hidden, be inacessible or forgotten due to the passage of time, inability to access because it is written in different original languages or use jargon. These situations can make the information unusable to subsequent scientific communities or even the lay public. For example, some may have experienced a situation where a software program, language, or specific computer script expires or becomes obsolete, which makes the data it may hold the key to, likewise, inaccessible. However, in many societies, the public has already paid a price (in taxes, subscriptions, etc.) for reasonable access to the research results. These potential breaks in knowledge transmission across generations should be a major concern.

We have previously referred to *databases*, especially in the context of genome sequencing discussed earlier, but a more formal definition is now warranted. A scientific database is a collection of scientific data that is organized and structured for easy access and analysis. This data can be in the form of research articles, datasets, patents, or other types of scientific information. Scientific databases are commonly used by researchers, scientists, and students to find and access relevant scientific information on specific topics. Like most popular databases, an effective data resource should be user-friendly, updated, content rich, and federated (interoperable) with other, more specialized databases. I will admit that database engineering and computer programming is not my 'forte', and I rarely code. However, I consider myself, like many, to be an avid database "user," and therefore would be qualified to recommend a few features that could constitute an optimal, operational, marine, biodiversity database for benthic assessment purposes.

Before diving into specifics, two well-known scientific database examples include: (1) Web of Science: a database of scholarly research articles, conference proceedings, and other scientific literature, developed by Clarivate Analytics and (2) the US National Center for Biotechnological Information or NCBI, which also holds at least 34 different databases such as "PubMed" with millions of biomedical literature records including abstracts and citations. Other distinct databases within NCBI include protein structures, reference genes (RefGENE), clinical variants, biological samples, chemical assays, Gene Expression Omnibus (GEO) and GenBank—the primary US repository of DNA/RNA, and protein sequences. NCBI was developed by the National Library of Medicine (NLM) at the National Institutes of Health (NIH) and ostensibly represents a mostly nonmarine example of a database par excellence. Focusing mostly on human biomedical applications, NCBI collects, curates, and makes available to the public a wide range of biological data, including nucleotide and protein sequences, gene-expression data, medical literature, and information on genetic variations (Sayers et al., 2021). Nonetheless, through its wide variety of online databases, useful nonhuman biodiversity related data can emerge. In addition to providing access to data, NCBI also develops and offers software tools for analyzing and manipulating biological data, such as the Basic Local Alignment Search Tool (BLAST) and the Clustal family of multiple sequence alignment programs. The "NCBI Handbook" (2012) is also available online (https://www.ncbi.nlm.nih.gov/books/NBK21091/) and gives a fairly robust explanation of NCBI's goals and tools available.

The integrated structure of NCBI databases, focused on biotechnology and its various related disciplines, warrants further discussion as a prime example and model of a versatile database at this juncture. Foremost in mind, having a comparable, sophisticated, and actively curated database for environmental and benthic research would most certainly benefit a wider community of scientists (Jetz et al., 2012).

Overall, a database remains a repository of data, not necessarily a concentration of knowledge. Databases may also contain mostly residual data after the original question and projects have been completed. The data, which is fortunate enough to be archived sufficiently, can then persist, possibly ad infinitum, as tools for a subsequent researcher to apply, synthesize, and create meaningful knowledge, and there lies the rub.

The backbone of NCBI's *interoperability* is the primary biological sequence (nucleotide or protein) itself. Although data is submitted by individual researchers (via https://www.ncbi.nlm.nih.gov/genbank/submit/), each entry is checked, curated, and identified by professional bioinformaticians at the NIH NLM. NCBI interoperability in turn rests on agreement between the major sequence database administrators (USA's National Center for Biotechnology Information—NCBI; European Molecular Biology Laboratory's European Bioinformatics Institute—EMBL EBI; and the DNA Data Bank of Japan—DDBJ), who all contribute to INSDC—the International Nucleotide Sequence Database Collaboration. Established in 1974, EMBL is an intergovernmental organization with 27 member states, headquartered in Heidelberg, Germany. Similarly, as nucleotide data began to burgeon, the

214   **Chapter 5** Conservation of benthic habitats and organisms

DDBJ was established in 1987 to handle and curate sequence data from various sources. The three main molecular sequence centers "provide freely available nucleotide sequence data and supercomputer systems, to support research activities in life science." The high level of cooperation between these centers with possibly more to be added in the future (China, Africa, etc.) contributes to the success and popularity of the databases. This collaboration covers the spectrum of data raw reads, through alignments and assemblies to functional annotation, enriched with contextual information relating to samples and experimental configurations.

In some ways, benthic researchers have missed deliberately creating a fully integrated and global benthic biodiversity repository, akin to an NCBI, and so our grasp of benthic data and knowledge remains incomplete or fragmented. On the other hand, nature has been accumulating species, fossil evidence, and mutations to its DNA code (a kind of "living" fossil and evolutionary database all its own) for millions of years. However, making sense and organizing this data for application and appreciation takes deliberate planning and structure. Rather, the existing databases of the benthos, their habitats, and related topics are mostly scattered and unfederated, although from a higher altitude, one could probably see how some could logically fit and link together. Envisioning such a tool could be fairly simple. Google Earth is a computer program, but also portal to an extensive database of satellite imagery. Google Maps is an extension of this capacity, and can theoretically remote sense every site visible to a satellite. A Google tool for the oceans already exists, although it is still fairly rudimentary and has room for expansion, e.g., after seabed mapping details are added.

These tools provide coordinates for specific locations on the map. A parallel can be drawn to the equally relevant and useful GenBank database of macromolecular sequences, which often represent the loci or "locations" on chromosomes. The BLAST tool provides ways to search this vast database quickly (Altschul et al., 1997) when it has a specific sequence (when at least 12 nucleotides long, the matching precision is increased).

Now imagine if researchers and the lay public (for example, see iNaturalist.org) had an interactive search engine for biodiversity geographic locales, perhaps based on specific latitudinal/longitude coordinates, and each coordinate had a specific identity and associated metadata akin to a chromosomal locus. This could be an enriching database that goes beyond Google Earth. Some of these aspirations have been implemented, as described in the following examples (Table 5.1). (Note that many of these are fairly recent, and parallel the rise of computing "big datasets").

First, one broadly used species database for biodiversity is the World Register of Marine Species (WoRMS) (https://www.marinespecies.org). Initiated in 2008, WoRMS is a taxonomic database derived from the European Register of Marine Species and the UNESCO-IOC Register of Marine Organisms (URMO). WoRMS approaches the goals to have a fully integrated biodiversity database, and even has links to NCBI's GenBank. However, the interoperability is not fully complete.

The Global Biodiversity Information Facility (GBIF), the Ocean Biogeographic Information System (OBIS), and the International Union for Conservation of Nature (IUCN) provide marine range maps, which are all very useful and ambitious (Otero, 2020). However, several sources have pointed out deficiencies in these biogeographic databases such as "duplicate records, positional uncertainty of records (Maldonado et al., 2015; Otegui et al., 2013), heterogeneity among taxa (i.e., poor knowledge of species distribution for many taxa), and spatially biased (i.e., uneven distribution of biodiversity information)" (Amano et al., 2016; Menegotto and Rangel, 2018; Meyer et al., 2015, 2016; Moudrý and Devillers, 2020). This could possibly be rectified by a concerted, multidisciplinary effort and requisite funding, which is easier said than done. For example, the eventual integration of high-resolution seafloor mapping data from the International Hydrographic Organization Data Centre for Digital Bathymetry (IHO DCDB) or the online Seabed 2030 project into biology rich databases would be a monumental task. However, if an integrated database was realized, it would provide essential baselines that could facilitate assessments.

However, establishing a unified benthic biology and biodiversity data network perhaps should continue to be a high priority item for various scientific circles (conservation, natural products, ecology, and evolution) in the coming years. A truly operational, effective, and accessible database dedicated to benthic biodiversity could broadly benefit both the habitat and its stakeholders (Troudet et al., 2018). The intentions are commendable. However, at least three major obstacles are recognized to be in the way: First, the lack of complete database compatibility allowing efficient information exchange between distinct sources, alongside with inconsistent file structures (Shanmughavel, 2007), would leave data frequently scattered, even for well-known taxa (Yesson et al., 2007). Second, the quality of several sources has been questioned regarding potential geographical data error. Third, there is the lack of a mutually agreeable universal identifier that can make all of the disparate databases relational and interoperable. Database interoperability is typically found across NCBI, perhaps with more ease, since the basis of many data entries stems from a macromolecular sequence (DNA, RNA, or protein) and the universal identification or accession number. This *identifier* relates the multiple databases to each other. A possible solution to link multiple benthic biodiversity databases together would be through the identifier. For example, perhaps this could be the agreed-upon species name or its synonyms.

In this context, a very useful tool within NCBI related to biodiversity is the "Taxonomy Browser," which has been installed since 1991 (Schoch et al., 2020; Federhen, 2003). This taxonomy database shows a hierarchical set of taxon names and classifications for all of the organisms that are represented in GenBank (Leray et al., 2019). NCBI taxonomy does not try to recapitulate an accurate phylogeny and thus should be contrasted with phylogenetic trees. Some prominent "tree of life" projects are discussed in more detail below. Taxonomy and classification mostly provide official, agreed-upon names of taxa in the Linnaean scheme from species to phyla (Ereshefsky, 2000). The NCBI resource also includes the disclaimer: "The NCBI taxonomy database is not an authoritative source for nomenclature or classification—please consult the relevant scientific literature for the most reliable information." An interesting feature of NCBI Taxonomy is that every taxon has been assigned a stable, *unique numerical identifier,* called the *Taxon ID (TaxID)*, which is universally recognized among the INSDC partners. Wider recognition of this number across all biodiversity databases could go a long way toward greater federation and interoperability.

By contrast, robust phylogenetic analyses and systematics aim to determine precise relationships among taxa at various levels, and thus can involve slightly more effort, computational resources, data, and philosophical and empirical underpinnings to resolve (Hillis et al., 1996; Nei and Kumar, 2000). This topic was previously discussed in Chapter 1. Over the last two decades, projects to share phylogenetic advances and its increasing breadth across biodiversity have multiplied online. Phylogenetic trees are crucial to biodiversity since they provide a way to visualize taxa that can be simultaneously quantitative (branch lengths), and hypothetical (placement of bifurcations and relationships), and yet intuitively comprehensible to the nonspecialist. Trees cannot be separated from biodiversity because they display evolution. The recent phylogeny projects have also taken advantage of burgeoning online resources for code sharing, such as the open-source Github server. For example, the "Open Tree of Life [project] aims to construct a comprehensive, dynamic and digitally-available tree of life by synthesizing published phylogenetic trees along with taxonomic data" (McTavish et al., 2015; Rees and

Cranston, 2017; https://tree.opentreeoflife.org/). The phylogenetic reconstructions are often collaborative, based on cumulative data, and are made public through Github (Hinchliff et al., 2015). Another similar initiative is the Interactive Tree of Life (iTOL) (Letunic and Bork, 2021). These various phylogenetic projects have similar aims to disseminate current view and understanding of biodiversity's relationship to itself.

As mentioned earlier, GBIF is a resource that has ambitiously attempted to provide the essential integration across various biodiversity sources. GBIF has an international scope, and "bridges research communities by providing the opportunity for synergy between museums, community science efforts, and ecology and evolution fields at large." GBIF was created from the Megascience Forum in 1999 sponsored by the International Organization for Economic Co-operation and Development (OECD), which is an international organization (https://www.oecd.org/science/inno/2105199.pdf).

The GBIF web portal appears user-friendly, updated, and well-curated with several live links and has illustrative graphics. As mentioned earlier, the GBIF database has the one major deficiency, like other biodiversity databases with similar goals, of lacking a unifying accession number or unifying identifier, as found in NCBI. Moreover, standardization will also remain a challenge because so many diverse disciplines are involved in biodiversity studies. DarwinCore, working with other databases such as WorMS helps provide some important criteria and stability within a flexible framework to store all fields from original data sources that can lead to standardization (Wieczorek et al., 2012).

Heberling et al. (2021) carried out a quantitative text analysis and bibliographic synthesis of >4000 studies published from 2003 to 2019 from GBIF to characterize its efficacy. The authors found studies that had been curated or tagged with GBIF use in mind a priori. In the following, I will discuss the drawbacks of this approach plus useful conclusions in this study. For example, certain disciplinary foci decreased or increased in importance. For example, taxonomic expertise has declined during this period, and has long been a red flag among practitioners. Clear linkages were demonstrated between diverse disciplines such as chemistry, math, earth sciences, infectious disease, and of course biology. Social sciences and the humanities were on the periphery. Even with this survey, more than 4000 biodiversity studies have been carried out, and so GBIF cannot capture them all, unless biodiversity publications can be curated on the scale of PubMed. When GBIF is queried for specific taxa that are known to have genetic/genomic entries in NCBI, the linkages are not always made (Waller, 2022).

Examples of specialized biodiversity databases include those focused on specific taxon groups (e.g., FishBase; Froese and Pauly, 2021, AlgaeBase; Guiry and Guiry, 2008). FishBase (http://www.fishbase.org) is a web-based platform that is accessible to anyone with an Internet connection. It is maintained and updated by a team of scientists and technicians at IFM-GEOMAR, and new information is added on a regular basis. The database is funded by a variety of sources, including government agencies, foundations, and private donations. Indeed, FishBase was one of the first successful biodiversity databases, being a global information resource for fish species, including their taxonomy, distribution, biology, and ecology. Created in the 1980s by Rainer Froese and Daniel Pauly at the Leibniz Institute for Marine Sciences (IFM-GEOMAR) in Kiel, Germany. FishBase provides a fairly comprehensive resource that includes information on over 33,000 species of fish, as well as images and other multimedia resources. According to the FishBase website, the database receives over 1.5 million visits per month from users around the world.

AlgaeBase (http://www.algaebase.org), a database about algae, also includes cyanobacteria (formerly known as blue-green algae). Created by Michael D. Guiry and Mary Guiry at the National University of Ireland, Galway, the AlgaeBase contains taxonomic, ecological, and distributional data on algae, as well as images and information on nomenclature and taxonomy (Guiry and Guiry, 2008). The database assists researchers and students in fields such as environmental management, aquaculture, besides phycology.

For broader taxonomic identification, the Barcode of Life Database (BOLD) also provides helpful tools and insight that assist in biodiversity assessments. BOLD collects, stores, and disseminates DNA Barcode data for a wide range of organisms. The BOLD system uses DNA barcodes, which are short genetic markers, to accurately identify species and track their distribution (Hebert et al., 2003). BOLD provides a platform for researchers to access barcode data and use tools for data analysis and visualization. The utility of BOLD is proportional to its data content. The inception of BOLD was controversial because it relied upon a short set of DNA sequences found in the cytochrome c oxidase subunit I (COI) gene, which is found in the mitochondria of most animals. Since the inception of BOLD, multiple studies have successfully applied barcoding to identify species and conduct molecular surveys of the benthos (Dettaï et al., 2011; Tzafesta et al., 2021; Leray and Knowlton, 2015; Ochieng et al., 2019). However, to this day, taxonomic and geographic biases still exist in the BOLD (Weigand et al., 2019). The BOLD system also supports citizen science projects that involve the public in collecting and submitting barcode data (Irga et al., 2018).

From an ecosystem perspective, NSF and its US Long-Term Ecological Research (LTER) programs exemplify the success and potential for public ecosystem-wide projects and management (Karasti et al., 2006; Cusser et al., 2021). Established in 1980, as a whole, LTER networks of specific sites investigate and attempt to understand the ecological processes that operate across various spatial and temporal scales. The LTER program provides a unique opportunity for scientists to conduct research on long-term ecological phenomena as well as to study how ecosystems respond to natural and human-induced changes. This research helps to advance knowledge of the complex relationships between the natural world and human activities, and it is critical for informing the development of effective management strategies and policies that can help to protect and preserve the Earth's natural resources. In addition, the LTERs provide valuable training and educational opportunities for students and researchers, which can promote scientific literacy and understanding among the general public.

Unfortunately, a federated database for all funded LTERs was not conceived at the outset by NSF, although retrospective efforts are now being established. For example, Huang et al. (2020) constructed an LTER bibliographic database to analyze and generate a few lessons from LTERs from 1981 to 2018. They found that proximity between collaborating institutions and establishing "friendship factors" via familiarity and camaraderie favored cohesion within the LTER team. This subsequently increased the number of joint publications. Data sharing also enabled more cohesion across the LTER, which the Environmental Data Initiative (EDI) facilitates but had a late start. Somewhat related to LTERs are MPAs.

In one review of current biodiversity knowledge bases, Asaad et al. (2019) point out that online atlases are available for specific localities including Ireland's Marine Atlas (https://atlas.marine.ie/), the Oregon Coastal Atlas (https://www.coastalatlas.net/), and the European Atlas of the Seas (Barale et al., 2015). These provide valuable background for each location, but, of course, due to variable costs, expertise, and available resources, the same effort cannot consistently be applied to every location. Nonetheless, there is also a growing realization of the need to establish reliable metadata standards. For example, Australia has attempted to standardize metadata on marine imagery for identifying benthic habitats and organisms: individual studies may collect additional data for specific projects, but key observations from common measurements are facilitated for comparison (Althaus et al., 2015; Muller-Karger et al., 2018). In a current biodiversity genomics program, the Darwin Tree of Life

218   **Chapter 5** Conservation of benthic habitats and organisms

Project (DToL) has implemented several methods to make newly minted whole genome data as widely available. Lawniczak et al. (2022) have described the standards for genomic data and ensure best practices to place the data into their proper biological and ecological contexts. The standards follow the principles that scientific data should be "Findable, Accessible, Interoperable and Reusable or FAIR" (Sansone et al., 2012; Wilkinson et al., 2016; McQuilton et al., 2020). An extension of DToL applies both NCBI taxonomy data and resources from GenomeHubs software to create a metadata portal for the burgeoning genomics community via the Genomes on a Tree (GoAT) resource (Table 5.3). Challis et al. (2023) describe how GoAT applies a denormalized, noSQL data structure implemented in Elasticsearch, and that metadata is approached from either a taxonomic or geographic angle, though perhaps less with the latter. Again, NCBI's hierarchical taxonomy database is closely integrated. Metadata connections are emphasized in the assertion to draw metadata graphs using more user friendly tools such as JSON-LD (Page, 2016).

Continuing on this theme, COPO (Collaborative Open Plant Omics) is an open source computational platform (https://copo-project.org/), originally conceived to aggregate and curate data for repositories from diverse plant researchers (Shaw et al., 2020). The platform aims to facilitate the use of data from initial generation to repository to the preparation of processed data for public dissemination in the context of FAIR. COPO, as well as other data managers, assert that this tool is *"fundamental to allow the application of modern data integration and interrogation approaches and tools to help unlock the value of knowledge held in disparate research groups, institutions, and communities."* For the above reasons, the COPO framework has now been adopted or emulated by various consortia, including the Darwin Tree of Life (DToL), to organize and process metadata associated with large projects (Johnson et al., 2021; Lawniczak et al., 2022).

The GRIIDC database addresses more marine data, both pelagic and benthic in scope (Gibeaut, 2016), and was established after the DWHOS mentioned in Chapter 1. From the outset, GOMRI administration emphasized the need to preserve all types of data—primary, geographic, and metadata related to the funded projects. GRIIDC thus stands as a robust archival resource that can serve many generations to convey the large amount of data on the Gulf of Mexico.

A relatively new WebGIS database called ABYSSAL (Assessment of the deep Benthic ecosYstemS of the Southern hemisphere AtLantic sector) has been published during the writing of this book (Bergamo et al., 2023). The authors assert that this public-private partnership from Brazil to be the first online multidisciplinary geodatabase of deep sea information for the region that contains biological, microbiological, geological, chemical, and physical data. These are stored as tables in .xlsx format with information of macrofauna reports, ROV footage in .mov or.mp4 formats, and geophysical surveys (e.g., multibeam echosounder) in GeoTIFF (georeferenced.tiff.files).

A broad database called GEOTRACES was implemented as an international program with US NSF funding to the Scientific Committee on Oceanic Research (SCOR) (Table 5.3). This resource increases the understanding of biogeochemical cycles and large-scale distribution of trace elements and their isotopes in the marine environment.

Education and engagement (for one and all)—"The Gathering of Tribes"

In the best case scenario, the knowledge bases discussed earlier are constructed and available for browsing, download, dissemination, and ingestion by the scientific community and perhaps lay public. "We can lead a horse to water, but we can't make her drink." The knowledge could be generated and

Education and engagement (for one and all)—"The gathering of Tribes" **219**

organized, and much of the public are avid readers, but would they consume information which is esoteric, and stemming from dark, cold, and remote places which they cannot see? This is the dilemma.

We should avoid making the mistake of trying to sugarcoat the global biodiversity problem. Putting live animals into the zoo or streaking a few microorganisms on a petri dish, hoping for some of them to survive and thrive, cannot replace the loss of whole or even partial ecosystems and symbiotically integrated habitats. In addition, having statistics of benthic organisms in a knowledge or data bank cannot match the real thing. If we do not expend our best efforts to protect even what we cannot see, then the data sets collected above, will ring hollow and the living organisms will disappear under the momentum of the Anthropocene's effects.

To risk another quote, musician Pete Townshend, the leader of the Who rock and roll group, once said that "one has to be young, to really be in touch." This statement was made during the 1960s and reflected the growing power and influence of that era's Baby Boomer youth culture, as it came of age. However, the appeal to the young may hold relevant lessons for contemporary millenials and GenZ generations as they begin to impinge their will and desires for the future, or on how to engage the public at large. Environmental issues may still motivate action, like climate change and the protection of biodiversity. The idealism of past eras led to positive initiatives such as the Peace Corps, which preceded the AmeriCorps in the 1990s (Ward, 2014). Perhaps civic duty and volunteerism for environmental and conservation concerns could again be instilled in younger generations, to create a genuine nonmilitary "marine corps."

Recent signs and evidence support the idea of growing impacts of youthful zeal. For example, in Montana a group of youthful plaintiffs (ages 9 to 22) won a decision in their environmental suit, *Held v. State of Montana*, in 2023. The judge declared that the state violated their constitutional rights to equal protection, dignity, liberty, health and safety, and public trust—i.e., the right to live in a clean and healthful environment (https://www.jurist.org/news/2023/08/montana-court-rules-in-favor-of-youth-plaintiffs-in-landmark-climate-trial/). Exemplary action was previously and prominently portrayed by the Swedish environmental activist Greta Thunberg, who began a "school strike for climate" *Skolstrejk för klimatet*) outside the gates of the Swedish Parliament in 2018 at the tender age of 15. Although she has been diagnosed with Asperger Syndrome, her resolute passion, determination, and candid statements, such as "I want you to act as if our house is on fire. Because it is" (Thunberg, 2019), earned her recognition as *Time* magazine's 2019 Person of the Year. Of course, not all of the passion and energy for environmental issues resides in the youth, but Thunberg has attempted to crystallize climate conclusions in simple terms for her demographic cohorts who are likely to inherit environmental problems for the rest of the twenty-first century (Murphy, 2021; Sabherwal et al., 2021). Thunberg's activism in recent years is deservedly poignant because her voice represents the *future* in both body and spirit. Her frankness and gravitas stunned much of the world. The successful Montana case had followed Thunberg's lead. In the same context, and as an event that began in 1992, World Ocean Day (June 8) is typically steered by a council comprised of young people. In 2022 the members ranged from ages 16 to 23 from 22 countries around the world (https://worldoceanday.org/wp-content/uploads/2022/08/Annual-Report-2022.pdf).

Also with respect to future generations, many formal science education programs exist across most developing countries in an effort to expand Science Technology Engineering and Math (STEM) training. I do not need to elaborate on all of those programs here, except that STEM hopes to train experts for the future. Unfortunately, too many countries may view these as a competitive endeavor instead of advanced education programs that can enlighten individual students to novel ideas and methods, possibly for application to benthic related science and conservation. In the context of genomics training, GIGA, and affiliate genome organizations initiated workshops and training in the genomics and bioinformatics concentrations (GIGA, 2017). An innovative EU consortium called IGNITE-ITN (Innovative Training Network) was initiated in 2017 after the second GIGA workshop by Gert Wörheide and multiple EU collaborators and has now trained 15 doctoral students (early stage researchers or ESRs). The program's stated goals of "generating high-quality reference genomes of under sampled animal lineages, developed new analytical tools and software, and provide novel insights into genomes in the context of an organism's function, environment and ecology, as well as how invertebrate genomes were shaped throughout the Earth's history" have been realized (https://cordis.europa.eu/project/id/764840/reporting) (Juravel et al., 2023). In one IGNITE research product, Guiglielmoni et al. (2021) have described different assembly strategies on novel genomes of the non-model invertebrate rotifer *Adineta vaga*. Another IGNITE trainee has conceived a workflow to produce novel omics data papers and facilitate the import and integration of metadata for genomics and biodiversity based projects (Penev et al., 2011; Dimitrova et al., 2021).

Two particular public scientific meetings focus on early science careers: the well-established annual US Benthic Ecology meeting (BEM) and the relatively new "Reef Futures." The BEM remains one very appropriate setting for finding and fostering optimism, start the networking process, as for many years it has been student focused. Also, the science of ecology aims to understand the distribution and abundance of species, and thus fundamentally related to biodiversity issues and conservation. The context of the BEM often helps highlight and spur further questions to drive biodiversity research. Reef Futures is a quadrennial meeting established by the Coral Restoration Consortium (https://www.crc.world/about) following a November 2016 "Workshop to Advance the Science and Practice of Caribbean Coral Restoration." CRC is also a part of the International Coral Reef Initiative (ICRI). Two Reef Futures conferences have been held with noticeable participation in student and early career researchers. A scan of the latest program from the 2022 meeting with attendees and their presentation titles reveals the urgency, focused energy, and wide geographic scope of restoration and conservation projects; "Developing a holistic stakeholder-driven coral reef restoration action plan for Guam"; "Eco-health Focused Coral Reef Conservation and Restoration in the Mesoamerican Reef"; "Collaborative Coral Restoration Planning in the USVI"; "Novel coral – sea urchin restoration protocol in La Parguera Natural Reserve, Puerto Rico"; and "Sixteen Geese, Four Sheep, or One Cow? The importance of diversity in enhanced invertebrate grazer assemblages," etc. The abovementioned are just a few examples of talks from the first day of the meeting!

I have already made references to several innovative expeditions by NOAA, HBOI, Bob Ballard, and other individual pioneers to the benthos, as well as educational entities that produce detailed videos of benthic habitats and expeditions. These may not be as polished or on par with professionally produced nature documentaries, which carry out impressive efforts to show the hidden beauty of deep-sea benthic habitats and fauna. However, raw footage of manned and unmanned automonous underwater vehicles (AUVs) or remotely operated vehicles (ROVs) provide tangible views of unique habitats, that may otherwise be viewed only by a score of scientists or never be seen at all. This compilation should continue to be funded and grow. Although some of the links may expire, I now list the current URLs to some of the most interesting videos and movies of the benthos in Table 5.4. They are meant to entice and sustain the momentum to protect what we have under the surface. The stunning beauty of many coral reefs has gained attention from divers and explorers and should be shared. For example, the Ocean Exploration Trust organization through the E/V Nautilus has captured multiple understudied fauna via their cruises, some of which are shared on YouTube—https://www.youtube.com/@EVNautilus/shorts.

Table 5.4 Examples of diverse benthos-related video links with public access.

Benthic habitat, location, or organism	Source	URL
Brine Pools	Beyond Blue	https://www.youtube.com/watch?v=NFjJJEHiirg
Varied	E/V Nautilus	https://www.youtube.com/watch?v=woTi–GCzwM&t=22s
		https://www.youtube.com/@EVNautilus/shorts
Varied	HBOI/Johnson Sea Link	https://hboi-marinebio-reference-collection.fau.edu/app/data-portal
Benthic juveniles	MBARI	https://www.youtube.com/watch?v=Q94zbh1g1No
Cuban coral reefs	HBOI/Johnson Sea Link	https://www.youtube.com/watch?v=GM4BFfwznnY
Varied	NOAA	https://www.ncei.noaa.gov/access/ocean-exploration/video/
Global	Time evolution of the Marine Protected Areas declared around the world.	https://static-content.springer.com/esm/art%3A10.1038%2Fs41586-020-2146-7/MediaObjects/41586_2020_2146_MOESM2_ESM.mp4
Pescadero Basin, JaichMaa 'ja'ag Vents	Schmidt Ocean	https://www.youtube.com/watch?v=VtrfuLAkPY8
Roatan, Honduras	WPBT	https://www.changingseas.tv/season-6/601/
Deepwater and Arctic reefs	TRACKS	https://www.youtube.com/watch?v=9I-OjB8cmeI
Florida Reef Tract	NSU	https://www.youtube.com/watch?v=z1QLmi2Isdg
Florida Reef Track	GIS and Geospatial Ecology Lab	https://www.youtube.com/channel/UCFyDIqtjGJ-eABamukrURxA
Johnston Atoll	E/V Nautilus	https://www.youtube.com/watch?v=wV8wKv-8fSU
Gulf of California	WHOI	https://www.whoi.edu/multimedia/seafloor-live/
Siphonophores	E/V Nautilus	https://www.youtube.com/watch?v=PLUVbYXmyhM
Deep Sea Rover Expeditions	Monterey Bay Aquarium Research Institute	https://www.youtube.com/watch?v=Nqe6tKIn628

The present and future (blue) bioeconomy

Many positive outcomes have been produced by the so-called biologically based commerce, or *bioeconomy* over decades. Indeed, most economies around the world are biological. One of the oldest and clearest examples of a bioeconomy is the harnessing wild plants and species into farms, systemic agriculture and sustainable yields of crops (Graeber and Wengrow, 2021). In essence, farming and agriculture can be viewed as one of the first biotechnologies, primitive as it may have been when it began at several gathering sites around the world such as the Fertile Crescent or Mesoamerica 5400–10,000 years ago (Benz, 2001). Crops were grown as products for consumption. Livestock were also raised for food or as tools to help maintain the fields. Commerce likely occurred at large ancient

222 **Chapter 5** Conservation of benthic habitats and organisms

marketplaces such as the 402 acre Poverty Point, a little known archeological dig and the first World Heritage site in Louisiana (Milner, 2004). Although not fully verified, indigenous Mississippian hunter-gatherer cultures in the area, dated at 1600 BCE, probably met at this site to either trade or practice religious rites on large concentric mounds (Graeber and Wengrow, 2021). Animal husbandry and breeding various stocks represented genetics practiced without Punnet squares or plasmid vectors on the laboratory bench. So, we can spring forward millennia to the "green revolution" (approximately 1960–2000) which was an extraordinary era of increased global food production and security (Pingali, 2012). Before this revolution, most agricultural practices barely changed over centuries— e.g., simple tools, or plows with beasts of burden sustained much of civilization across multiple continents and cultures. We should also acknowledge that different macroeconomic models exist in the world, but for most practical purposes, western style capitalism with its free market and incentives basis underpins this section and discussion.

Hidden or unassessed ecosystem services provided by wildlife in natural habitats should also be included in the broader bioeconomy. The concept of a "Blue" or "Oceans Economy" arose from the United Nations Conference on Sustainable Development in Rio de Janeiro in 2012 (United Nations Conference on Sustainable Development, 2012) and essentially encompasses the sustainable use of ocean resources for as many people as possible (Smith-Godfrey, 2016). The UN Rio document touches upon economic topics such as bioprospecting and bio-tourism previously discussed, while society's dependence on many ocean products (e.g., various fisheries for food) can be easily intuited. Barbier et al. (2011) reviewed some of the deficiencies in society's valuation of various estuarine and coastal ecosystems, their spatial and temporal heterogeneity, and some solutions for preserving the services. Oyster reefs discussed in Chapter 1 provide ecosystem services (excluding oyster harvesting) such as enhancing water quality through filtration and shoreline stabilization, connecting the pelagic to the benthos, all of which have an economic value between $5500 and $99,000 per hectare per year (Grabowski et al., 2012). Similar estimates could be made for seagrass beds and coral reefs. In 2017, diving tourism (e.g., diving, snorkeling, and glass bottom-boat tours) on reefs was calculated to be worth ca. US$35.8 billion globally per annum (international and domestic visitors) (Spalding et al., 2017).

At times, there does seem to be a push and pull for monetizing biodiversity. Putting a specific price on a species or a community or the process of evolution is not trivial and perhaps should include as many voices from the public as possible. However, most of the time, we may yield these important decisions to the business sectors. Innovation with business and public policies will be required, such as measuring and integrating "natural capital," the value and benefits of resources that ecosystems provide to humans, into future strategies (Freeman et al., 2014; Dasgupta, 2021; Luisetti and Schratzberger, 2023). Economies and underlying biology often have a tacit interdependency. Thus, biology should enter into business education. The fields of economics and ecology share the etymology of their root "eco" (Greek oikos "house, household"). The former focuses on the distribution and consumption of resources, while the latter studies the relationships of living organisms that compose the environment. Thus, serious business school or economics curriculum should consider including a required course in basic ecology that teaches the principles of organismal and community dynamics. This background would expose business students, capitalists and prospective ministers of finance that their future profits and welfare greatly depend on nature and often healthy, functioning ecosytems.

With the advent of genetic engineering and then biotechnology spinoffs in the 1970s, another multibillion dollar bioeconomy was born. We can list scores of successful companies that have profited and produced useful biological products that have benefitted society: Genentech, AmGen, Analym, Second Genome, and big pharma subsidiaries. Many more companies probably did not succeed. For example, we watched a startup company called Diversa Inc. form with much fanfare by aiming to "mine" biodiversity and its metabolic pathways through deep sequencing in the first promising years of this century. But ultimately this company did not bring a natural product to market and ultimately folded (Wolfson, 2005). Perhaps this enterprise was too far ahead of its time. The previously mentioned enterprises of extractive methods and bioprospecting of marine products have obvious commercial interests, which can also meet societal needs. As briefly discussed in Chapter 3, marine habitats can yield a cache of diverse genes, enzymes, and metabolic products courtesy of evolution and adaptation (Ghattavi and Homaei, 2023). However, these enterprises rely upon biodiversity, but some may not always be sustainable. Therefore, the biological and policy gatekeepers should remain wary to assure that biological resources remain protected. The NP is one such safeguard. Discovery, innovation, creativity, and application of best practices will assist in the conservation of benthic resources.

As one example of the current bioeconomy's breadth with regard to healthcare, over the last decade, microbiome enterprises have increased, creating dozens of companies. The European Union funded 216 projects under the Seventh Framework Programme (2007–2013) and Horizon 2020 (2014–2020) to promote metagenomics (the study of genetic material recovered directly from environmental samples) and advance knowledge of microbes. These projects studied the role of the microbiome in chronic inflammatory diseases (inflammatory bowel disease, systemic lupus erythematosus, and rheumatoid arthritis) as well as mechanisms of microbiome resilience and disruption. Together, these projects invested more than €498 million (Hadrich, 2018). Since the rise of the COVID-19 pandemic, the expenditures reached over $18 billion for the development of the vaccines through Operation Warp Speed, distributed among nine different companies (Kim et al., 2021; Neumann et al., 2021). Another $2.5 billion was required for equipment, including storage vials and syringes needed to administer the vaccines. For example, Moderna Inc. received $955 million in US federal grants for R&D and testing, but the funds did not include purchases of the vaccine. Johnson & Johnson subsidiary received $456 million in R&D funding (D'Alcott, USA Today, https://www.gobankingrates.com/money/economy/much-covid-19-vaccines-cost-taxpayers/). These examples technically represent the bioeconomy.

I have already devoted a large amount of space to describing current genetic methods, including whole genome sequencing efforts, so there is no need to reiterate. Bugge et al. (2016) have reviewed bioeconomy views and contrast a "bio-ecology vision that highlights the importance of ecological processes that optimize the use of energy and nutrients, promote biodiversity, and avoid monocultures and soil [habitat] degradation" to more technological approaches to a bioeconomy. Renewable energy could help drive this vision as well (Jacobson et al., 2017).

With the above in mind, we can ask—"Can and how would biodiversity, and more specifically marine and benthic diversity, fit into a modern bioeconomy? What is the future of biodiversity in the applications, and how creative do we have to be to ensure the fit?" Past history and current research provide several affirmative answers and promising examples. Carrageenans, linear sulfated polysaccharides, have long been extracted from red edible seaweeds, such as the family Rhodophyceae.

224 **Chapter 5** Conservation of benthic habitats and organisms

These polysaccharides are now a mainstay in the food industry and used for gelling, thickening, and stabilizing properties. The Caribbean sea whip (Gorgonian) *Pseudopterogorgia elisabethae* has yielded a special extract that led to Resilience, a line of skin care products from the Estée Lauder company in the 1990s. The active ingredient of this cosmetic was a pseudopterosin, which are potent anti-inflammatory and analgesic agents that inhibit eicosanoid biosynthesis by inhibition of both phospholipase A2 (PLA2) and 5-lipoxygenase (Martins et al., 2014). Going back to some of the biomedical applications discussed for marine natural products, PharmaMar is part of Spanish pharmaceutical industry, which is in the top 10 worldwide in terms of market share with 26.3 billion dollars in sales in 2020. PharmaMar produces anti-neoplastic compounds called Trabectedin, also referred to as ecteinascidin 743 and ET-743 and marketed as Yondelis. This compound was first discovered in the Caribbean tunicate *Ecteinascida turbinata*. More recently, Thraustochytrids, an organism once viewed as a nuisance in some experimental systems, are oleaginous protists which have the potential to produce some high-value biochemicals, such as nutritional supplements (omega-3 fatty acids), squalene, exopolysaccharides (EPSs), enzymes, aquaculture feed, and biodiesel and pigment compounds (Gupta et al., 2022). Although not marine, the worm *Zophobas atratus* (Family: Tenebrionidae) has the ability to mineralize and biodegrade polystyrene (PS) found in styrofoam (Yang et al., 2020). PS is a common petroleum-based plastic used in various commercial forms since 1930 and thus commonly litters all of our environments. Since the worm trait is based on its gut microbial symbionts, the trait could potentially be found in other organisms, similar to the many other incompletely characterized marine symbiotic and chemosynthetic systems. These diverse biological traits represent just a few examples where the utility of biodiversity can be used as leverage to conserve and better understand the organisms and mechanisms that generate wealth and "natural capital."

An official "Ocean sampling day" was conceived by the European Union 7FP project Micro B3 (Marine Microbial Biodiversity, Bioinformatics, Biotechnology) (Kopf et al., 2015). These are excellent activities that mobilize the lay public to address the monumental goal of "sampling the oceans." However, sampling the ocean bottom, as previously discussed, has inherent difficulties of going underwater. Not everyone scuba dives. There have been many urgent calls for all of us to become better stewards of our environment (Mengerink et al., 2014). We should continue to heed these calls, which are summarized below:

(1) Prioritize benthic sampling, which will allow species identifications and within species genotyping.
(2) Organize broad coalitions with sufficient authority and resources for implementation and enforcement.
(3) Continue outreach projects to raise awareness among a wider public.
(4) Attempt to integrate databases that focus on benthic habitats, their assessment and conservation.

The interdisciplinary field of biological (marine) conservation represents a crossroads between traditional approaches (taxonomy, monitoring, and environmental policy), instinctive impetuses to stewardship and cooperation, and ever-burgeoning technologies such as artificial intelligence, CRISPR-based genetic engineering. The results could be a well ordered merging and free flow of information to create constructive new ideas and large scale campaigns driven by new maps and methods. Or it could result in just another frustrating traffic accident, causing a backup.

However, when the habitats are out of immediate sight, such as on the deep seabed, which begin to lose light at two hundred meters or so below the water's surface, it becomes literally more difficult to see the services and the values. This conundrum of distancing and maintaining connections has relevance to our economic activities—proximity to ports or rivers, which can facilitate commercial transactions and shipping of materials, etc. We should evaluate and integrate these economic interests, which do not appreciate the oceans in the same way as a biologist, recreational diver, or even a lobster fisherman, into conservation strategies that could help the benthic habitats. A farmer probably has little care for what a sailor knows by heart, and vice versa.

Nonetheless, the dearth of details and studies relative to the immense breadth and scale of benthic habitats leaves much to our imagination. The diversity of benthic habitats stems from the topography and rugosity, sometimes created biogenically or via abiotic processes, including the movement of deep underwater currents, eddies and downdrafts that can affect the resident inhabitants. And yet to love and protect something, we must first learn to fully know and understand the object. And so, part of the goal of this book is to make readers not only understand the ocean's seafloor and benthos, but also to love it, without the visceral experience of immersion or seeing through the darkness for the first time.

Coda

What will it mean to lose some of the benthic habitats described herein? The old adage "out of sight, out of mind" pertains mostly to human relationships. Could this apply to close relationships that humans sometimes have with their surrounding environments? Perhaps. Because many environments and the organisms that perform ecosystem services are out of our direct sights, yet still have a profound effect on our health and well-being. For example, weather patterns are generally created far away from where they may actually most affect us (e.g., dust storms and series of low pressure systems leave the coast of Africa every summer riding west on the trade winds with the possibility of later transforming into hurricanes that threaten N. America) (Shinn et al., 2000).

Throughout this book, I have described various examples of methods and expeditions set out to discover and characterize life on the seafloor. Besides sometimes requiring specific oceanic vessels, R/Vs from the *Challenger* to *Alvin* to R/V *Edwin Link* and *Ronald Brown*, each has in common their home ports and safe harbors to sail from. The expedition actually begins with months of planning, budgeting, and packing of equipment and personal belongings to last 1 week or 3 months. The feeling of elation and high expectations culminate upon launching the voyage. Gazing back to see the home port can be a unique experience to mariners going out on long expeditions (Fig. 5.3). They venture forth into our last and largest earthly frontier.

I have tried to present or point to multidisciplinary, cross-sectional, and contemporary views of assessing the benthos and hope that I have accomplished conveying some useful scientific information that can be applied at both local and wider regional levels. The perils of losing biodiversity can no longer remain understated. At the same time, I have also tried to weave a thread of wonder, started by many other benthic scientists who preceded me. The thread connects the various benthic regions, shallow and deep, across the second largest habitat on the planet, so that these messages are both informational and potentially inspirational.

226 Chapter 5 Conservation of benthic habitats and organisms

FIG. 5.3

(A) *Sea Change* by Agnes Pelton, 1931. (B) A safe harbor at sunset in South Florida.

Photo by the author.

I have suggested that humanity find ways to visit, monitor, and support the seafloor exploration efforts more, not less, at the risk of tiring ourselves or causing too much disruption to sensitive habitats. The primary goal that can be applied to the benthos, and all other natural places worth saving, however, is to maintain the awe and appreciation whether from near or afar. And if we are lucky enough to visit these typically inaccessible depths, more than once or routinely during our lives, hopefully we will find and appreciate the valuable treasures found there.

We shall not cease from exploration
And the end of all our exploring
Will be to arrive where we started
And know the place for the first time.
Through the unknown, remembered gate
When the last of earth left to discover
Is that which was the beginning;
At the source of the longest river
The voice of the hidden waterfall
And the children in the apple-tree
Not known, because not looked for
But heard, half-heard, in the stillness
Between two waves of the sea.

Eliot (1943), from "Little Gidding," Four Quartets

References

Abson, D.J., Fischer, J., Leventon, J., Newig, J., Schomerus, T., Vilsmaier, U., et al., 2017. Leverage points for sustainability transformation. Ambio 46, 30–39.

Allemand, D., Tambutté, É., Zoccola, D., Tambutté, S., 2011. Coral calcification, cells to reefs. In: Dubinsky, Z., Stambler, N. (Eds.), Coral Reefs: An Ecosystem in Transition. Springer, pp. 119–150.

Althaus, F., Hill, N., Ferrari, R., Edwards, L., Przeslawski, R., Schönberg, C.H., et al., 2015. A standardised vocabulary for identifying benthic biota and substrata from underwater imagery: the CATAMI classification scheme. PLoS One 10 (10), e0141039.

Altschul, S.F., Madden, T.L., Schäffer, A.A., Zhang, J., Zhang, Z., Miller, W., Lipman, D.J., 1997. Gapped BLAST and PSI-BLAST: a new generation of protein database search programs. Nucleic Acids Res. 25 (17), 3389–3402.

Amano, T., Lamming, J.D.L., Sutherland, W.J., 2016. Spatial gaps in global biodiversity information and the role of citizen science. Bioscience 66, 393–400.

Ambler, J., Dearden, P.K., Wilcox, P., Hudson, M., Tiffin, N., 2021. Including digital sequence data in the Nagoya protocol can promote data sharing. Trends Biotechnol. 39 (2), 116–125.

Anbleyth-Evans, J., 2018. Local Ecological Knowledge, the Benthos and the Epistemologies of Inshore Fishing (Doctoral dissertation). University of Brighton.

Asaad, I., Lundquist, C.J., Erdmann, M.V., Costello, M.J., 2019. An interactive atlas for marine biodiversity conservation in the coral triangle. Earth Syst. Sci. Data 11 (1), 163–174.

Asimov, I., 1966. Foundation, revised ed. Bantam Spectra Books (1991).

Atlas, R.M., 2012. One health: its origins and future. In: One Health: The Human-Animal-Environment Interfaces in Emerging Infectious Diseases: The Concept and Examples of a One Health Approach. Springer, pp. 1–13.

Balog-Way, D.H., McComas, K.A., 2022. COVID-19: reflections on trust, tradeoffs, and preparedness. In: Covid-19. Routledge, pp. 6–16.

Barale, V., Assouline, M., Dusart, J., Gaffuri, J., 2015. The European atlas of the seas: relating natural and socio-economic elements of coastal and marine environments in the European Union. Mar. Geod. 38 (1), 79–88.

228 **Chapter 5** Conservation of benthic habitats and organisms

Barbier, E.B., Hacker, S.D., Kennedy, C., Koch, E.W., Stier, A.C., Silliman, B.R., 2011. The value of estuarine and coastal ecosystem services. Ecol. Monogr. 81 (2), 169–193.

Benz, B.F., 2001. Archaeological evidence of teosinte domestication from Guilá Naquitz, Oaxaca. Proc. Natl. Acad. Sci. 98 (4), 2104–2106.

Bergamo, G., Carrerette, O., Souza, B.H., Banha, T.N., Nagata, P.D., Corrêa, P.V., et al., 2023. ABYSSAL database: an integrated WebGIS platform for deep-sea information from the South Atlantic. Ocean Coast. Res. 70. https://doi.org/10.1590/2675-2824070.22076gb.

Bonilla-Aldana, D.K., Dhama, K., Rodriguez-Morales, A.J., 2020. Revisiting the one health approach in the context of COVID-19: a look into the ecology of this emerging disease. Adv. Anim. Vet. Sci. 8 (3), 234–237.

Borgstein, N., Beltra, D.M., Prada, C., 2020. Variable growth across species and life stages in Caribbean reef octocorals. Front. Mar. Sci. 7, 483. https://doi.org/10.3389/fmars. 2020.00483.

Boyd, P.W., Claustre, H., Levy, M., Siegel, D.A., Weber, T., 2019. Multi-faceted particle pumps drive carbon sequestration in the ocean. Nature 568 (7752), 327–335.

Bugge, M.M., Hansen, T., Klitkou, A., 2016. What is the bioeconomy? A review of the literature. Sustainability 8 (7), 691.

Caporaso, J.G., Lauber, C.L., Walters, W.A., Berg-Lyons, D., Huntley, J., Fierer, N., et al., 2012. Ultra-high-throughput microbial community analysis on the Illumina HiSeq and MiSeq platforms. ISME J. 6 (8), 1621–1624.

Casey, S., 2023. The Underworld: Journeys to the Depths of the Ocean. Doubleday.

CBD (Convention on Biological Diversity), 2010. Decision Adopted by the Conference of the Parties to the Convention on Biological Diversity at its Tenth Meeting. X/2. Strategic Plan for Biodiversity 2011–2020 and the Aichi Biodiversity Targets. http://www.cbd.int/doc/decisions/cop-10/cop-10-dec-02-en.pdf.

Challis, R., Kumar, S., Sotero-Caio, C., et al., 2023. Genomes on a tree (GoaT): a versatile, scalable search engine for genomic and sequencing project metadata across the eukaryotic tree of life [version 1; peer review: awaiting peer review]. Wellcome Open Res. 8, 24. https://doi.org/10.12688/wellcomeopenres.18658.1.

Che-Castaldo, J.P., Grow, S.A., Faust, L.J., 2018. Evaluating the contribution of North American zoos and aquariums to endangered species recovery. Sci. Rep. 8 (1), 1–9.

Chivian, D., Jungbluth, S.P., Dehal, P.S., Wood-Charlson, E.M., Canon, R.S., Allen, B.H., 2022. Metagenome-assembled genome extraction and analysis from microbiomes using KBase. Nat. Protoc. https://doi.org/10.1038/s41596-022-00747-x.

Cigliano, J.A., Meyer, R., Ballard, H.L., Freitag, A., Phillips, T.B., Wasser, A., 2015. Making marine and coastal citizen science matter. Ocean Coast. Manag. 115, 77–87.

Collins, J., Boenish, R., Kleisner, K., Rader, D., Fujita, R., Moritsch, M., 2022. Coastal Natural Climate Solution: An Assessment of Scientific Knowledge Surrounding Pathways for Carbon Dioxide Removal & Avoided Emissions in Nearshore Blue Carbon Ecosystems. https://oceanrep.geomar.de/id/eprint/57687/1/Coastal%20Natural%20Climate%20Solutions.pdf.

Corballis, M.C., 2019. Evolution of cerebral asymmetry. Prog. Brain Res. 250, 153–178.

Cusser, S., Helms IV, J., Bahlai, C.A., Haddad, N.M., 2021. How long do population level field experiments need to be? Utilising data from the 40-year-old LTER network. Ecol. Lett. 24 (5), 1103–1111.

Dasgupta, P., 2021. The Economics of Biodiversity: The Dasgupta Review. Hm Treasury.

De Lisle, 2021. Art Gets us through. W & M Magazine. https://magazine.wm.edu/issue/2021-spring/art-gets-us-through.php.

Demarin, V., Roje Bedeković, M., Bosnar Puretić, M., Bošnjak Pašić, M., 2016. Arts, brain and cognition. Psychiatr. Danub. 28 (4), 343–348.

Dettaï, A., Adamowizc, S.J., Allcock, L., Arango, C.P., Barnes, D.K., Barratt, I., et al., 2011. DNA barcoding and molecular systematics of the benthic and demersal organisms of the CEAMARC survey. Polar Sci. 5 (2), 298–312.

Dimitrova, M., Meyer, R., Buttigieg, P.L., Georgiev, T., Zhelezov, G., Demirov, S., et al., 2021. A streamlined workflow for conversion, peer review, and publication of genomics metadata as omics data papers. Giga-Science 10 (5), giab034.

Done, T., Roelfsema, C., Harvey, A., Schuller, L., Hill, J., Schläppy, M.L., et al., 2017. Reliability and utility of citizen science reef monitoring data collected by Reef Check Australia, 2002–2015. Mar. Pollut. Bull. 117 (1–2), 148–155.

Duarte, C.M., Agusti, S., Barbier, E., Britten, G.L., Castilla, J.C., Gattuso, J.P., Fulweiler, R.W., Hughes the NSU GIS and Spatial Ecology Lab TP, Knowlton, N., Lovelock, C.E., Lotze, H.K., Predragovic, M., Poloczanska, E., Roberts, C., Worm, B., 2020. Rebuilding marine life. Nature 580 (7801), 39–51. https://doi.org/10.1038/s41586-020-2146-7.

Dullo, W.C., 2005. Coral growth and reef growth: a brief review. Facies 51 (1–4), 33–48.

Eliot, T.S., 1943. Little Giddings, main ed. Gardners Books. 30 April 2001.

Ereshefsky, M., 2000. The Poverty of the Linnaean Hierarchy: A Philosophical Study of Biological Taxonomy. Cambridge University Press.

Fauville, G., Dupont, S., von Thun, S., Lundin, J., 2015. Can Facebook be used to increase scientific literacy? A case study of the Monterey Bay Aquarium Research Institute Facebook page and ocean literacy. Comput. Educ. 82, 60–73.

Federhen, S., 2003. The taxonomy project. *The NCBI Handbook*. http://www.icgeb.res.in/whotdr/cd1/PreCourseReading/NCBI_Handbook2/ch4d1.pdf. (Accessed 11 August 2023).

Folke, C., Carpenter, S.R., Walker, B., Scheffer, M., Chapin, T., Rockström, J., 2010. Resilience thinking: integrating resilience, adaptability and transformability. Ecol. Soc. 15 (4) (9 pp.).

Freeman III, A.M., Herriges, J.A., Kling, C.L., 2014. The Measurement of Environmental and Resource Values: Theory and Methods. Resources for the Future Press.

Froese, R., Pauly, D., 2021. FishBase. World Wide Web Electronic Publication. Available from: http://www.fishbase.org.

Gaines, S.D., White, C., Carr, M.H., Palumbi, S.R., 2010. Designing marine reserve networks for both conservation and fisheries management. Proc. Natl. Acad. Sci. U. S. A. 107 (43), 18286–18293.

Ghattavi, S., Homaei,, A., 2023. Marine enzymes: classification and application in various industries. Int. J. Biol. Macromol., 123136.

Gibeaut, J., 2016. Enabling data sharing through the Gulf of Mexico Research Initiative Information and Data Cooperative (GRIIDC). Oceanography 29 (3), 33–37.

GIGA, 2017. Advancing genomics through the global invertebrate genomics alliance (GIGA). Invertebr. Syst. 31 (1), 1–7. http://www.publish.csiro.au/is/Fulltext/IS16059.

Gilbert, N., 2022. Troubled global biodiversity plan gets billion-dollar boost. Nature 610, 19.

Gilbert, J.A., Field, D., Swift, P., Thomas, S., et al., 2010. The taxonomic and functional diversity of microbes at a temperate coastal site: a 'multi-omic' study of seasonal and diel temporal variation. PLoS One 5 (11), e15545.

Gilbert, J.A., O'Dor, R., King, N., Vogel, T., 2011. The importance of metagenomic surveys to microbial ecology: or why Darwin would have been a metagenomic scientist. Microb. Inf. Exp. 1, 5.

Gilbert, J.A., Jansson, J.K., Knight, R., 2014. The earth microbiome project: successes and aspirations. BMC Biol. 12, 1–4.

Gonzalez, A., Navas-Molina, J.A., Kosciolek, T., McDonald, D., et al., 2018. Qiita: rapid, web-enabled microbiome meta-analysis. Nat. Methods 15 (10), 796–798.

Goodsell, D.S., Jenkinson, J., 2018. Molecular illustration in research and education: past, present, and future. J. Mol. Biol. 430 (21), 3969–3981.

Górska, A., Przystupski, D., Niemczura, M.J., Kulbacka, J., 2019. Probiotic bacteria: a promising tool in cancer prevention and therapy. Curr. Microbiol. 76, 939–949.

230 Chapter 5 Conservation of benthic habitats and organisms

Gould, S.J., 1983. Hen's Teeth and Horse's Toes. W. W. Norton & Company, New York.

Gould, S.J., Eldredge, N., 1993. Punctuated equilibrium comes of age. Nature 366 (6452), 223–227.

Gould, S.J., Lewontin, R.C., 1979. The spandrels of San Marco and the Panglossian paradigm: a critique of the adaptationist programme. Proc. R. Soc. Lond. B Biol. Sci. 205 (1161), 581–598.

Grabowski, J.H., Brumbaugh, R.D., Conrad, R.F., Keeler, A.G., Opaluch, J.J., Peterson, C.H., et al., 2012. Economic valuation of ecosystem services provided by oyster reefs. Bioscience 62 (10), 900–909.

Graeber, D., Wengrow, D., 2021. The Dawn of Everything: A New History of Humanity. Penguin UK.

Guiglielmoni, N., Houtain, A., Derzelle, A., Van Doninck, K., Flot, J.F., 2021. Overcoming uncollapsed haplotypes in long-read assemblies of non-model organisms. BMC Bioinform. 22 (1), 1–23.

Guiry, M.D., Guiry, G.M., 2008. AlgaeBase. In: AlgaeBase. World-wide Electronic Publication, National University of Ireland, Galway.

Gupta, H., Singh, N.K., 2023. Climate change and biodiversity synergies: a scientometric analysis in the context of UNFCCC and CBD. Anthropocene Sci. 2, 1–14.

Gupta, A., Barrow, C.J., Puri, M., 2022. Multiproduct biorefinery from marine thraustochytrids towards a circular bioeconomy. Trends Biotechnol. 40 (4), 448–462.

Habibi, A., Setiasih, N., Sartin, J., 2007. A Decade of Reef Check Monitoring: Indonesian Coral Reefs, Condition and Trends. The Indonesian Reef Check Network. 32 p.

Hadrich, D., 2018. Microbiome research is becoming the key to better understanding health and nutrition. Front. Genet. 9, 212.

Haeckel, E., 1904. Kunstformen der Natur. Leipzig und Wien Verlag des Bibliographischen Instituts. https://www.biodiversitylibrary.org/item/182319#page/7/mode/1up.

Haywood, B.K., 2014. A "sense of place" in public participation in scientific research. Sci. Educ. 98 (1), 64–83.

Heberling, J.M., Miller, J.T., Noesgaard, D., Weingart, S.B., Schigel, D., 2021. Data integration enables global biodiversity synthesis. Proc. Natl. Acad. Sci. 118 (6), e2018093118.

Hebert, P.D.N., Cywinska, A., Ball, S.L., deWaard, J.R., 2003. Biological identifications through DNA barcodes. Philos. Trans. R. Soc. Lond. Ser. B Biol. Sci. 358 (1433), 603–625. https://doi.org/10.1098/rstb.2003.1331.

Hillis, D.M., Moritz, C., Mable, B.K., 1996. Molecular Systematics. Sinauer Associates, Sunderland, MA.

Hinchliff, C.E., Smith, S.A., Allman, J.F., Burleigh, J.G., Chaudhary, R., Coghill, L.M., et al., 2015. Synthesis of phylogeny and taxonomy into a comprehensive tree of life. Proc. Natl. Acad. Sci. 112 (41), 12764–12769.

Hodgson, G., Stepath, C.M., Seas, S.O., 1998. Using reef check for long-term coral reef monitoring in Hawaii. In: Proceedings of the Hawaii Coral Reef Monitoring Workshop: a Tool for Management. 9–11 June.

Hopwood, N., 2022. 'Not birth, marriage or death, but gastrulation': the life of a quotation in biology. Br. J. Hist. Sci. 55 (1), 1–26.

Hotez, P., Batista, C., Ergonul, O., Figueroa, J.P., Gilbert, S., et al., 2021. Correcting COVID-19 vaccine misinformation: Lancet Commission on COVID-19 vaccines and therapeutics task force members. EClinicalMedicine 33.

Hu, C., Wang, M., Lapointe, B.E., Brewton, R.A., Hernandez, F.J., 2021. On the Atlantic pelagic Sargassum's role in carbon fixation and sequestration. Sci. Total Environ. 781, 146801.

Huang, T.Y., Downs, M.R., Ma, J., Zhao, B., 2020. Collaboration across time and space in the LTER network. Bioscience 70 (4), 353–364.

Irga, P.J., Barker, K., Torpy, F.R., 2018. Conservation mycology in Australia and the potential role of citizen science. Conserv. Biol. 32 (5), 1031–1037.

Jackson, J.B., Kirby, M.X., Berger, W.H., Bjorndal, K.A., Botsford, L.W., Bourque, B.J., et al., 2001. Historical overfishing and the recent collapse of coastal ecosystems. Science 293 (5530), 629–637.

Jacobson, M.Z., et al., 2017. 100% clean and renewable wind, water, and sunlight all-sector energy roadmaps for 139 countries of the world. Joule 1 (1), 108–121. https://doi.org/10.1016/j.joule.2017.07.005.

Jetz, W., McPherson, J.M., Guralnick, R.P., 2012. Integrating biodiversity distribution knowledge: toward a global map of life. Trends Ecol. Evol. 27 (3), 151–159.

Johnson, D., Batista, D., Cochrane, K., Davey, R.P., 2021. ISA API: an open platform for interoperable life science experimental metadata. GigaScience 10 (9), giab060.

Jones, J.P., Thomas-Walters, L., Rust, N.A., Veríssimo, D., 2019. Nature documentaries and saving nature: reflections on the new Netflix series our planet. People Nat. 1 (4), 420–425.

Juravel, K., Porras, L., Höhna, S., Pisani, D., Wörheide, G., 2023. Exploring genome gene content and morphological analysis to test recalcitrant nodes in the animal phylogeny. PLoS One 18 (3), e0282444.

Kahng, S.E., Akkaynak, D., Shlesinger, T., Hochberg, E.J., Wiedenmann, J., Tamir, R., Tchernov, D., 2019. Light, Temperature, Photosynthesis, Heterotrophy, and the Lower Depth Limits of Mesophotic Coral Ecosystems. Springer International Publishing, pp. 801–828.

Karasti, H., Baker, K.S., Halkola, E., 2006. Enriching the notion of data curation in e-science: data managing and information infrastructuring in the long term ecological research (LTER) network. Comput. Supported Coop. Work 15 (4), 321–358.

Kaufman, A.B., Bashaw, M.J., Maple, T.L., (Eds.),, 2019. Scientific Foundations of Zoos and Aquariums: Their Role in Conservation and Research. Cambridge University Press.

Khoury, C.K., Kisel, Y., Kantar, M., Barber, E., Ricciardi, V., Klirs, C., et al., 2019. Science–graphic art partnerships to increase research impact. Commun. Biol. 2 (1), 295.

Kim, J.H., Hotez, P., Batista, C., Ergonul, O., et al., 2021. Operation Warp Speed: implications for global vaccine security. Lancet Glob. Health 9 (7), e1017–e1021.

Kittelmann, U., Wertheim, C., 2022. Christine and Margaret Wertheim: Value and Transformation of Corals: Catalogue for the exhibition at Museum Frieder Burda 2022. Wienand Verlag.

Kjær, K.H., Winther Pedersen, M., De Sanctis, B., De Cahsan, B., Korneliussen, T.S., Michelsen, C.S., et al., 2022. A 2-million-year-old ecosystem in Greenland uncovered by environmental DNA. Nature 612 (7939), 283–291.

Knight, R., Jansson, J., Field, D., Fierer, N., Desai, N., 2012. Unlocking the potential of metagenomics through replicated experimental design. Nat. Biotechnol. 30, 513–520.

Knowlton, N., 2010. Citizens of the Sea. National Geographic Partners.

Knowlton, N., 2017. Doom and gloom won't save the world. Nature 544 (7650), 271.

Knowlton, N., 2021. Ocean optimism: moving beyond the obituaries in marine conservation. Annu. Rev. Mar. Sci. 13, 479–499.

Knowlton, N., Di Lorenzo, E., 2022. The search for ocean solutions. PLoS Biol. 20 (10), e3001860.

Kohler, S., Dietrich, T.C., 2021. Potentials and limitations of educational videos on YouTube for science communication. Front. Commun. 6, 581302.

Kopf, A., Bicak, M., Kottmann, R., Schnetzer, J., Kostadinov, I., Lehmann, K., et al., 2015. The ocean sampling day consortium. Gigascience 4 (1), s13742-015. https://doi.org/10.1186/s13742-015-0066-5.

Kritzer, J.P., DeLucia, M.B., Greene, E., Shumway, C., Topolski, M.F., Thomas-Blate, J., et al., 2016. The importance of benthic habitats for coastal fisheries. Bioscience 66 (4), 274–284.

Lapointe, N.W., Tremblay, M.A., Barna, H., 2016. Tools for improving the effectiveness of academic partnerships in informing conservation practices. Nat. Areas J. 36 (1), 93–101.

Lasker, H.R., Bramanti, L., Tsounis, G., Edmunds, P.J., 2020. The rise of octocoral forests on Caribbean reefs. In: Riegl, B. (Ed.), Advances in Marine Biology, first ed. vol. 87. Academic Press, pp. 361–410.

Lawniczak, M.K., Durbin, R., Flicek, P., Lindblad-Toh, K., Wei, X., Archibald, J.M., et al., 2022. Standards recommendations for the earth BioGenome project. Proc. Natl. Acad. Sci. 119 (4), e2115639118.

Leray, M., Knowlton, N., 2015. DNA barcoding and metabarcoding of standardized samples reveal patterns of marine benthic diversity. Proc. Natl. Acad. Sci. 112 (7), 2076–2081.

Leray, M., Knowlton, N., Ho, S.L., Nguyen, B.N., Machida, R.J., 2019. GenBank is a reliable resource for 21st century biodiversity research. Proc. Natl. Acad. Sci. 116 (45), 22651–22656.

Leslie, H.M., Goldman, E., Mcleod, K.L., Sievanen, L., Balasubramanian, H., Cudney-Bueno, R., et al., 2013. How good science and stories can go hand-in-hand. Conserv. Biol. 27 (5), 1126–1129.

232 **Chapter 5** Conservation of benthic habitats and organisms

Letunic, I., Bork, P., 2021. Interactive tree of life (iTOL) v5: an online tool for phylogenetic tree display and annotation. Nucleic Acids Res. 49 (W1), W293–W296.

Lough, J.M., Cantin, N.E., Benthuysen, J.A., Cooper, T.F., 2016. Environmental drivers of growth in massive *Porites* corals over 16 degrees of latitude along Australia's northwest shelf. Limnol. Oceanogr. 61 (2), 684–700.

Lubchenco, J., Grorud-Colvert, K., 2015. Making waves: the science and politics of ocean protection. Science 350 (6259), 382–383.

Luisetti, T., Schratzberger, M., 2023. Including biological diversity in natural capital accounts for marine biodiversity conservation and human well-being. Biodivers. Conserv. 32 (1), 405–413.

Mace, G.M., Collar, N.J., Gaston, K.J., Hilton-Taylor, C.R.A.I.G., Akçakaya, H.R., Leader-Williams, N.I.G.E.L., et al., 2008. Quantification of extinction risk: IUCN's system for classifying threatened species. Conserv. Biol. 22 (6), 1424–1442.

Mackenzie, J.S., Jeggo, M., 2019. The one health approach—why is it so important? Trop. Med. Infect. Dis. 4 (2), 88.

Maestro, M., Pérez-Cayeiro, M.L., Chica-Ruiz, J.A., Reyes, H., 2019. Marine protected areas in the 21st century: current situation and trends. Ocean Coast. Manag. 171, 28–36.

Magris, R.A., Andrello, M., Pressey, R.L., Mouillot, D., Dalongeville, A., Jacobi, M.N., Manel, S., 2018. Biologically representative and well-connected marine reserves enhance biodiversity persistence in conservation planning. Conserv. Lett. 11 (4), e12439.

Maldonado, C., Molina, C.I., Zizka, A., Persson, C., Taylor, C.M., Albán, J., Chilquillo, E., Rønsted, N., Antonelli, A., 2015. Estimating species diversity and distribution in the era of big data: to what extent can we trust public databases? Glob. Ecol. Biogeogr. 24, 973–984.

Manzer, L.E., 1990. The CFC-ozone issue: progress on the development of alternatives to CFCs. Science 249 (4964), 31–35.

Martins, A., Vieira, H., Gaspar, H., Santos, S., 2014. Marketed marine natural products in the pharmaceutical and cosmeceutical industries: tips for success. Mar. Drugs 12 (2), 1066–1101.

McAfee, D., Doubleday, Z.A., Geiger, N., Connell, S.D., 2019. Everyone loves a success story: optimism inspires conservation engagement. Bioscience 69 (4), 274–281.

McCauley, D.J., Pinsky, M.L., Palumbi, S.R., Estes, J.A., Joyce, F.H., Warner, R.R., 2015. Marine defaunation: animal loss in the global ocean. Science 347 (6219), 1255641.

McClure, E.C., Sievers, M., Brown, C.J., Buelow, C.A., Ditria, E.M., Hayes, M.A., et al., 2020. Artificial intelligence meets citizen science to supercharge ecological monitoring. Patterns 1 (7), 100109.

McLuhan, M., 1962. The Gutenberg Galaxy: The Making of Typographic Man.

McQuilton, P., Batista, D., Beyan, O., et al., 2020. Helping the consumers and producers of standards, repositories and policies to enable FAIR data. Data Intell. 2 (1–2), 151–157.

McTavish, E.J., Hinchliff, C.E., Allman, J.F., Brown, J.W., Cranston, K.A., Holder, M.T., et al., 2015. Phylesystem: a git-based data store for community curated phylogenetic estimates. Bioinformatics 31, 2794–2800. https://doi.org/10.1093/bioinformatics/btv276.

Meadows, D.H., 1999. Leverage Points: Places to Intervene in a System. https://tinyurl.com/ycx75r52.

Menegotto, A., Rangel, T.F., 2018. Mapping knowledge gaps in marine diversity reveals a latitudinal gradient of missing species richness. Nat. Commun. 9, 4713.

Mengerink, K.J., Van Dover, C.L., Ardron, J., Baker, M., Escobar-Briones, E., Kristina Gjerde, J., Koslow, A., et al., 2014. A call for deep-ocean stewardship. Science 344 (6185), 696–698.

Messing, C.G., 2003. Three new species of Comasteridae (Echinodermata, Crinoidea) from the tropical western Pacific. Zoosystema 25 (1), 149–162.

Messing, C.G., 2020. A revision of the unusual feather star genus Atopocrinus with a description of a new species (Echinodermata: Crinoidea). Zootaxa, 4731: 471–491. Zootaxa 4772 (3), 600.

Meyer, C., Kreft, H., Guralnick, R., Jetz, W., 2015. Global priorities for an effective information basis of biodiversity distributions. Nat. Commun. 6, 8221.

Meyer, C., Jetz, W., Guralnick, R.P., Fritz, S.A., Kreft, H., 2016. Range geometry and socio-economics dominate species-level biases in occurrence information. Glob. Ecol. Biogeogr. 25, 1181–1193.

Miller, D.C., Agrawal, A., Roberts, J.T., 2013. Biodiversity, governance, and the allocation of international aid for conservation. Conserv. Lett. 6 (1), 12–20.

Milner, G.R., 2004. The Moundbuilders: Ancient Peoples of Eastern North America. Thames & Hudson Ltd, London.

Morely, D.M., Sherman, R.L., Jordan, L.K., Banks, K.W., Quinn, T.P., Spieler, R.E., 2008. Environmental enhancement gone awry: characterization of an artificial reef constructed from waste vehicle tires. In: Environmental Problems in Coastal Regions. 7. WIT Press, pp. 73–87.

Moudrý, V., Devillers, R., 2020. Quality and usability challenges of global marine biodiversity databases: an example for marine mammal data. Ecol. Inform. 56, 101051.

Muka, S., 2022. Taking hobbyists seriously: the reef tank hobby and knowledge production in serious leisure. Stud. Hist. Phil. Sci. 93, 192–202.

Muller-Karger, F.E., Miloslavich, P., Bax, N.J., Simmons, S., et al., 2018. Advancing marine biological observations and data requirements of the complementary essential ocean variables (EOVs) and essential biodiversity variables (EBVs) frameworks 211. Front. Mar. Sci. 211.

Murphy, P.D., 2021. Speaking for the youth, speaking for the planet: Greta Thunberg and the representational politics of eco-celebrity. Pop. Commun. 19 (3), 193–206.

NASEM, 2019. A Decision Framework for Interventions to Increase the Persistence and Resilience of Coral Reefs. National Academy Press, Washington DC.

National Academies of Sciences, Engineering, and Medicine (NASEM), 2021. A Research Strategy for Ocean-Based Carbon Dioxide Removal and Sequestration.

Nayak, S., Iwasa, J.H., 2019. Preparing scientists for a visual future: visualization is a powerful tool for research and communication but requires training and support. EMBO Rep. 20 (11), e49347.

Nedimyer, K., Gaines, K., Roach, S., 2011. Coral Tree Nursery©: an innovative approach to growing corals in an ocean-based field nursery. Aquac. Aquar. Conserv. Legis. 4 (4), 442–446.

Nei, M., Kumar, S., 2000. Molecular Evolution and Phylogenetics. Oxford University Press, USA.

Neumann, P.J., Cohen, J.T., Kim, D.D., Ollendorf, D.A., 2021. Consideration of value-based pricing for treatments and vaccines is important, even in the COVID-19 pandemic: study reviews alternative pricing strategies (cost-recovery models, monetary prizes, advanced market commitments) for COVID-19 drugs, vaccines, and diagnostics. Health Aff. 40 (1), 53–61.

Neumann, B., Vafeidis, A.T., Zimmermann, J., Nicholls, R.J., 2015. Future coastal population growth and exposure to sea-level rise and coastal flooding-a global assessment. PLoS One 10 (3), e0118571.

NOAA Fisheries, 2019. Restoring seven iconic reefs: a mission to recover the coral reefs of the Florida keys. In: Mission: Iconic Reefs–Summary. NOAA Fisheries.

Nusrat, S., Harbig, T., Gehlenborg, N., 2019. Tasks, techniques, and tools for genomic data visualization. Comput. Graphics Forum 38 (3), 781–805.

O'Brien, S.J., 2003. Tears of the Cheetah. Thomas Dunne Books.

O'Brien, S.J., 1987. The ancestry of the Giant panda. Sci. Am. 257 (5), 102–107.

O'Brien, S.J., Wildt, D.E., Bush, M., 1986. The cheetah in genetic peril. Sci. Am. 254 (5), 84–92.

Ochieng, H., Okot-Okumu, J., Odong, R., 2019. Taxonomic challenges associated with identification guides of benthic macroinvertebrates for biomonitoring freshwater bodies in East Africa: a review. Afr. J. Aquat. Sci. 44 (2), 113–126.

O'Donoghue, S.I., Baldi, B.F., Clark, S.J., Darling, A.E., Hogan, J.M., Kaur, S., et al., 2018. Visualization of biomedical data. Annu. Rev. Biomed. Data Sci. 1, 275–304.

Ortega, A., Geraldi, N.R., Alam, I., Kamau, A.A., Acinas, S.G., Logares, R., et al., 2019. Important contribution of macroalgae to oceanic carbon sequestration. Nat. Geosci. 12 (9), 748–754.

234 **Chapter 5** Conservation of benthic habitats and organisms

Otegui, J., Ariño, A.H., Encinas, M., Pando, F., 2013. Assessing the primary data hosted by the Spanish node of the Global Biodiversity Information Facility (GBIF). PLoS One 8, e55144.

Otero, M.M., International Union for the Conservation of Nature, 2020. Conservation overview of Mediterranean deep-sea biodiversity: a strategic assessment. https://www.researchgate.net/publication/340584052_Conservation_overview_of_Mediterranean_deep-sea_biodiversity_a_strategic_assessment#fullTextFileContent.

Overmann, J., Scholz, A.H., 2017. Microbiological research under the Nagoya protocol: facts and fiction. Trends Microbiol. 25 (2), 85–88.

Page, R., 2016. Towards a biodiversity knowledge graph. Res. Ideas Outcomes 2, e8767. https://doi.org/10.3897/rio.2.e8767.

Palumbi, A.R., Palumbi, S.R., 2021. The Extreme Life of the Sea. Princeton University Press.

Patterson, K., Terrill, B., Dorfman, B.S., Blonder, R., Yarden, A., 2022. Molecular animations in genomics education: designing for whom? Trends Genet. 38, 517–520.

Peixoto, R.S., et al., 2017. Beneficial microorganisms for corals (BMC): proposed mechanisms for coral health and resilience. Front. Microbiol. 8, 341.

Peixoto, R.S., Voolstra, C.R., Sweet, M., Duarte, C.M., Carvalho, S., Villela, H., et al., 2022. Harnessing the microbiome to prevent global biodiversity loss. Nat. Microbiol. 7 (11), 1726–1735.

Penev, L., Mietchen, D., Chavan, V., Hagedorn, G., Remsen, D., Smith, V., Shotton, D., 2011. Pensoft Data Publishing Policies and Guidelines for Biodiversity Data. Pensoft Publ.

Pingali, P.L., 2012. Green revolution: impacts, limits, and the path ahead. Proc. Natl. Acad. Sci. 109 (31), 12302–12308.

Politico, 2020. Coronavirus Will Change the World Permanently. Here's how. Politico. 19 March 2020 https://www.politico.com/news/magazine/2020/03/19/coronavirus-effect-economy-life-society-analysis-covid-135579.

Pratchett, M.S., Anderson, K.D., Hoogenboom, M.O., Widman, E., Baird, A.H., Pandolfi, J.M., et al., 2015. Spatial, temporal and taxonomic variation in coral growth—implications for the structure and function of coral reef ecosystems. Oceanogr. Mar. Biol. Annu. Rev. 53, 215–295.

Reed, J.K., 1980. Distribution and structure of deep-water Oculina varicosa coral reefs off central Eastern Florida. Bull. Mar. Sci. 30 (3), 667–677.

Reed, J.K., 2002a. Deep-water Oculina coral reefs of Florida: biology, impacts, and management. Hydrobiologia 471 (1–3), 43–55.

Reed, J.K., 2002b. Comparison of deep-water coral reefs and lithoherms off southeastern USA. Hydrobiologia 471, 57–69.

Reed, J.K., Shepard, A.N., Koenig, C.C., Scanlon, K.M., Gilmore, R.G., 2005. Mapping, habitat characterization, and fish surveys of the deep-water Oculina coral reef marine protected area: a review of historical and current research. In: Freiwald, A (Ed.), Cold-Water Corals and Ecosystems. Springer, pp. 443–465.

Rees, J., Cranston, K., 2017. Automated assembly of a reference taxonomy for phylogenetic data synthesis. Biodivers. Data J. 5, e12581. https://doi.org/10.3897/BDJ.5.e12581.

Rees, S.E., Foster, N.L., Langmead, O., Pittman, S., Johnson, D.E., 2018. Defining the qualitative elements of Aichi Biodiversity Target 11 with regard to the marine and coastal environment in order to strengthen global efforts for marine biodiversity conservation outlined in the United Nations Sustainable Development Goal 14. Mar. Policy 93, 241–250.

Rice, M.E., 1993. Sipuncula. In: Microscopic Anatomy of Invertebrates, Vol. 12: Onychophora, Chilopoda, and Lesser Protostomata. Wiley-Liss.

Rohwer, F., 2010. Corals in a Microbial Sea. Plaid Press.

Rome, A., 2013. The Genius of Earth Day: How a 1970 Teach-in Unexpectedly Made the First Green Generation. MacMillan.

Rosado, P.M., Leite, D.C., Duarte, G.A., Chaloub, R.M., Jospin, G., Nunes da Rocha, U., et al., 2019. Marine probiotics: increasing coral resistance to bleaching through microbiome manipulation. ISME J. 13 (4), 921–936.

Rubenzer, R., 1979. The role of the right hemisphere in learning & creativity implications for enhancing problem solving ability. Gift. Child Q. 23 (1), 78–100.

Rubin, E.S., Davison, J.E., Herzog, H.J., 2015. The cost of CO_2 capture and storage. Int. J. Greenhouse Gas Control 40, 378–400. https://doi.org/10.1016/j.ijggc.2015.05.018.

Sabherwal, A., Ballew, M.T., van Der Linden, S., Gustafson, A., Goldberg, M.H., Maibach, E.W., et al., 2021. The Greta Thunberg effect: familiarity with Greta Thunberg predicts intentions to engage in climate activism in the United States. J. Appl. Soc. Psychol. 51 (4), 321–333.

Sansone, S.A., Rocca-Serra, P., Field, D., Maguire, E., Taylor, C., Hofmann, O., et al., 2012. Toward interoperable bioscience data. Nat. Genet. 44 (2), 121–126.

Sayers, E.W., Beck, J., Bolton, E.E., Bourexis, D., Brister, J.R., Canese, K., et al., 2021. Database resources of the national center for biotechnology information. Nucleic Acids Res. 49 (D1), D10.

Schoch, C.L., Ciufo, S., Domrachev, M., Hotton, C.L., Kannan, S., Khovanskaya, R., et al., 2020. NCBI taxonomy: a comprehensive update on curation, resources and tools. Database 2020, baaa062.

Shanmughavel, P., 2007. An overview on biodiversity information in databases. Bioinformation 1, 367–369.

Shaw, F., Etuk, A., Minotto, A., Gonzalez-Beltran, A., Johnson, D., Rocca-Serra, P., et al., 2020. COPO: a metadata platform for brokering FAIR data in the life sciences. F1000Research 9 (495), 495.

Shinn, E.A., Smith, G.W., Propero, J.M., Betzer, P., Hayes, M.L., Barber, R.T., 2000. African dust and the demise of Caribbean coral reefs. Geophys. Res. Lett. 27 (19), 3029–3032.

Shrivastava, P., Smith, M.S., O'Brien, K., Zsolnai, L., 2020. Transforming sustainability science to generate positive social and environmental change globally. One Earth 2 (4), 329–340.

Small, C., Nicholls, R.J., 2003. A global analysis of human settlement in coastal zones. J. Coast. Res., 584–599.

Smith-Godfrey, S., 2016. Defining the blue economy. Marit. Aff. 12 (1), 58–64.

Sosa-Gutierrez, R., Jouanno, J., Berline, L., Descloitres, J., Chevalier, C., 2022. Impact of tropical cyclones on pelagic Sargassum. Geophys. Res. Lett. 49 (6), e2021GL097484.

Sousa, D.A., Pilecki, T., 2013. From STEM to STEAM: Using Brain-Compatible Strategies to Integrate the Arts. Corwin Press.

Spalding, M., Burke, L., Wood, S.A., Ashpole, J., Hutchison, J., Zu Ermgassen, P., 2017. Mapping the global value and distribution of coral reef tourism. Mar. Policy 82, 104–113.

Strupp, B.J., Levitsky, D.A., 1995. Enduring cognitive effects of early malnutrition: a theoretical reappraisal. J. Nutr. 125 (Suppl 8), 2221S–2232S.

Tàbara, J.D., Frantzeskaki, N., Hölscher, K., Pedde, S., Kok, K., Lamperti, F., et al., 2018. Positive tipping points in a rapidly warming world. Curr. Opin. Environ. Sustain. 31, 120–129.

Thomas, T., Moitinho-Silva, L., Lurgi, M., Easson, C.G., Björk, J., Astudillo, C., Erpenbeck, D., Gilbert, J., Knight, R., Lopez, J.V., Taylor, M., Thacker, R., Montoya, J.M., Hentschel, U., Webster, N., 2016. The global sponge microbiome: symbiosis insights derived from a basal metazoan phylum. Nat. Commun. 7 (11870), 1–12. https://doi.org/10.1038/ncomms11870. http://www.nature.com/articles/ncomms11870.

Thompson, L.R., Sanders, J.G., McDonald, D., Amir, A., Ladau, J., Locey, K.J., et al., 2017. A communal catalogue reveals Earth's multiscale microbial diversity. Nature 551 (7681), 457–463.

Thunberg, G., 2019. No One Is Too Small to Make a Difference. Penguin.

Tichenor, P., Donohue, G.A., Olien, C.N., 1970. Mass media flow and differential growth in knowledge. Public Opin. Q. 34 (2), 159. https://doi.org/10.1086/267786.

Tonkiss, J., Galler, J., Morgane, P.J., Bronzino, J.D., Austin-Lafrance, R.J., 1993. Prenatal protein malnutrition and postnatal brain function a. Ann. N. Y. Acad. Sci. 678 (1), 215–227.

236 **Chapter 5** Conservation of benthic habitats and organisms

Troudet, J., Vignes-Lebbe, R., Grandcolas, P., Legendre, F., 2018. The increasing disconnection of primary biodiversity data from specimens: how does it happen and how to handle it? Syst. Biol. 67, 1110–1119.

Tzafesta, E., Zangaro, F., Specchia, V., Pinna, M., 2021. An overview of DNA-based applications for the assessment of benthic macroinvertebrates biodiversity in Mediterranean aquatic ecosystems. Diversity 13 (3), 112.

United Nations, 2011. Secretariat of the Convention on Biological Diversity, Nagoya Protocol on Access to Genetic Resources and the Fair and Equitable Sharing of Benefits Arising from their Utilization to the Convetion on Biological Diversity.

United Nations, 2020. Make Bold Environmental Action Central Focus of Post Pandemic Economic Recovery, Speakers Urge as General Assembly Holds First Ever Global Biodiversity Summit. https://press.un.org/en/2020/ga12274.doc.htm. (Accessed 5 August 2023).

United Nations, 2017. https://www.un.org/sustainabledevelopment/wp-content/uploads/2017/05/Ocean-fact-sheet-package.pdf.

United Nations Conference on Sustainable Development, 2012. Blue Economy Concept Paper. United Nations Conference on Sustainable Development, s.l. http://unctad.org/en/PublicationsLibrary/ditcted2014d5_en.pdf.

Van Bonn, W., Oliaro, F.J., Pinnell, L.J., 2022. Ultraviolet light alters experimental aquarium water microbial communities. Zoo Biol., 1–9. https://doi.org/10.1002/zoo.21701.

Vaughan, D.E., 2021. Active Coral Restoration: Techniques for a Changing Planet. J. Ross Publishing.

Vaughan, D.E., Nedimyer, K., 2021. Emerging technologies. In: Active Coral Restoration: Techniques for a Changing Planet. Ross Publishing, Plantation, FL.

Waller, J., 2022. Finding data gaps in the GBIF backbone taxonomy. Biodivers. Inf. Sci. Stand. 6, e91312.

Ward, K.D., 2014. Cultivating public service motivation through AmeriCorps service: a longitudinal study. Public Adm. Rev. 74 (1), 114–125.

Ware, M., Garfield, E.N., Nedimyer, K., Levy, J., et al., 2020. Survivorship and growth in staghorn coral (Acropora cervicornis) outplanting projects in the Florida Keys National Marine Sanctuary. PLoS One 15 (5), e0231817.

Wee, S.Y.C., Sam, S.Q., Sim, W.T., Ng, C.S.L., Taira, D., Afiq-Rosli, L., et al., 2019. The role of in situ coral nurseries in supporting mobile invertebrate epifauna. J. Nat. Conserv. 50, 125710.

Weigand, H., Beermann, A.J., Čiampor, F., Costa, F.O., Csabai, Z., Duarte, S., et al., 2019. DNA barcode reference libraries for the monitoring of aquatic biota in Europe: Gap-analysis and recommendations for future work. Sci. Total Environ. 678, 499–524.

Wells, H.G., 1921. The Outline of History: Being a Plain History of Life and Mankind. Cassell.

Widder, E., 2021. Below the Edge of Darkness: A Memoir of Exploring Light and Life in the Deep Sea. Random House.

Wieczorek, J., Bloom, D., Guralnick, R., Blum, S., Döring, M., Giovanni, R., et al., 2012. Darwin Core: an evolving community-developed biodiversity data standard. PLoS One 7 (1), e29715.

Wilkinson, M.D., Dumontier, M., Jan Aalbersberg, I.J., et al., 2016. The FAIR guiding principles for scientific data management and stewardship. Sci. Data 3, 160018.

Witze, A., 2022. Astronomers forced to rethink early Webb telescope findings. Nature 610, 243.

Wolfson, W., 2005. Diversa builds a business with designer bacteria. Chem. Biol. 12 (5), 503–505.

World Meteorological Organization (WMO), 2022. Executive Summary. Scientific Assessment of Ozone Depletion: 2022, GAW Report No. 278. WMO, Geneva. 56 pp https://ozone.unep.org/system/files/documents/Scientific-Assessment-of-Ozone-Depletion-2022-Executive-Summary.pdf.

Yang, Y., Wang, J., Xia, M., 2020. Biodegradation and mineralization of polystyrene by plastic-eating superworms Zophobas atratus. Sci. Total Environ. 708, 135233.

Yesson, C., Brewer, P.W., Sutton, T., Caithness, N., Pahwa, J.S., Burgess, M., et al., 2007. How global is the global biodiversity information facility? PLoS One 2 (11), e1124.

Zheng, M., Waller, M.P., 2017. ChemPreview: an augmented reality-based molecular interface. J. Mol. Graph. Model. 73, 18–23.

Index

Note: Page numbers followed by *f* indicate figures and *t* indicate tables.

A

Abyssal zone, 3–5, 155
Abyssoprimnoa gemina, 148–149
Acetyl-coenzyme A pathway, 51
Acoustic mapping, 90–91, 95
Acropora millepora, 111, 156
Acropora tenuis, 46–47, 155–156
Acyl-CoA oxidase, 155
Adhesion phenotype, 153
Aequorea victoria, 154
AlgaeBase, 216–217
Alvinocaris longirostris, 154
Amazon plume, 14
Amphibalanus amphitrite, 153
Amphipoda species, 31–32
Amplified sequence variants (ASVs), 44–45
Andros Island reef, 34
Anthosigmella varians, 138
Anthozoan Notch ligand-like (AnNLL) molecule, 155–156
Anticancer agents, 102–105
Antifreeze proteins (AFPs), 153–154
Antillogorgia americana, 138
Antiviral drugs, 103
Apex predators, 21–22
Aquatic Symbiosis Genome (ASG) Project, 44, 47–48
Arenicola marina, 179–180
Argus/Jason ROV, 89–90
Artemis program, 88
Artificial intelligence (AI), 97–98, 197
Artificial soft-bodied robot, 96
Arts, 201–203
Asteroids, 32
Atlantic and Gulf Rapid Reef Assessment Program (AGRRA), 93, 211*t*
Autochemosynthetic pathways, 97
Autonomous reef monitoring structures (ARMSs), 112–113
Autonomous underwater vehicles (AUVs), 8–9, 82, 89–91, 150, 220
Avirulent strains, 77
Axinella corrugata, 10

B

Baltic Sea, 16
Barcode of Life Database (BOLD), 211*t*, 217
Bathyal zone, 3–5, 133
Bathymetric maps, 81

Bathymodiolus thermophilus, 154
Bdelloid rotifers, 36
Benthic diversity, threats to
   climate change, 173–177
   destructive extraction methods, 177–178
   disturbance, 172–173
   eutrophication, 178–180
   habitat loss, 177–178
   marine diseases, 180–184
   pollution, 178–180
   stability, 172–173
Benthic microalgae (BMA), 17
Benthic-pelagic coupling (BPC), 2–3, 5–6
   Baltic Sea, 16
   biological gravitational pump (BGP), 10–11
   biomineralization, 16
   carcasses, 14–15
   coral mass spawning, 17
   DOM–sponge–fauna pathway, 15–16
   foraminiferans, 15
   marine microorganisms, 13
   marine snow, 12–13
   meridional overturning circulation (MOC), 17–18
   microplastics, 15
   nitrification, 13
   nutrients, 17
   organic matter decomposition, geochemical and biological
      views of, 11, 12*f*
   oyster bed reefs, 16
   particulate organic carbon (POC), 10–13
   particulate organic material (POM), 10–11
   particulate sediments, 12–13
   physical and biological parameters, 11
   phytoplankton, 15
   reef health, turbidity effects on, 13–14
   river plumes, effects of, 14
   sedimentation, 12–14
   sediment biota, 14
   thermo haline circulation (THC), 17–18
   underwater technologies, invention of, 15
Benthos, 1–3, 10
   abyssal zone, 3–5
   assessments, wedge technologies for, 86
   bathyal zone, 3–5
   bathymetric maps, 81
   benthic-pelagic coupling (BPC), 2–3, 5–6
      Baltic Sea, 16

237

238 Index

Benthos *(Continued)*
  biological gravitational pump (BGP), 10–11
  biomineralization, 16
  carcasses, 14–15
  coral mass spawning, 17
  DOM–sponge–fauna pathway, 15–16
  foraminiferans, 15
  marine microorganisms, 13
  marine snow, 12–13
  meridional overturning circulation (MOC), 17–18
  microplastics, 15
  nitrification, 13
  nutrients, 17
  organic matter decomposition, geochemical and biological
    views of, 11, 12*f*
  oyster bed reefs, 16
  particulate organic carbon (POC), 10–13
  particulate organic material (POM), 10–11
  particulate sediments, 12–13
  physical and biological parameters, 11
  phytoplankton, 15
  reef health, turbidity effects on, 13–14
  river plumes, effects of, 14
  sedimentation, 12–14
  sediment biota, 14
  thermo haline circulation (THC), 17–18
  underwater technologies, invention of, 15
 biodiversity, 37–41, 80
 continuity and cosmopolitanism
  apex predators and pelagic taxa, 21–22
  ascidian species, 21–22
  deep-water corals, 24–25
  extant benthic single-celled taxa and metazoan phyla,
    examples of, 22–24, 23–24*t*
  feather duster worms, 24–25
  interstitial meiofaunal annelid, 24–25
  intertidal mussels, 21–22
  meiofanua, 22–24
  microbiomes, 24
  species distribution modeling (SDM), 25
 diversity hotspots, 132–133
  Census of Marine Life (CML), 133–134
  Clarion-Clipperton Fracture Zone (CCZ), 148–150
  continuum, 150–152
  Florida Reef Tract (FRT), 137–146
 expeditions, 8
 habitat conservation
  biodiversity databases, 211*t*, 212–218
  governmental organizations involved in, 197–200
  marine-protected areas (MPA), 207–208
  nonprofit organizations and projects involved in, 197–200,
    200*t*

 hadal zone, 3–5
 heterogeneity and patchiness of, 2
 large-scale deep benthic experiments, 79
 littoral and neritic zone, 3–5
 locales, 8–9
 macrofauna, diversity of
  cnidaria, 33–37
  Echinodermata, 32–33
  metazoan diversity, 26–27
  molluscs and crustaceans, 31–32
  Porifera, 28–31
 monitoring, 92–95
 Rho-associated kinase (ROCK) gene homologs, 152
 species, connectivity within
  Caribbean spiny lobster, 20
  "deep reef refugia" hypothesis, 21
  demographic connectivity, 19
  endemism, 19
  gene flow, 18–19
  genetic connectivity, 18, 20
  in nature preserves, 19
  physical and biological processes, 18
  reef organisms, 20–21
  vertical connectivity, 19–20
  vicariance events, 20
 stressor, 5
 sublittoral zone, 3–5
 video links, 220, 221*t*
Big experiments, 76–81
"Big Science" framework, 2–3
"Big Science" projects, 75–77
Bilaterian phylogeny reconstruction, 100, 101*f*
Bioacoustics, 95–96
Biodiversity, 26, 75
 abyssal, 5
 benthic, 37–41
 databases, 211*t*, 212–218
 disruption of ecosystems, 173
 Coral Triangle, 143, 144*f*
 definition, 25
 generation, 40–41
 genes and functions, 152–157
 2030 goals for, 198–200
 hotspot, 131
 issues, 136–137
 origins of, 49–51
 policies, 197–200
 strategic plan for, 75–76
Biodiversity ecosystems functions (BEF) approach, 92–93
Bioeconomy, 107–108, 221–225
Bioerosion, 28–29, 93, 176–177
Biological cementing substances, 153

Index   239

Biological gravitational pump (BGP), 10–11
Bioluminescence, 26, 43, 92, 96, 154, 209–210
Bioluminescent bacteria, 13
Biomass, 5, 15–17, 22, 27–28, 37, 80–81, 90, 94, 179–180, 206
Biomineralization, 15–16, 28–29
Biomolecule-based electrode materials, 97
Bioprospecting, 101–105, 107, 209, 222–223
Biosynthetic gene clusters (BGCs), 106, 157, 173
Biotechnology, 75–76, 99–100, 223
BLAST tool, 109–110, 213–214
Blue economy, 221–225
Bottrylids, 30–31
Brevetoxin, 105–106
Bryostatin, 103
Bryostatin-1, 103

C

California Conservation Genomics Project (CCGP), 114
Calyptrophora persephone, 148–149
Cambrian benthic communities, 50–51
Candid critter camera, 91
Carbon sequestration, 206
Carcasses, 14–15
Carcinus maenas, 179–180
Caribbean reefs, 138
Caribbean spiny lobster, 20
Caulophacus sponges, 148–149
cDNA sequences, 7
Cell cultures, 114–115
Cell freezing, 114–115
Cement proteins, 153
Census of Marine Life (CML), 133–134
ChatGPT, 97–98
Cinachyrella alloclada, 155
Clarion-Clipperton Fracture Zone (CCZ), 31–32, 79, 148–150, 177–178
Climate change, 173–177
Cliona delitrix, 136–137
Clustered regularly interspersed palindromic repeats (CRISPR)-CAS methods, 99
Cnidaria, 23–24*t*, 25, 27, 30, 33–37, 40–41, 46, 79–80, 153, 156
Coaxing organisms, 106
Coccolithus braarudii, 177
Collaborative Open Plant Omics (COPO), 211*t*, 218
Combinatorial biosynthesis, 106–107
Commercial cruise lines, 9–10
Committee on Space Research (COSPAR), 75–76
Conference of parties (COP), 75–76, 198
Congo River plume, 14
Conservation genomics, 107–115, 224
Conus snails, 35

Convention for Biological Diversity (CBD), 107, 197–198, 207–208
Convolutional neural networks (CNN), 98
Cooperative Ecosystem Studies Units (CESU) Network, 199–200, 200*t*
Coral diseases, 180–182, 181*f*, 184
Coral mass spawning, 17
Coral nurseries, 205–207
Coral point count, 94
Coral Reef Conservation Program (CRCP), 93–95
Coral Reef Evaluation and Monitoring Project (CREMP), 94
Coral Reef Information System (CoRIS), 93–94
Coral Reef Restoration, Assessment & Monitoring (CRRAM), 94
Coral reefs, 11, 27, 45–49
    ocean acidification (OA) effects on, 78–79
    profiles of, 136–137
Coral Restoration Consortium (CRC), 220
Corals
    and algal symbionts, nutritional interactions between, 45
    Aquatic Symbiosis Genome (ASG) Project, 47–48
    chemosymbiotic associations, 47
    dinoflagellates associated with, 46
    microalgal-coral symbiosis, 46
    Symbiodinium, 45–46
    symbiotic bacterial microbiomes, 46–47
    whole genome sequencing, 47–48
Corexit 9500, 80
Crowdsourced bathymetry, 197
Crustaceans, 1, 21–22, 26, 31–32, 36, 155, 177
Crustose coralline algae (CCA), 78–79, 94
Ctenophores, 29–30, 100, 151, 156
Cuban reef, 11
Curacao reef, 11
C6-zinc cluster proteins, 106

D

Dampier Ridge, 9
Darwin Tree of Life (DToL) project, 47–48, 211*t*, 217–218
Deep Diver submersible, 87
DEEPEND/RESTORE project, 5–7
Deep learning algorithms, 98, 175–176*f*
Deep-Ocean Genomes Program, 197–198
"Deep reef refugia" hypothesis, 21
Deep sea mining (DSM), 79–80, 177–178, 178*f*
Deepwater Horizon oil spill (DWHOS), 5–6, 31, 80–81, 209–210
Defensomes, 85
Denaturing gradient gel electrophoresis (DGGE), 46
Deoxyribonucleic acid (DNA), 99
    pyrosequencing technology, 109
Destructive extraction methods, 177–178

240 Index

Diadema antillarum, 136–137, 182
Dideoxy chain termination sequencing technique, 109
Discodermolide, 104–105
Dissolved inorganic nitrogen (DIN), 17
Dissolved organic matter (DOM), 12–16, 28–29, 48, 50–51
Dissolved organic nitrogen (DON), 17
Disturbance, 172–173
Disturbance and recolonization (DISCOL) experiment, 79–80, 177–178
Diversity hotspot, 21–24, 131–133, 137, 148
DNA metabarcoding, 112
DNA sequencing methods, 7, 24–25, 29–30, 84–86, 107–109, 112
Double-digest RadSeq (ddRadSeq), 111
Drosophila, 49, 84, 152, 156
Drug discovery program, 104
DSV Alvin submersible, 5, 47, 87–88, 96
Dynamic random access memory (DRAM), 84

E

Earth Biogenome Project (EBP), 84–86
Echinodermata, 23–24t, 27, 32–33, 40–41, 149–150, 156
Echo sounders, 82
Eco-acoustics, 95–96
Ecological field experiments, 78
Ecotoxicology experiments, 77–78
Education, 191–197, 218–220
Emiliania huxleyi, 177
Endemism, 5, 19–22, 109, 131, 149–150
Engagement, 191–197, 218–220
Environmental DNA (eDNA), 112–113, 197
Essential Biodiversity Variables (EBVs), 93
Essential Ocean Variables (EOVs), 93
Eukaryotic nuclear genomes, 108, 110–111
Eunicea flexuosa, 138
Eunicella gazella, 173
Euprymna scolopes, 43
European Reference Genome Atlas (ERGA), 84–85
Eutrophication, 173, 178–180
Exclusive Economic Zone (EEZ), 80, 91, 135–136
Explorer of the Seas, 9–10
Expressed sequence tags (ESTs), 29–30

F

Feather duster worms, 24–25
FishBase, 216
Flavin adenine dinucleotide (FADH2), 97
Florida Department of Environmental Protection-Coral reef Conservation Program (FDEP-CRCP), 95
Florida manatees (*Trichechus manatus latirostris*), 180
Florida Reef Track (FRT), 94–95, 136–146, 184, 196, 221t
Fluorescence in situ hybridization (FISH) methods, 48

Foraminiferans, 1, 15, 149
Framework of Ocean Observing (FOO), 93

G

Galathea, 8
Gastropods, 34–35, 133
Gastrulation, 42
GenBank database macromolecular sequences, 214
Gene flow, 18–22, 41, 146–148
Genetic and genomic maps, 82–86
Genetic barcoding, 112
Genetic engineering, 99, 152–153, 223
Genetic imprinting, 113–114
Genome 10,000 (10K) Project, 84–86
Genome sequencing, 107–115
Genome-wide association studies (GWAS), 111
Geographic information system (GIS), 81, 98
Geo-orbital satellites, 81
GEOTRACES, 218
Gephyrocapsa oceanica, 177
Global Biodiversity Information Facility (GBIF), 211t, 214–216
Global Coral Reef Monitoring Network, 133
Global Explorer ROV, 89–90
Global Invertebrate Genomes Project (GIGA), 27, 85–86
Global Ocean Observing System (GOOS), 93
Glutathione S-transferase, 154
Gorgonia ventalina, 37, 94, 138
Great Barrier Reef (GBR), 78, 92, 148, 196
Green fluorescent protein (GFP) gene, 154
GRIIDC database, 211t, 218
Group on Earth Observations Biodiversity Observation Network (GEO BON), 93
Gulf of Mexico (GOM), 20
    cold seep communities, 47
    DEEPEND/RESTORE project in, 5–6
    Deepwater Horizon oil spill in, 80
    Loop Current of, 10
    shrimp species, 31
Gulf of Mexico Research Initiative (GOMRI), 5–6, 218

H

Habitat heterogeneity hypothesis, 2, 10, 40–41, 135, 143
Habitat loss, 16, 177–178
Habitat of Particular Concern (HAPC), 210
Hadal zone, 3–5, 135–136, 148
Harbor Branch Oceanographic Institution (HBOI), 39, 87–88, 104–105, 209–210
Harpacticoids, 79
Heat Shock factor 1 (hsf1), 155
Heat Shock Proteins (HSPs), 155
Hermodice carunculata, 172
High-definition (HD) video, 91

Index **241**

High microbial abundance (HMA) sponges, 44–45
High-performance computing (HPC), 98
High Seas Treaty, 80
High-throughput (HT) DNA sequencing, 7, 44–45, 84, 99, 112
High-throughput sequencing (HTS), 44–45, 99, 106–107, 112
HMS Challenger expedition, 7, 86
Homeobox (HOX) genes, 155–156
Hsp90, 155
Human Microbiome Project (HMP), 42–43
Hybrid remotely operated vehicles (HROVs), 90–91
Hydras, 36
Hydrothermal vents, 36, 51, 87, 97, 150, 154–155

I

IGNITE-ITN (Innovative Training Network), 219–220
Indian Deep-sea Environment Experiment (INDEX), 79
Information age, 98, 170–172
Informative maps, 76–77, 81–82
International Organization for Economic Co-operation and Development (OECD), 216
International Seabed Authority (ISA), 75–76, 148, 177–178, 197–198, 200*t*
International Space Station (ISS), 75–76, 192–193
International Union for Conservation of Nature (IUCN), 199–200, 214–215
Inter Ocean Metal Benthic Impact Experiment (IOM BIE), 79

J

Japanese spider crab, 30–31
Johnson Sea-Link (JSL) submersible, 87, 96, 104

K

Kunming-Montreal global biodiversity framework, 199

L

'Lab-on-a-chip' systems, 96–97
Light detection and ranging (LIDAR) systems, 81, 92
Long interspersed nuclear elements (LINES), 110–111
Loop current, 10, 80
Loriciferans, 36, 47
Low microbial abundance (LMA) sponges, 44–45
Lugworms (*Arenicola marina*), 179–180
LuxA, 154–155
LuxB, 154–155
Lysyl oxidase (LOX), 153

M

Machine learning, 97–98
Macroalgae, 32, 44, 94, 104, 112, 136–138, 180, 182
Macrobenthos, 1
Madracis, 34
Marine Biodiversity Observation Network (MBON), 93

Marine Biotechnology Reference Collection (MBRC), 104, 211*t*
Marine-protected areas (MPA), 95–96, 207–208
Marine snow, 12–13, 15, 17, 88, 97
Mars exploratory programs, 49
Meiobenthos, 1
Meiofanua, 22–24
Meridional overturning circulation (MOC), 17–18
Mesonychoteuthis hamiltoni, 30–31
Mesophotic coral ecosystems (MCEs), 90–91, 146–148, 147*f*
Mesophotic reefs, 21, 33, 41, 90–91, 138–143
Metabarcoding, 112–113, 148–149
Metagenomic datasets, 7–8, 14
Metagenomics, 39, 43, 107–108, 155, 223
Metatranscriptomics, 47–48, 108, 155
Methylation-sensitive amplified polymorphism (MSAP), 113–114
Microbe-associated molecular patterns (MAMPs), 32–33, 43
Microbenthos, 1
Microbial diversity, 37–40, 44–45
Microbial secondary metabolites, 105–107
Microbiome stewardship, 207
Microplastics, 15, 179–180
Miocene ocean circulation model, 24–25
Miri-Sibuti Coral Reef National Park (MSCRNP), 13–14
Mitochondrial sulfur detoxification pathway, 47
Molecular ecology, 107–115
Molluscan proteins, 16
Molluscs, 16, 22–24, 26, 28–29, 31–32, 35–36, 44
Montastraea cavernosa, 138, 146–148
Montipora capitata, 78–79
Montreal Protocol, 192
Multibeam echosounders (MBES), 82

N

Nagoya Protocol (NP), 107, 198
NanoSIMS, 48
National Coral Reef Monitoring Program (NCRMP), 93–94, 138, 211*t*
National Oceanic and Atmospheric Administration (NOAA), 9–10, 31, 88, 91, 93–94, 135–136, 183, 211*t*, 212, 220, 221*t*
Natural products, bioprospecting for, 101–105
Near reef complex (NRC), 138
Nematodes, 1, 21–22, 35–36, 77–78, 84–85, 90, 133
Nitrification, 13, 183
Nutrient overloading, 178–180
Nutrients, 1–5, 11, 13–18, 22, 26, 28, 40–41, 45–46, 51, 98, 102, 143, 148–149, 173, 178–180, 223

O

Observational cameras, 91
Ocean acidification (OA), 78–79, 94, 137, 172, 176–177

242 Index

Ocean Biogeographic Information System (OBIS), 133–134, 214–215
Ocean conveyor belt (OCB), 17–18
Ocean Exploration and Research (OER) program, 91
Ocean optimism, 195–196
Ocean seabed
  average global depth, 2
  benthic habitats (*see* Benthos)
  biochemical and ontogenetic perspective, 1–2
  biodiversity, origins of, 49–51
  depth zones of, 3, 4*f*
  expeditions, 5–10
  flow patterns, 2
  microbial diversity, 37–40
  symbiosis, 48–49
    corals, 45–49
    deep-sea sponge symbioses, 44–45
    definition, 42
    Euprymna scolopes, 43
    gastrulation, 42
    between macrosymbionts and microsymbionts, 42
    microbiomes, 42–43
    photosymbiotic traits, 44
    research, 42–43
    sponge-microbial symbiosis systems, 44
    symbiotic cells, origin of, 42
    Vibrio fischeri, 43
Octopus, 14–15, 34–35, 85, 153
Oculina reefs, 208–212
Okeanos Explorer ROV, 88, 90–91, 135–136
One Health Initiative (OHI), 82–84, 193
Open reading frame (ORF), 49, 153–154
Operational taxonomic units (OTU), 39–40, 44–45, 149, 183–184
Optimism, 191–197
Osedax worms, 14–15
Osmotic stress response (OSR) genes, 155
Overturning, 17–18
Oysters, 16, 27, 152, 173, 205–206, 222
Ozone-depleting substances (ODSs), 192

P

Pacific Islands Ocean Observing System, 94
Paralvinella pandorae irlandei, 154
Pardaliscidae amphipods, 32
Particulate nitrogen (PN), 17
Particulate organic carbon (POC), 10–13, 15, 112
Particulate organic material (POM), 10–11, 15–16, 28, 48
Particulate organic nitrogen (PON), 12–13
Particulate sediments, 12–13
Patatin-like proteins (PLPs), 155
Pathogen-associated molecular patterns (PAMPS), 32–33, 43

Pattern recognition receptors (PRRs), 32–33
PAX genes, 153
PharmaMar, 223–224
Phelliactis anemone, 148–149
Phosphotransferase system (PTS), 107
Photosynthesis, 13, 46–47, 208–209
Photo/video image systems, 79–80
Phylogenetic trees, 38, 100, 215–216
Physical maps, of ocean, 81
Phytoplankton, 3–5, 7–8, 11, 14–15, 17, 21–22, 206
Pisum sativum, 150–151
Planktonic microbiomes, 7–8
Pleurobrachia pileus, 156
Plumes, 13–15, 79–80, 177–178
Pocillopora damicornis, 78–79, 115, 207
Pollution, 16, 78–79, 170, 173, 178–182
Polychaetes, 1, 28, 36, 153
Polyketides, 103–107
Polymerase chain reaction (PCR), 86, 99–100
Polystyrene (PS), 179–180, 223–224
Porifera, 23–24*t*, 27–31, 40–41, 44, 50–51, 79–80, 149–150, 182
Porifera Tree of Life (PorToL), 29
Predictive mapping, 82, 83*f*
Protocadherin genes, 85, 153
Pseudoliparis swirei, 96
Pyrosequencing technology, 39–40, 109, 183–184

Q

Quinones, 97

R

Raman spectroscopy, 48
Random Forests methods, 98
Reduction-oxidation (redox) reactions, 97
Reef fish species, 20–21, 34, 95, 138–143
Remotely operated vehicles (ROV), 36, 87–92, 148–149, 218
Remote sensing, 81, 92, 98
Remote Sensing Object-Based Image Analysis (RSOBIA), 95
Restriction associated DNA (RadSeq), 111
Rho-associated kinase (ROCK) gene homologs, 152
Riboflavin, 97
Ribosomal RNA (rRNA), 38–40, 44–46, 100, 112–113
RNAseq sequencing approaches, 108
Rosenstiel School of Marine and Atmospheric Science (RSMAS), 9–10
Royal Caribbean (RC), 9–10
Ruditapes philippinarum, 155
Russian-American Long Term Census of the Arctic (RUSALCA), 133–134

Index **243**

S

Seabed. *See* Ocean seabed
Seafloor mapping, 82, 214–215
Seagrasses, maps of, 82
Seagrass habitats, 180
Seamounts, 9, 20, 81, 134, 170
Sea star-associated densovirus (SSaDV), 182
Sea star wasting disease (SSWD), 182
Sea surface temperatures (SST), 13–14, 17, 44, 78–79, 137, 174, 175–176f, 184
Secondary metabolites (SMs)
  bioprospecting for, 101–105
  from marine microbes, 105–107
Sedimentation, 5, 12–14, 17, 79–80, 98, 136, 176–178, 184
Sediment community oxygen consumption (SCOC) rates, 80–81
Sequencing, 7
Shinkaia crosnieri, 154
Shore crab *(Carcinus maenas)*, 179–180
Short interspersed nuclear elements (SINES), 110–111
Short tandem repeats (STRs), 111
"Shotgun" cloning approach, 7
Siderastrea siderea, 138
Silicatein, 152–153
Silicon microchips, 84
Single-nucleotide polymorphism (SNP) analyses, 111
Skepticism, 171
Slope-abyss source-sink (SASS) hypothesis, 148–149
Smithsonian Marine Station (SMS), 212
Snail fish, 96
Sonar sounders, 82
Sorcerer II expeditions, 7
Southeast Florida Coral Reef Evaluation and Monitoring Project (SECREMP), 94, 138
Southeast Florida Coral Reef Initiative (SEFCRI), 95
Southeast Florida Reef Tract (SEFRT), 98
South Florida Association of Environmental Professionals-Community Coral Nursery Project (SFAEP CCNP), 205–206
Sox gene, 156
Spawning, 17, 19–20, 34, 78–79, 111
Species distribution modeling (SDM), 25
Spheciospongia vespariuum, 138
Spiny sea urchin, 85, 182
"Sponge Loop" hypothesis, 15–16, 28
Sponge orange band (SOB) disease, 183–184
Sponges, 1, 15–16, 28–30, 37, 44–47, 77–78, 99, 101–102, 104–105, 136–138, 156, 182–184
Squid, 30–31, 43
Stability, 172–173
Stony coral tissue loss disease (SCTLD), 137–138, 184

Strategies and Techniques for Analyzing Microbial Population Structures (STAMPS), 40
Stronglyocentrotus purpuratus, 156
Sub-bottom profilers (SBP), 91
Sublittoral zone, 3–5
Submerged patch reefs, 13–14
Submersible, 5, 8, 87–92, 104, 133–134, 150, 209–210, 212
Sustainable Seabed Knowledge Initiative (SSKI), 197–198
Symbiodiniaceae, 45–46, 136, 174
Symbiosis, 2–3, 42–43, 45–49, 136, 155–156, 174
  ciliates as models for, 48
  corals, 45–49
  deep-sea sponge symbioses, 44–45
  definition, 42
  Euprymna scolopes, 43
  gastrulation, 42
  between macrosymbionts and microsymbionts, 42
  marine, 42–45
  microbiomes, 42–43
  photosymbiotic traits, 44
  research, 42–43
  sponge-microbial symbiosis systems, 44
  symbiotic cells, origin of, 42
  Vibrio fischeri, 43
Synchronous laboratory experiments, 78
Synthetic bacterial genome, 7

T

Tara Oceans, 7–8
Taxonomy Browser, 211t, 215
Telepresence methods, 91
Tetraploid rotifer, 36
Thermo haline circulation (THC), 17–18
Tongue of the Ocean (TOTO), 34
Total nitrogen (TN), 17, 178–179
Total phosphorous (TP), 17
Trenches, 3–5, 8–9, 34, 96
Trichechus manatus latirostris, 180
Trimethylamine N-oxide (TMAO), 155
Tube worms, 47
Tunicata, 30–31
Turbidity, 12–14, 21–22, 143, 153, 173, 178–180, 184
Turf algae, 180

U

Underwater maps, 82
Underwater soundscapes, 95–97
Underwater Visual Censuses (UVC), 89–90, 94

244 Index

United Nations (UN), 75–76
  biodiversity policies, 197–198
  conference of parties (COP), 75–76
  Conference on Sustainable Development in Rio de Janeiro, 222
  Decade of Ocean Science for Sustainable Development (DOSSD), 197–198
  Earth Summit, 197–198
  High Seas Treaty, 80
  Intergovernmental Panel on Climate Change (IPCC), 173–174
  Oceans Day, 194
United Nations Environment Program, 75–76
Urchins, 32–33, 85, 136, 156, 182
Uronema marinum, 179–180
US Benthic Ecology meeting, 220
US Long-Term Ecological Research (LTER), 211*t*, 217
US National Center for Biotechnological Information (NCBI) databases, 213, 215

V

Verrucomicrobia, 183–184
Vibrio coralliilyticus, 207
Vibrio fischeri, 43, 154–155

W

Wand'rin' Star, 6–7
WebGIS database, 218
Web of Science, 146, 201, 213
Wedge technologies, for benthic assessments, 86
Whole genome sequencing, 30, 44, 47–48, 84–86, 111, 150–151, 223
World Meteorological Organization (WMO), 93
World Register of Marine Species (WoRMS), 214

X

Xenophyophores, 15
Xestospongia, 30–31
Xestospongia muta, 30, 94, 138, 183

Printed in the United States
by Baker & Taylor Publisher Services